DISCARDED

SEDIMENTATION ON THE MODERN
CARBONATE TIDAL FLATS
OF NORTHWEST ANDROS ISLAND, BAHAMAS

The Johns Hopkins University Studies in Geology No. 22

Sedimentation on the Modern Carbonate Tidal
Flats of Northwest Andros Island, Bahamas

Sedimentation on the Modern Carbonate Tidal Flats of Northwest Andros Island, Bahamas

EDITED BY LAWRENCE A. HARDIE

The Johns Hopkins University Press · Baltimore and London

Copyright © 1977 by The Johns Hopkins University Press
All rights reserved. No part of this book may be reproduced
or transmitted in any form or by any means, electronic or
mechanical, including photocopying, recording, xerography, or
any information storage or retrieval system, without permission
in writing from the publishers.
Manufactured in the United States of America

The Johns Hopkins University Press, Baltimore, Maryland 21218
The Johns Hopkins Press Ltd., London

Library of Congress Catalog Card Number 76-47389
ISBN 0-8018-1895-8

Library of Congress Cataloging in Publication data
will be found on the last printed page of this book.

CONTENTS

ACKNOWLEDGMENTS — xiii

1. INTRODUCTION/Lawrence A. Hardie — 1
2. METHODS/Peter Garrett and Lawrence A. Hardie — 4
3. EXPOSURE INDEX: A QUANTITATIVE APPROACH TO DEFINING POSITION WITHIN THE TIDAL ZONE/Robert N. Ginsburg, Lawrence A. Hardie, Owen P. Bricker, Peter Garrett, and Harold R. Wanless — 7
4. GENERAL ENVIRONMENTAL SETTING/Lawrence A. Hardie and Peter Garrett — 12
5. LAYERING: THE ORIGIN AND ENVIRONMENTAL SIGNIFICANCE OF LAMINATION AND THIN BEDDING/Lawrence A. Hardie and Robert N. Ginsburg — 50
6. BIOLOGICAL COMMUNITIES AND THEIR SEDIMENTARY RECORD/Peter Garrett — 124
7. ALGAL STRUCTURES IN CEMENTED CRUSTS AND THEIR ENVIRONMENTAL SIGNIFICANCE/Lawrence A. Hardie — 159
8. DISTINCTIVE FEATURES OF A RAINY, LOW-ENERGY, TROPICAL CARBONATE TIDAL FLAT: A SUMMARY/Lawrence A. Hardie — 178
9. SOME MISCELLANEOUS IMPLICATIONS AND SPECULATIONS/Lawrence A. Hardie and Peter Garrett — 184
10. A CONCLUDING NOTE: SENSITIVITY OF THE RECORD/Lawrence A. Hardie — 188

REFERENCES — 190

INDEXES — 197

TABLES

1. Weather data for Nassau, Bahamas. *16*
2. Range and height of tides on Three Creeks tidal flats. *18*
3. Types of layering in the Three Creeks tidal flat sediments. *53*
4. Comparison of features of smooth flat lamination of different sub-environments. *64*
5. Rate of deposition observations at selected sites on the Three Creeks tidal flats. *94*
6. Suspended sediment in the waters of the tidal flats and offshore of northwest Andros Island. *95*

7. Major onshore storms and storm flooding at Three Creeks tidal flats in the period March 1968 to May 1971. *98–99*
8. Characteristic species of the nearshore community of the Three Creeks tidal flat complex. *125*
9. Characteristic species of the pond community of the Three Creeks tidal flats. *126*
10. Characteristic species of the levee community of the Three Creeks tidal flats. *126*
11. Pelleting rate of *Cerithidea costata* and *Batillaria minima* on wet supratidal mat. *141*
12. Pelleting rate of *Marphysa sanguinea*. *143*
13. Basic features of cemented crusts of Three Creeks and Pumpion Cay areas. *160*
14. Catalog of major sedimentary structures in the Three Creeks tidal flat sediments. *178–79*
15. Summary of structures and textures in the sediments of different subenvironments of the Three Creeks tidal flats. *180–81*
16. Environmental elements and their specific sedimentary record at Three Creeks. *182*

ILLUSTRATIONS

1. Map showing location of study areas on Andros Island. *2*
2. Map of Three Creeks study area, showing locations of observation, etc., sites. *5*
3. Graph of mean Exposure Index vs. surface elevation for levee-pond subenvironments at Three Creeks. *8*
4. Relationship between sedimentary features and Exposure Index at Three Creeks. *10*
5. Graphs showing seasonal variation of Exposure Index at Three Creeks. *11*
6. A. Aerial photograph of Three Creeks area. B. Map showing environments at Three Creeks. *14*
7. Two oblique aerial views of parts of Three Creeks tidal flats. *15*
8. Cross-section through "palm hammock" at Pumpion Cay. *16*
9. A. Levelled profile across a levee at Three Creeks. B. Levelled profile across a beach ridge at Three Creeks. *17*
10. Map of levee-pond complex, showing areal distribution of subenvironments. *21*
11. View of knobby, crab-burrowed surface of a channel bank. *22*
12. A. General view of smooth-surfaced, vegetation-covered levee crest. B. Close-up view of levee crest surface, showing current lineations. *23*
13. A. Scars in levee crest surface where algal mat-bound sediment has been ripped out. B. "Air bubble" dome in algal mat-bound sediment of levee crest surface. *24*
14. View of mud-cracked levee backslope surface. *25*
15. Chipped surface of levee backslope, showing intraclast pocket. *26*
16. A. View of surface of wrinkled *Scytonema* mat covering high algal marsh zone of levee-pond complex. B. View of upper edge of high algal marsh, showing growth of *Scytonema* on crests of current lineations. *27*

CONTENTS

17. A. Raised polygons of *Scytonema*-bound sediment of high algal marsh zone of levee-pond complex. B. Patchy mound-like growth of *Scytonema* mats of high algal marsh zone of levee-pond complex. *28*

18. A. Typical view of *Scytonema* "pincushions" of the low algal marsh zone at the edge of an intertidal pond during wet summer months. B. Desiccated *Scytonema* "pincushions" of the low algal marsh zone during dry winter months. *30*

19. A. View of a pond at low tide. B. Close-up view of surface of a pond, showing pellets of gastropods and polychaetes. *32*

20. A. Pellet mounds of the polychaete *Marphysa* on a pond surface. B. "Prism-cracks" at the surface of a pond. *33*

21. Map showing distribution of subenvironments in shoreline area of Three Creeks. *35*

22. A. General view of beach "mound-and-hollow" surface at Three Creeks near low tide. B. Close-up view of a beach "mound-and-hollow" area. C. *Schizothrix* algal mat covering crest of a beach mound. *36*

23. A. View of older pond sediment exhumed by erosion to seaward of beach ridge at Three Creeks. B. *Penicillus* exposed on a channel mouth bar during an exceptionally low tide at Three Creeks. *38*

24. A. View looking onshore toward beach ridge at Three Creeks, showing eroded beach cliff and rippled beach surface at low tide. B. Close-up view of eroded beach cliff. *39*

25. A. View of *Schizothrix*-bound surface of beach terrace at Three Creeks, showing flat-pebbles of cohesive sediment welded to the surface by algal mats to produce LLH-S protostromatolites. B. View of eroding algal-bound laminated sediment at the seaward edge of the beach terrace at Three Creeks. *40*

26. A. Imbricate flat-pebble trains on the beach-ridge washover crest at Three Creeks. B. Close-up view of beach-ridge washover crest surface, showing patchy cover of rippled sand. *41*

27. Chipped, desiccated algal mat of beach-ridge backslope surface at Three Creeks. *42*

28. Intertidal, prism-cracked, channel bar in the axis of a small distributary channel at Three Creeks. *43*

29. Sediment core profile across a channel bar. *44*

30. Slumped blocks along a small channel. *45*

31. Levelled profile of inland marsh surface at Three Creeks. *46*

32. A. *Scytonema* "pincushions" covering low depressions in the surface of the inland algal marsh at Three Creeks. B. View of flat *Scytonema* mat that covers the "high" ground of the inland algal marsh at Three Creeks. *47*

33. A. General view of patchy, lightly cemented *Scytonema* mats of the high "crowns" along seaward edge of inland algal marsh at Three Creeks. B. Close-up of *Scytonema* mounds of high "crowns" along seaward edge of inland algal marsh at Three Creeks. *48*

34. A. View of channel bank surface showing *Schizothrix*-draped knobs. B. Vertical section through plastic-impregnated channel-bank algal knob. C. Vertical section through plastic-impregnated channel-bank algal knob which accreted over a laminated intraclast pebble. *54*

35. Photomicrograph of thin section of part of a channel-bank algal knob. *56*

36. A. Slab of levee crest sediment, showing typical uniformly thin,

smooth, flat lamination. B. Slab of beach-ridge washover sediment, showing destruction of lamination by oligochaete burrowing. *58*

37. A. Thin section photomicrograph of smooth flat lamination of levee crest sediment. B. Thin section photomicrograph of smooth flat lamination of beach-ridge washover crest sediment. *60*

38. A. Thin section photomicrograph of smooth flat lamination of beach terrace. B. Thin section photomicrograph of clotted, peloidal fabric typical of muddy laminae of all lamination types on the Andros Island tidal flats. *61*

39. A. Thin section photomicrograph of smooth flat lamination, showing vertical algal filament molds in a muddy lamina. B. Thin section photomicrograph of large ovoid grain with peloid core and muddy rim found in both smooth and disrupted flat laminated sediment. *62–63*

40. Surface and vertical views of disrupted flat lamination of levee backslope sediment. *65*

41. A. Slab showing laminated flat chips in levee backslope sediment. B. Thin section of photomicrograph of levee backslope sediment showing peloid-filled erosion scar. *66*

42. A. Thin section photomicrograph of levee backslope sediment, showing mud-cracks, filament molds, oligochaete burrows. B. Thin section photomicrograph of levee backslope sediment, showing sharp separation of muddy from sandy laminae. *67*

43. A. Slab of sediment from high algal marsh zone of levee-pond complex, showing typical crinkled fenestral lamination. B. Slab of high algal marsh sediment treated with acid to reveal *Scytonema* filament complex. *69*

44. A. Thin section photomicrograph of lamination of high algal marsh, showing wavy organic partings between clotted fenestral sediment laminae. B. Close-up photomicrograph view of crinkled laminae of of high algal marsh sediment, showing clotted fabric and algal filaments. *70*

45. A. Slab of wavy fenestral laminae draping *Scytonema* clots in sediment of the upper edge of high algal marsh. B. Draping of laminae over fenestral clots in sediment from the high algal marsh-beach-ridge washover backslope boundary. *72*

46. A. Cemented surface crust from levee backslope-high algal marsh boundary, showing lamination with columnar "palisade" structure. B. Lithified mushroom-shaped algal "head" with fibrous "palisade" structure from inland algal marsh. *73*

47. A. Bedding-plane view of "palisade"-structured crust, showing head- and sheet-like mounds with honeycomb surface. B. Bedding-plane view of "palisade"-structured crust after the uncemented sediment had been washed from between the cemented columns. *74*

48. Thin section photomicrograph of fibrous "palisade"-structured lamination. *75*

49. A. Slab showing fibrous "palisade"-structured domal lamination with lithified fibrous tubes piercing unlithified domal sediment laminae. B. Thin section photomicrograph of fibrous "palisade"-structured domal lamination shown in Fig. 49A. *76*

50. A. Slab of a cemented crust from high algal marsh-levee backslope boundary, showing fibrous "palisade"-structured domes overlying flat lamination with "palisade"-structure. B. Thin section photomicrograph of cemented crust shown in Fig. 50A. *77*

51. A. View of surface of living *Scytonema* algal mat, showing honey-

	comb-like structure of the filament tufts. B. Thin section photomicrograph of living *Scytonema* mat, showing the columnar structure of the algal tufts. *79*
52.	A. Slab showing typical thin bedding to thick lamination of the inland algal marsh sediment. B. Slab of layered inland algal marsh sediment, showing thin paper-crusts and intraclasts. *80*
53.	A. Thin section photomicrograph of inland algal marsh sediment, showing a cycle of well-sorted peloidal sediment, capped by paper-crust which in turn is overlain by algal tufa. B. Thin section photomicrograph of inland algal marsh sediment, showing the same cycle as in Fig. 53A, but with well-defined filament molds in the tufa. *81*
54.	A. Slab of sediment from the low algal marsh at a pond edge, showing disrupted fenestral bedding. B. Thin section photomicrograph of slab shown in Fig. 54A. *83*
55.	A. Flat-pebble gravel on the levee backslope, showing curled algal-bound sediment clasts derived from levee crest. B. Flat-pebble gravel from base of a small erosion ledge on beach terrace. *86*
56.	A. Flat-pebble gravel on the beach, showing large clasts made of exhumed cemented crust layers (*top & bottom*). B. Flat-pebble gravel made of cemented crust fragments on levee in Pumpion Cay area. *87*
57.	Flat-pebble gravel of curled, algal-bound clasts derived from the levee, collecting at pond edge. *89*
58.	A. Slab of unlayered sediment from 4 km offshore of Three Creeks. B. Thin section photomicrograph of unlayered sediment from the offshore. *91*
59.	A and B. Slabs of sediment of a pond. *92*
60.	Slab of beach-terrace sediment, showing pigment layers used to measure rate of deposition of sediment. *97*
61.	Graph of wind speed against storm surge generated by wind. *101*
62.	A. Slab from pond on Crane Key, Florida, showing hurricane sediment layers and algal layers. B. Mud-cracked hurricane layer exposed on Snake Hammock in the Pumpion Cay area. C. Accumulation of sediment in the lee of bushes on a levee at Three Creeks. *104–5*
63.	Patches of peloid sand filling depressions in levee crest after an onshore storm at Three Creeks. *107*
64.	Map of Three Creeks area showing distribution of layering in surface sediment. *118*
65.	Schematic diagram of cores, showing vertical distribution of layering at Three Creeks. *119*
66.	Schematic cross-section through Three Creeks tidal flats to show "laminite" cap. *120*
67.	Schematic drawing showing what a vertical core through the Andros tidal flats might look like at the completion of a regressive cycle. *122*
68.	Sketches of common pond organisms. *127*
69.	Graph of distribution of pond and levee organisms against Exposure Index. *128*
70.	Sketches of common levee and algal marsh organisms. *129*
71.	Diagrammatic illustration of the relationship between degree of exposure and salinity variation. *130*
72.	Diagrammatic illustration of the relationship between degree of exposure and temperature variation. *131*

73. Diagrammatic illustration of the relationship between degree of exposure and desiccation. *131*
74. A. Photomicrograph of sand fraction of pond sediment, showing smooth porcellaneous peloids. B. Photomicrograph of test of the foraminifer *Peneroplis proteus*. *133*
75. A. Compacted pond mud, showing faint outlines of pellets preserved near pores but destroyed away from pores. B. Thin section photomicrograph of peloidal sediment of an abandoned channel. *134*
76. Profile and abundance of cerithid gastropods across a pond-levee complex. *135*
77. Profile and abundance of cerithid gastropods across a pond-levee complex. *135*
78. A. Slab of channel sediment, showing cerithid and *Peneroplis* tests in pelleted muddy matrix. B. Burrows, mostly of polychaete worms, in mud of an intertidal channel bar. *137–38*
79. A. Thin section photomicrograph of endolithic algal borings in *Geloina* shell. B. Stages in the disintegration of cerithid shells by endolithic algal borings. *140*
80. A. Pellet-strewn surface of a subtidal pond. B. Slab of pond sediment, showing lack of layering and pelleted texture. *142*
81. Schematic cross-section through a pond, showing bioturbating animals and the depths of their effect. *144*
82. A. Beach-mound sediment with burrows of polychaete worms. B. Resin casts of polychaete burrows. *147–48*
83. A. *Apseudes* burrows in pond sediment. B. *Panopeus* burrow entrance in channel bank. *149*
84. A. Surface view of complex burrow entrance holes of *Alpheus*. B. Resin cast of *Alpheus* burrow. *150*
85. A. Earthworm (*Pontodrilus*) exposed within its burrow. B. Tunnels of an ant burrow in levee sediment. *152*
86. A. Territory of a fiddler crab (*Uca*). B. Resin cast of *Uca* and *Sesarma* burrows. *153*
87. A. Slab of algal marsh sediment near a pond edge, showing *Uca* burrows. B. Bonefish feeding pits on a beach. *154–55*
88. Map of Pumpion Cay area, showing distribution of cemented crusts. *161*
89. Schematic cross-section of a levee-channel complex, showing distribution of buried, exposed, and eroded cemented crusts. *163*
90. Schematic cross-section through a partly lithified algal tufa mound of the marsh zone of a levee-pond complex. *164*
91. S.E.M. photo of aragonite wall cement lining a vug in a cemented crust from Three Creeks. *166*
92. Thin section photomicrograph of a cemented crust from Three Creeks, showing aragonite cement lining floors of vugs. *166*
93. Plot of relative surface elevation of a levee-pond complex against maximum length of time the surface was submerged in seawater. *168*
94. A. Bedding-plane and vertical views of laminated algal mounds covering mud-cracks in the Ordovician St. Paul's Group of Western Maryland. B. Slab of Precambrian dolomite from Great Slave Lake, Canada, showing palisade structure. *176–77*

CONTRIBUTORS

Owen P. Bricker (Ph.D., Harvard), Geologist, Maryland Geological Survey, The Johns Hopkins University, Baltimore, Md. 21218.

Peter Garrett (Ph.D., Johns Hopkins), Research Associate, Dept. of Geology, University of California, Santa Barbara, California 93106.

Robert N. Ginsburg (Ph.D., Univ. of Chicago), Professor of Geology and Director of the Comparative Sedimentology Laboratory, Rosenstiel School of Marine and Atmospheric Science, University of Miami, Fisher Island, Miami Beach, Florida 33139.

Lawrence A. Hardie (Ph.D., Johns Hopkins), Professor of Geology, Dept. of Earth and Planetary Sciences, The Johns Hopkins University, Baltimore, Md. 21218.

Harold R. Wanless (Ph.D., Johns Hopkins), Assistant Professor, Rosenstiel School of Marine and Atmospheric Science, University of Miami, Virginia Key, Florida 33149.

ACKNOWLEDGMENTS

The study was supported by NSF Grant GA 1345, and we thank the Foundation for its generous aid. Some of the field observations, sediment cores, and rate-of-deposition experiment data were collected by our (then) graduate students, while Paul Hoffman initiated the first mapping project. To Paul, and the other students we offer our special thanks—without their willing help, advice, and companionship this study could never have been so thoroughly and happily carried out. Benjamin Lewis of Andros and Monty Dyer were excellent field assistants and good companions. Dr. Hans Eugster and Dr. Hans Füchtbauer read the manuscripts and made many valuable suggestions; we appreciate their advice and counsel. We would like to thank J. R. Clackson, chief meteorologist at Nassau Airport, for providing numerous weather records for the Bahama area, and also the RCA Service Company for providing weather and sea-state data gathered at the AUTEC station at Fresh Creek, Andros. Kathleen Shannon's skill and patience in typing the several drafts of the manuscript is saluted. Finally, we would like to express our appreciation to the Bahamas government for allowing us the opportunity to carry out our work on Andros.

SEDIMENTATION ON THE MODERN CARBONATE TIDAL FLATS OF NORTHWEST ANDROS ISLAND, BAHAMAS

1 INTRODUCTION

Lawrence A. Hardie

Our ability to recognize the general depositional environment of ancient carbonate rocks has advanced significantly through studies of modern carbonate deposits (e.g., the work of Black 1933; Cloud 1962; Illing et al. 1965; Logan et al. 1969; Logan et al. 1970; Newell et al. 1959; Purdy 1963; Shinn et al. 1969; among many others). The advance has been most spectacular in the shallowest environments, the tidal[1] flats and their lagoonal connections to the open sea. Tidal flats, being partly land and partly sea, are subject to the widest spectrum of environmental conditions, and their sediments consequently offer the richest variety of sedimentary and diagenetic features found in carbonate deposits, features such as flat lamination, undulating lamination, thin beds, stromatolites, mud-cracks, sheet-cracks, flat-pebble gravels, intraclastic sands, rippled and cross-bedded sands, pelleted lime mud, fenestral pores, burrows, evaporite mineral casts and molds, caliche crusts, root casts and molds. Analogous macro-structures, beautifully preserved, are found in carbonate rocks throughout the geologic column (Lucia 1972; Ginsburg 1975), but are especially characteristic of the great thicknesses (up to 5,000 meters) of carbonates that dominate the North American record from late Precambrian to middle Ordovician as well as the Permo-Triassic of the Alps and the southwestern U.S.A., so that it is with great confidence that we can identify these ancient carbonates as deposits of tidal flat-lagoon complexes.

The next step in our advance must be toward very refined reconstructions of the tidal flat environment—we want to be able to read with precision the climate, weather patterns, tidal regime, water circulation, physiography, water chemistry, etc. There is little doubt that there is a wealth of as yet unexploited environmental information stored in carbonate tidal deposits: a simple comparison of modern deposits shows that no two tidal flats are quite the same (compare, for example, Shinn et al. 1969; Logan et al. 1970; and Kendall & Skipwith 1969). And this is certainly true for ancient carbonates; there seem to be as many varieties of tidal flat as there are ancient tidal flat deposits (compare, for example, Fischer 1964; Bosellini 1967; Laporte 1967; Matter 1967; Roehl 1967; Schenk 1967; Bosellini & Hardie 1973). The implication is clear, tidal deposits are very sensitive to variations in environmental parameters. We need to know, therefore, which environmental parameters leave their record in tidal flat sediments and how precisely variations in these parameters are reflected in sedimentary and diagenetic features. This information would

---

1. Tidal is meant here in the broadest sense, that is, short-period fluctuations in water level due to either astronomical or meteorological causes, or both.

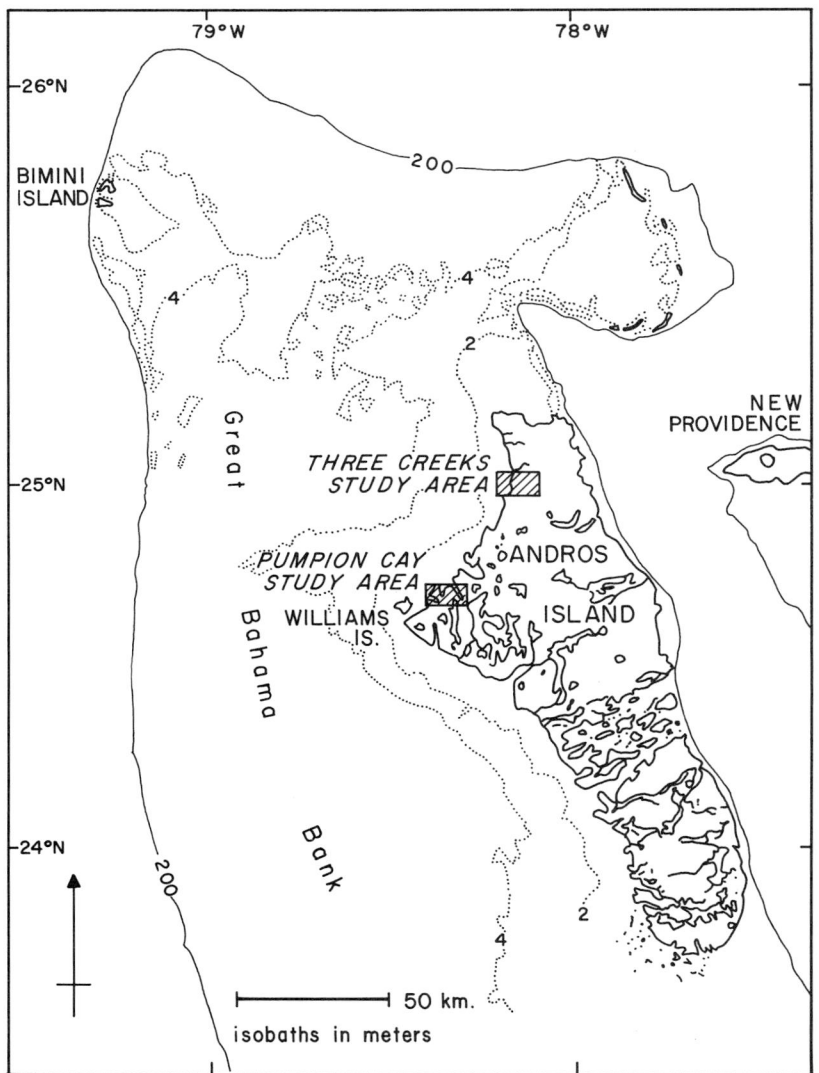

Fig. 1. Map of northern part of Great Bahama Banks, showing the location of the study areas of Three Creeks and Pumpion Cay on Andros Island.

allow us to make very precise interpretations of tidal flat deposits and to recognize the proper significance of the many different varieties.

As a small start in this direction we studied in detail a small part of the modern carbonate tidal flat belt of northwest Andros Island, Bahamas (Fig. 1). Our aim was to determine as precisely as possible what information on climate, weather, tides, water chemistry, and indigenous life is recorded in the sediments of this one particular kind of tropical carbonate tidal flat type—the low-energy, high-rainfall, nonevaporitic, totally marine type. This publication presents our modest progress toward this goal.

Shinn, Lloyd, and Ginsburg (1969) have worked out the major physiographic subdivisions of the northwest Andros tidal flats and have described the essential sedimentary features and stratigraphic relations. Their work has provided a solid foundation for us to build on. In our study we have placed special emphasis on understanding the origin of layering (Chapter 5) and on understanding animal-sediment relations (Chapter 6), but providing, at the same time, detailed descriptions of those dominant features that would help the geologist recognize analogous features in the geologic column. We have also examined the zonation of algae and their "instant" preservation in cemented surface crusts (Chapter 7). To provide a framework in which each of these particular studies can be properly viewed we have described near the

INTRODUCTION

beginning of this book (Chapter 4) the overall environmental setting, with particular emphasis on the descriptions of the surface features of the numerous subenvironments[2] that make up this tidal flat complex. To place each of the subenvironments and their features in a quantitative position within the tidal zone, we have introduced a new approach—the "Exposure Index" (Chapter 3). The Exposure Index is used extensively throughout the book where precision is required, but we have used the arbitrary terms "subtidal" (below mean low water), "intertidal" (between mean low water and mean high water), and "supratidal" (above mean high water) where a general sense of position is all that is needed. All the significant features that characterize the northwest Andros Island tidal flat complex and that have been discussed in detail in the foregoing chapters are brought together in summary form in one chapter (Chapter 8). Here we have tried to emphasize the relationship between environmental parameters and sedimentary and organic features and how this relationship puts its special stamp on the Andros Island deposit, making it different from other tidal flat deposits. Finally, we end with two chapters (9 and 10) that offer some miscellaneous speculations that we thought were worth while.

---

[2] "Subenvironment" is used here to categorize any part of the tidal flat that has a distinctive physiography and distinctive assemblage of surface features, both physical and organic. Such subenvironments should be of mappable size (see, for example, Figs. 10 and 21).

2 METHODS

Peter Garrett and Lawrence A. Hardie

1. General

We made many short visits to the field area during the period March 1968 to May 1971, each visit being from a few days to a month or more in length. Intervals between visits varied from one to five months, averaging three months. The first two visits were based on board a 65-foot boat which traversed the Great Bahama Bank from Miami via Bimini and was anchored several kilometers offshore. All other visits were based in a custom-built (!) palm-thatched "chickee" on stilts, which protected our researchers from floodwaters but not from violent weather. The chickee was sited on a levee crest at the mouth of Three Creeks, allowing easy access by two 14-foot outboard skiffs to a large area of the tidal flats.

2. Weather and Tide Records

Local weather was monitored for two and a half years by a battery-driven, portable, continuous weather recorder (Science Associates model 400B) which marked off wind speed and direction, temperature, and rainfall on a carbonized roll of graph paper. The record, though useful, had gaps in it due to the difficulty of keeping the battery contacts clean. However, supplementary weather data from the AUTEC station at Fresh Creek, Andros, 50 km to the SE, kindly provided by the RCA Service Company, and from Nassau Airport 80 km to the E, proved to be sufficiently similar to our own records so that gaps could be confidently filled.

For the tide data we set up two long-run (four months), constant-recording strip chart tide gauges (Stevens type A 35). These provided two and a half years of continuous records from two separate physiographic locations. Both were installed on tubular supports seated on firm bedrock, one in the south channel of Three Creeks near its mouth and the other in a small distributary channel where it terminates within a large pond (see Fig. 2 for locations).

3. Mapping

A base map was prepared by dye-line enlargement of vertical airphotos (Hunting Surveys nos. BAH 35–96 and BAH 48–64), which provided sufficient detail for location and sufficient contrast for the differentiation and mapping of many small physiographic divisions. Our mapping method was to traverse across the physiographic trend, noting characteristic associations of surface features and differentiating these as subenvironments on the map.

The maps presented in this book (for example, Fig. 10) are simplifications of the maps prepared in the field.

4. Levelling

In order to tie in the surface features, junctions between mapped units, and tidal levels as indicated on the guages, we established a levelled network of surface elevations, using a Dumpy level and a surveyors rod marked in feet

METHODS

Fig. 2. Map of Three Creeks area, showing local place names and locations of levelled profiles, pigment patches for rate of deposition experiments, tide gauges, and weather station.

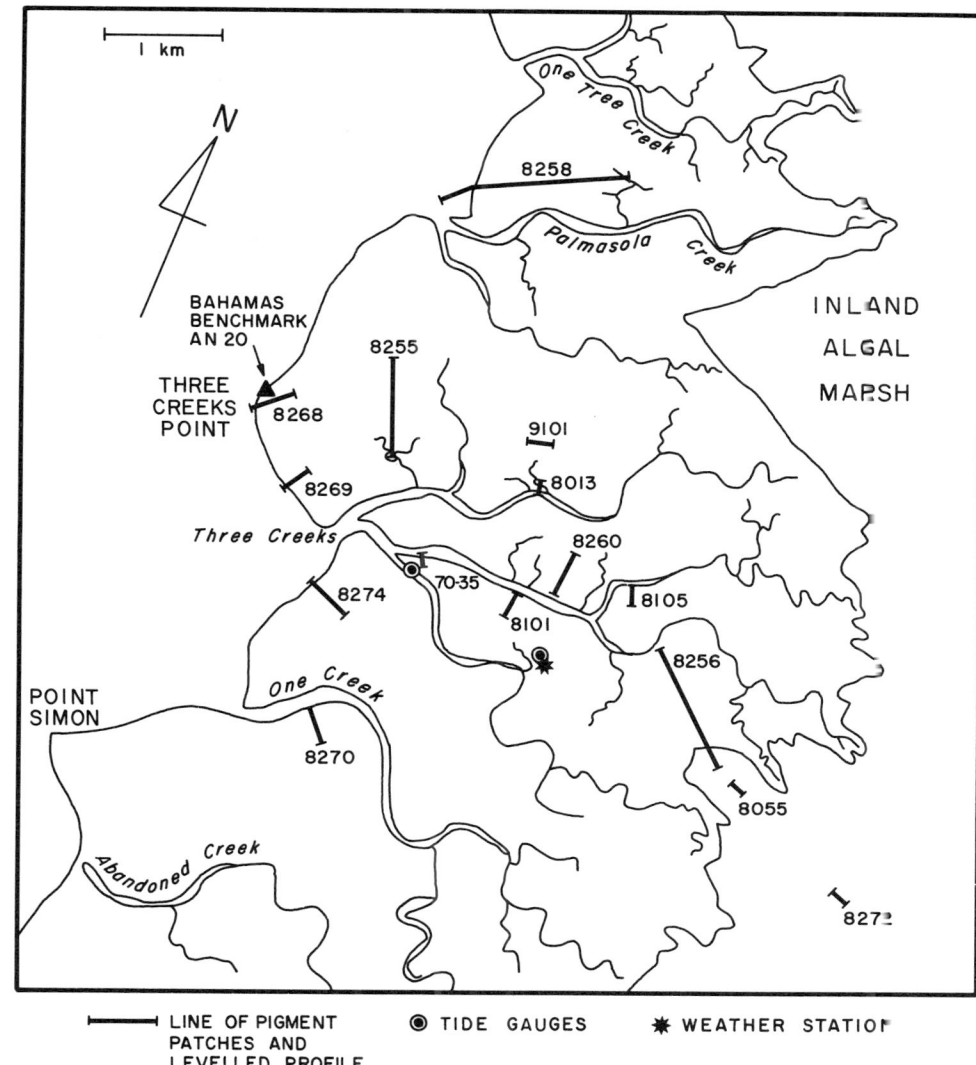

and tenths of feet. The network was tied in to benchmark AN20 of the Bahamas survey, located on Three Creeks Point (Fig. 2) and extended all the way to the pine forests on the Pleistocene bedrock of Andros. Our technique was to survey along closed traverses, with radial shots being taken around certain positions. Each position was reversed and the results of a traverse were discarded if the accumulated error was greater than 0.1 ft.

5. Biological Studies

We made as complete a collection of organisms as was possible, which we identified by standard manuals or with the help of taxonomists at the U.S. National Museum. We also observed the habits, distribution, and abundance of these organisms in the field.

Habits were determined by careful observation and by field experiments, as on the bioturbation rates of cerithid gastropods and certain polychaetes, these experiments being described in the text.

Distributions were mapped and tied in to the levelled network of surface features, so that upper and lower limits of distribution could be correlated with water levels.

Abundance was estimated from averaged counts of individuals per unit area (as for gastropods, burrow entrances of polychaetes, shrimps and crabs and certain plants, etc.) or from numbers of polychaetes or bivalves picked from a unit volume of sediment (the sediment being too cohesive to sieve effectively).

6. Water Sampling We collected water samples seasonally both from open water bodies and from interstitial pore waters, the latter being obtained in the field with a nitrogen squeezer apparatus.

Alkalinity and pH were measured in the field, samples then being stored for later chemical analysis, and in the case of open water samples, weight determination of suspended sediment content was made by millepore filtering.

7. Sediment Cores We collected over 100 cores, in three-inch aluminum pipes or in oblong one-gallon gasoline cans. Some of these were collected specifically to determine the stratigraphy of the sediments, but many were short, intended only for correlation with the causative processes. Most cores were retrieved simply by pushing down the core barrel, capping it, and pulling it up, but longer cores required the method and equipment of Ginsburg and Lloyd (1956).

All cores were transported wet to Baltimore, where they were extruded, dried in a low temperature oven, impregnated with clear polyester resin (Ginsburg et al. 1966), slabbed, and mounted on boards for observation and description.

3 EXPOSURE INDEX: A QUANTITATIVE APPROACH TO DEFINING POSITION WITHIN THE TIDAL ZONE

Robert N. Ginsburg, Lawrence A. Hardie, Owen P. Bricker,
Peter Garrett, and Harold R. Wanless

1. Introduction

The criteria for identifying ancient tidal flat deposits—sedimentary structures, textures, fossils—come from the study of modern active tidal flats (Ginsburg 1975; Lucia 1972). Implicit in such interpretations is the recognition that individual sedimentary and organic features are restricted to specific zones within the tidal flat (Lucia 1972). These zones are determined largely by fluctuations in water level, due to either astronomic or meteorologic causes, tides in the most general sense. Accordingly, workers studying tidal flat deposits, both modern and ancient, describe and interpret sedimentary and organic features as subtidal, intertidal, or supratidal (Ginsburg 1975; Lucia 1972). This three-fold subdivision has become essential to the recognition and interpretation of ancient tidal flat deposits for three reasons: (1) recognizing one zone provides a reference point for interpreting overlying or underlying zones; (2) establishing a sequence of successive zones makes identification of the tidal flat environment more reliable, and it facilitates comparison of different deposits; (3) the succession of zones can be used to analyze the balance between relative rates of subsidence and sediment supply and to establish transgressive or regressive cycles of deposition.

A basic drawback of this three-fold subdivision of tidal flats is that there are substantial differences in the definition of the three zones as used by different workers. For example, some workers (Lucia 1972; Davies 1970a) use intertidal for the zone between mean low water and mean high water, some as the zone between normal low and normal high tide, without specifying what is meant by normal (Shinn et al. 1965); others use "extreme low water" to high water springs (Woods & Brown 1975); at least one worker (Fischer 1964, p. 131) uses intertidal "not in the sense of daily tides, but for the bottoms exposed and covered by the tidal extremes during the year." These differences in definition can produce apparent contradiction in interpretation, as, for example, when one worker describes algal stromatolites as intertidal and another classifies them as supratidal. Ecologists working in the tidal zone have expressed a similar dissatisfaction with arbitrary subdivisions like "lower littoral," noting that they have "become ambiguous by reason of ill and dissimilar definition" (Doty 1946, p. 316). A further complication is that on some modern tidal flats there are strong seasonal effects, for example, in the Bahamas the pond subenvironment that would be classified as subtidal during the wet, rainy summer months must be classed as intertidal over the long, dry winter (Hardie & Garrett, Chapter 4).

To avoid the existing confusion in the three-fold subdivision of tidal zones, with its inherent limitations, and to provide a more precise means of calibration of the sedimentary and organic features of tidal flat deposits, we propose

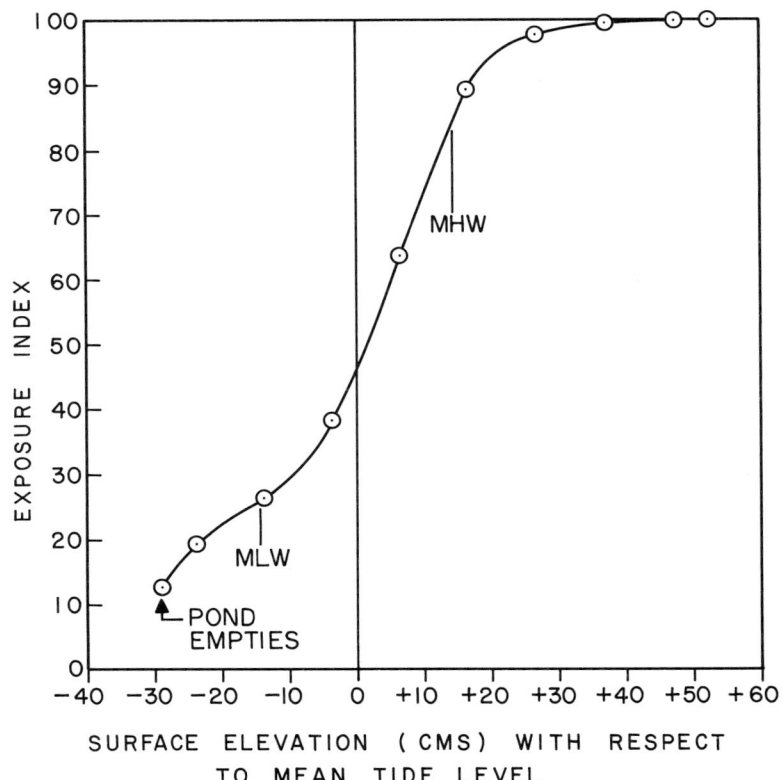

Fig. 3. Mean Exposure Index as a function of relative surface elevation for levee-pond subenvironment complex at Three Creeks. MLW = mean low water, MHW = mean high water.

a quite different approach, the "Exposure Index." A preliminary report on this "Exposure Index" was published by Ginsburg et al. (1970).

2. Exposure Index

Our approach follows the method of ecologists studying the distribution of intertidal plants and animals. This method is to determine the duration of exposure or submergence at successive elevations directly from tide guage records or from curves constructed from tide tables (Doty 1946). The results are presented either as a graph of total time or percent of time exposed or submerged plotted against elevation, or as the maximum or minimum period of a single exposure or submergence at successive elevations (see, for example, Doty 1946, Fig. 4). This kind of analysis has shown that changes in the distribution of tidal organisms on rocky shores can be closely correlated with breaks (termed critical tide factors by Doty 1946) in the length of submergence or exposure.

To get the necessary records of water-level fluctuations, we set up tide gauges in the study area. We began with several short-term recorders, but maintaining them proved so difficult that we replaced them with long-term recorders. These float-type, clock-driven recorders (Chapter 2) are sensitive to changes in water level of as little as 1 cm, and they run without rewinding for up to four months. One recorder was positioned at the mouth of the main channel in the Three Creeks area, the other in one of the nearby ponds (Fig. 2). These two recorders were maintained without significant interruption for over two years. From the continuous strip records of these two gauges, the duration of exposure was determined at vertical elevation intervals of 5 to 10 cm. An example of the results, converted to percentage of time exposed, is shown in Figure 3. The smooth curve fitted to these points defines what we term here the Exposure Index of all points within the range of tides. The Exposure Index of any one point on the tidal flat surface is simply the mean annual percentage of time that point is exposed. We have chosen exposure rather than submergence as an index because so many of the easily recognized and diagnostic sedimentary features are formed by exposure.

Now, such curves as that shown in Figure 3 will be different, both as regards shape and absolute elevation, from one tidal flat to another and even from one subenvironment to another within the same tidal flat. However, the major advantage of using a measure such as the Exposure Index is that it is a *relative but quantitative* measure that will allow precise comparison of the tide factors from one area to another, no matter what the topographic or hydrologic differences; there are no ambiguities with regard to position within the particular tide zone under consideration.

3. Elevations of Sedimentary and Organic Features: A Test of the Usefulness of the Exposure Index

We determined the elevations of most of the sedimentary and organic features in the vicinity of the two tide gauges using a Dumpy level. The precision of this instrument is ± 2 mm and reproducibility of individual shots by a single operator was found to be within 3 mm. Each shot with the level was reversed to ensure an accurate determination, and all traverses were closed to determine cumulative errors. The combined effect of all errors gave the determinations an overall precision of within 3 cm.

Figure 4 shows the positions of the various sedimentary and organic features with respect to the Exposure Index, as determined by correlating tide gauge levels with tidal flat surface elevations, using the Dumpy level. These sedimentary and organic features are described and illustrated in the chapters to follow, and their origin and significance will not be discussed here. The point that is to be made here is that the Exposure Index plot in Figure 4 brings out very forcefully the narrow ranges occupied by the different sedimentary structures and organic features within the overall tidal zone. Clearly, the Exposure Index is a sensitive indicator of the relationship between the tide factors and the sedimentary record, an indicator that cannot be matched for accuracy and flexibility by the arbitrary three-fold subdivisions currently in use. Of course, we recognize that other factors than duration of exposure and submergence influence the sedimentary record: climate, weather, physiography, rate and amount of sediment supply, the kind, variety, and abundance of organisms, taken individually and in concert, determine which kinds of sedimentary and organic features are developed in a given tidal flat. Nevertheless, the Exposure Index provides a useful way of precisely "calibrating" the existing record of modern tidal flat deposits.

4. Discussion: The Need for More Studies

The curve of the Exposure Index shown in Figure 3 is an overall average based on twenty-two months of continuous records of tidal level and is therefore considered quite reliable. However, it is important to note that there are strong seasonal effects on Andros Island, the summer months May through October are very rainy, while the winter months are very dry. Therefore, short-term measurements can be very misleading as the sequence of three-month period curves shown in Figure 5 demonstrates. For studies of the Exposure Index in other localities, we suggest that at least twelve months of record are needed to establish a reliable mean curve. In areas where seasonal effects are minimal then something like four one-month records should be made over a twelve-month period. We would also like to point out the danger of taking tide-table data from a nearby open water area and applying it to a tidal flat; the influence of the topography and weather (particularly wind and rain) on the tidal regimen in the shallow and constricted waters of tidal flats is so strong that no extrapolations are meaningful. It is essential to measure the tidal fluctuations *in situ*.

The approach we have presented here very much needs testing in other modern tidal flat areas. Such additional studies will show whether similar features in different settings respond to tide factors (exposure and submer-

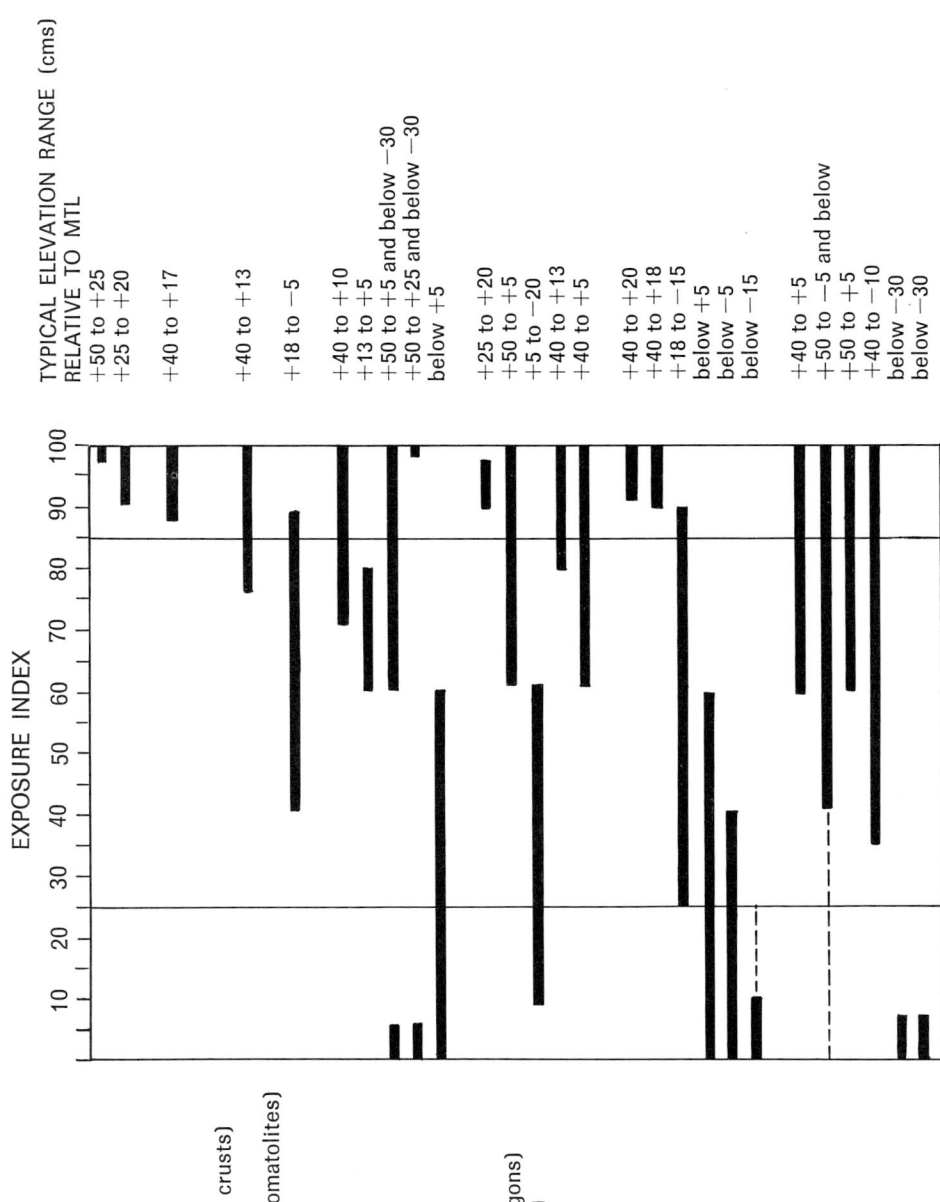

Fig. 4. Relationship between some sedimentary and organic features of the Three Creeks tidal flat and Exposure Index range of the surface environment in which the features, or the sediment containing them, occur. *Solid lines* indicate measured range, *dashed lines* indicate probable ranges (not measured but roughly estimated). Range of relative elevations are rounded values based on Dumpy level measurements.

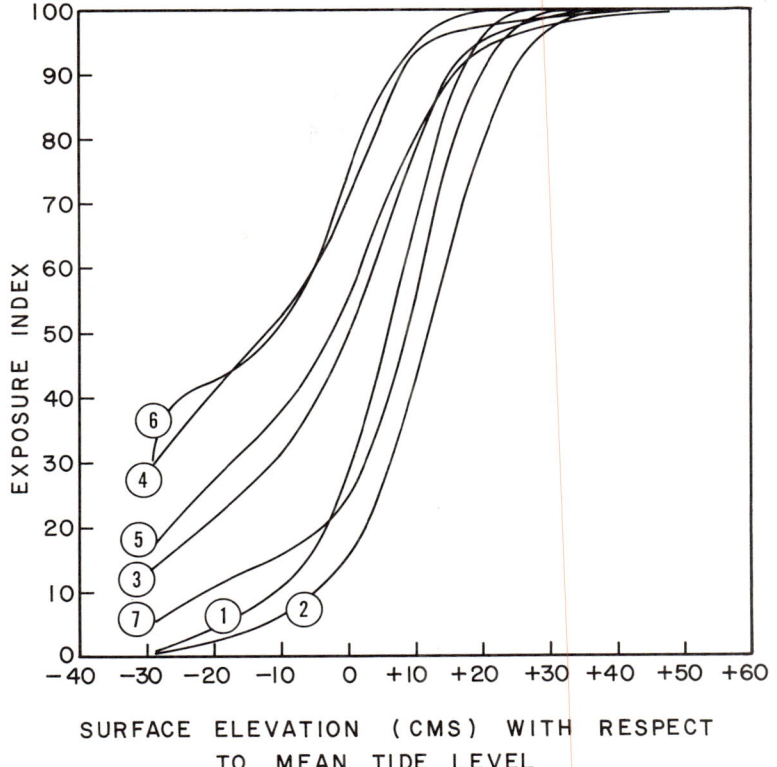

Fig. 5. Curves showing seasonal variation of Exposure Index for levee-pond surface subenvironments at Three Creeks: (1) June–Aug. 1969, (2) Aug.–Oct. 1969, (3) Oct.–Dec. 1969, (4) Jan.–Mar. 1970, (5) Mar.–Apr. 1970, (6) Apr.–June 1970, (7) June–Sept. 1970.

gence patterns) in the same way or not. If they do, then we would have a very sensitive and unambiguous calibration of the position of diagnostic sedimentary and organic features in the tidal zone, and this would open the way to precise reconstructions of ancient tidal flat environments.

We are not suggesting that the terms "subtidal," "intertidal," and "supratidal" be abandoned—they are useful as general descriptive terms when explicitly defined—but we are recommending for quantitative calibration of the sedimentary and organic record that a more sensitive measure, such as the Exposure Index, be used.

4 GENERAL ENVIRONMENTAL SETTING

Lawence A. Hardie and Peter Garrett

1. Introduction

This chapter is a short description of the northwest Andros Island tidal flat complex, its physiographic setting, climate and weather, tides, and surface features of the significant depositional environments.

We have tried here, as briefly as we can, to sketch a picture of the essential features of the environment that will provide for the reader both a clear "roadmap" and a working foundation for the several particular in-depth studies we have made of this modern carbonate tidal flat deposit.

2. Location

The main study area is located in the extensive carbonate tidal flat belt on the west side of Andros Island, Bahamas (Fig. 1). On navigational charts (e.g., U.S. Naval Oceanographic Office H. O. 5991) the area is identified as Point Simon (25°00′N, 78°10′W), on the northwest coast of Andros; locally it is known as "Three Creeks." Figure 2 gives the local place names and the locations of sampling sites, levelled profiles, and instrument sites referred to throughout this book. We made a less intensive study of another area, the Pumpion Cay area, near Williams Island (Fig. 1). Unless otherwise noted the descriptions below refer to the Three Creeks area.

We chose the Three Creeks area for study because the basic physiographic and sedimentary elements had already been worked out and described by Shinn, Lloyd, and Ginsburg (1969)[1] and so provided a well-documented deposit on which to try our approach.

3. Sediments, Age and Stage of Development

The sediment of the Andros tidal flat belt is mainly pelleted carbonate mud, consisting of aragonite aggregated into silt and sand-sized peloids which show varying degrees of induration. Skeletal material, mostly foraminiferal and molluscan, makes up a variable but small proportion of the sediment, usually less than 10%.

The peloids (described and discussed in some detail by Purdy 1963; Cloud 1962; Garrett 1971; and in Chapter 6) occur in a spectrum of induration states in which the two end members are those peloids which appear light colored in transmitted light and are soft enough to squeeze together on compaction, and those which appear dark grey-brown in transmitted light and retain their shape and identity even when surrounded by a compacted matrix. Both soft and indurated peloids are ellipsoidal or rounded-irregular in shape and may contain smaller peloids or recognizable skeletal material. They range in size from 30 to 500 μ.

---

1. The area described by Shinn *et al.* (1969) as Loggerhead Point (see their Fig. 2) is actually Point Simon at Three Creeks. Loggerhead Point lies some 9 km to the north.

GENERAL ENVIRONMENTAL SETTING

Offshore the sediment is sandy, with mostly indurated and semi-indurated peloids in the medium and fine sand fractions and less than 5% skeletal material. On the tidal flats, the peloids, which we believe to have been stirred up offshore and washed onshore by storms (Chapter 5), are finer, rarely exceeding 200 μ in diameter, except where clearly pelleted by tidal flat organisms. The indurated peloids occur either concentrated into fine sand laminae and thin beds or thoroughly mixed with a majority of soft peloids in a bioturbated sediment. When compacted, the soft peloids of this sediment squeeze together, lose their identity, and become matrix in which the indurated peloids and skeletal material "floats." Skeletal material, consisting mostly of small and large (5 mm diameter) foraminifera and whole and broken molluscs, is usually most abundant in its life-environments. However, the small forams, and other skeletal fragments, hydrodynamic equivalents of small peloids, are transported, sorted and deposited with the peloids. In a few places molluscs and larger forams are concentrated into coarse lag deposits. In some special situations, blue-green algae induce precipitation of small (5 μ) crystals of Mg-calcite around their filaments to produce an *in situ* algal dust sediment (see Chapter 7).

This pelleted lime mud sediment, with a maximum thickness of 3 m, forms a wedge that has accreted onto the west side of the island in the lee of the prevailing easterly tradewinds. It overlies unconformably a hard, lithified Pleistocene peloid and ooid limestone basement which, in the Three Creeks area, has an irregular but subdued karst surface. Freshwater peat from the base of the soft sediment wedge 24 km west of Three Creeks at a depth of 2.7 m below sea level gave a radiocarbon age of 4890 ± 200 years BP (Traverse & Ginsburg 1966, p. 422). Accumulation has occurred, therefore, during the latest Holocene rise of sea level.

Quite different stages of development occur in different parts of the tidal flat belt. To the south of Williams Island (Fig. 1) the tidal flats are apparently in a regressive stage that has produced a broad exposed coastal plain (4 to 8 km wide) with few or no active tidal channels, but with the scars of abandoned ones (Shinn et al. 1969, Fig. 33). The regression has left a complex of shallow ponds and lakes stranded inland. To the north of Williams Island the flats are undergoing a stage of trangression and are dissected by a very active system of branching tidal channels. Our study locality at Three Creeks is a typical example of this channeled area (Figs. 6 & 7). Connecting these two areas is yet a third tidal flat type. Here, directly east of Williams Island, the development of the tidal flats is strongly controlled by the Pleistocene bedrock topography. Sediment is collecting in the lows among long curving or short squat ridges of Pleistocene rock (Fig. 8). These "island" ridges are the palm hammocks described by Shinn et al. (1965, Figs. 2 & 3).[2] A few branching channel systems are present, but are not well developed. These narrow, shallow channels connect to wide, landward-tapering waterways that cut across and between the Pleistocene islands to end in wide shallow ponds (Shinn et al. 1969, Fig. 33, upper right hand corner).

Of these three tidal flats the northwest area is the most active area and therefore best suited for our approach. For stratigraphic studies the regressing southwest section below Williams Island should be most revealing. The area is presently under investigation by Conrad Gebelein of the Bermuda Biological Station.

4. General Physiographic Setting

Shinn et al. (1969) recognized three main geomorphic belts in the Three Creeks area: (1) on the landward side, a broad 4 to 8 km-wide almost feature-

---

2. It is worth pointing out here that not all the palm hammocks are modern sediment build-ups as implied by Shinn et al. (1965) in their Figure 4, but instead many are Pleistocene rock highs with a sediment fringe, as we show in our Figure 8.

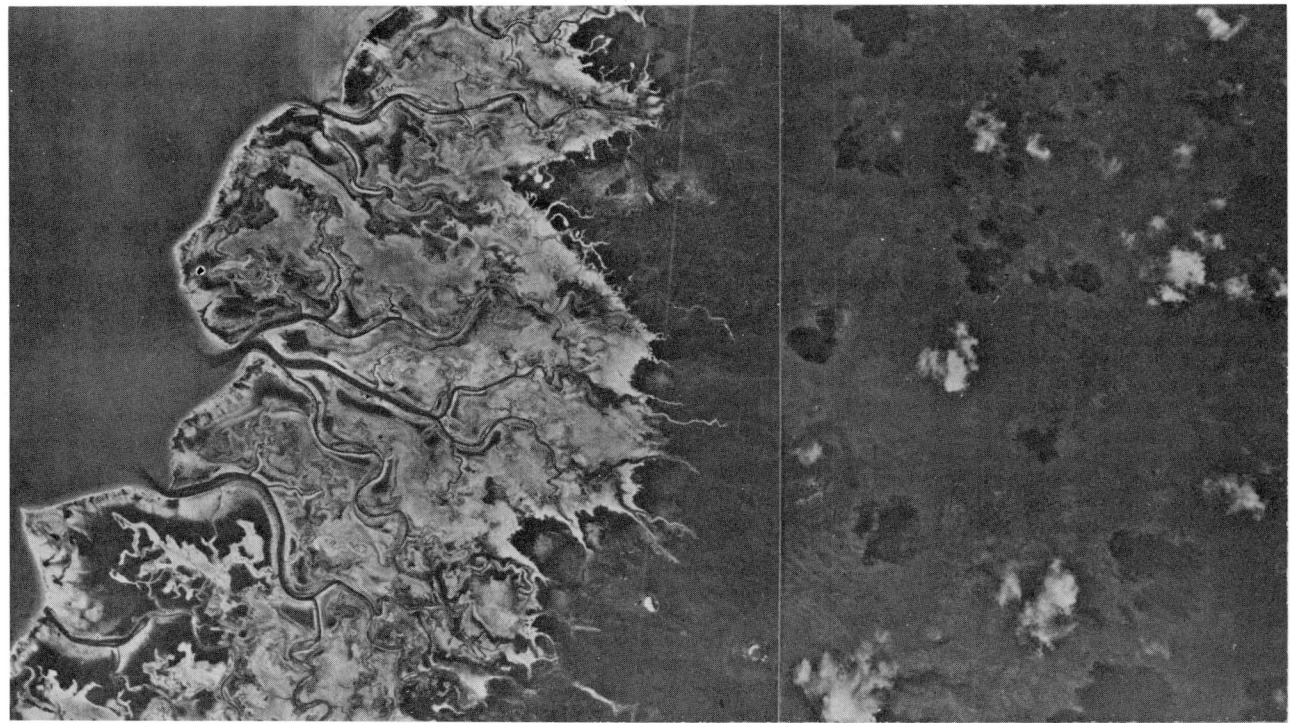

Fig. 6A. Aerial photograph of Three Creeks area. From left to right: offshore, channeled belt, inland algal marsh. Note sharp boundary between channeled belt ponds and inland marsh. See Fig. 6B below for details of channeled belt physiography.

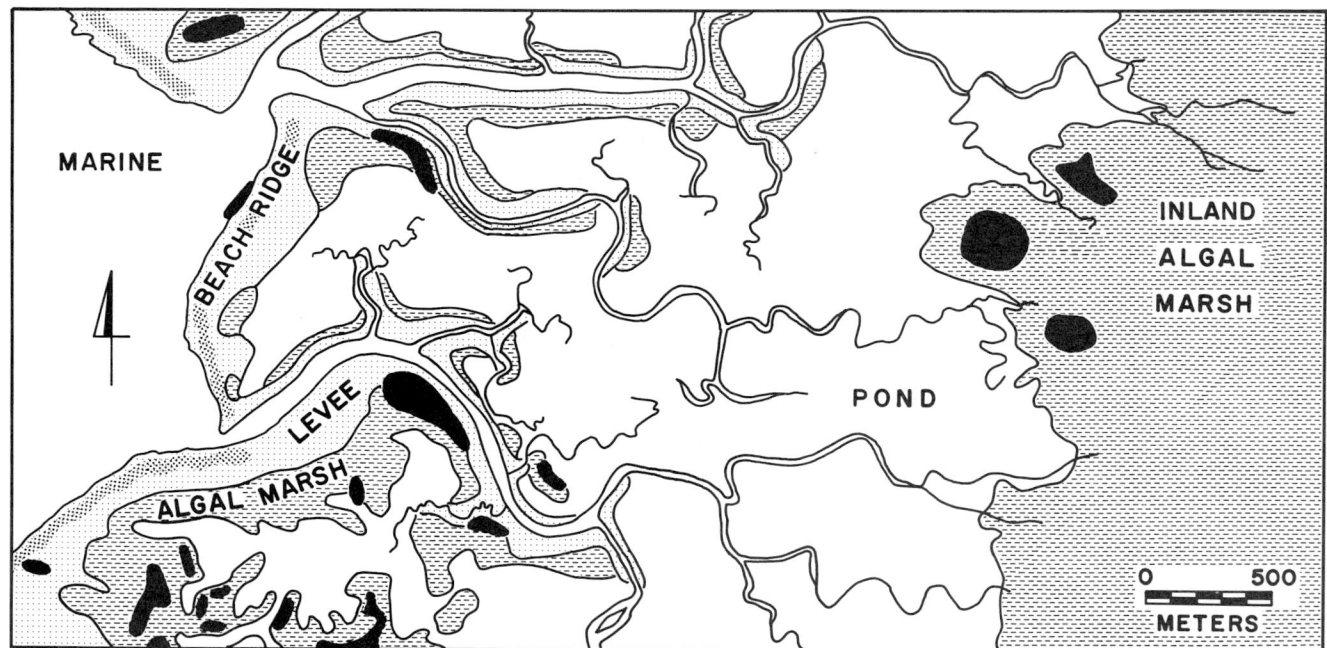

Fig. 6B. Map showing the complex of environments at Three Creeks; compare with aerial photo above. *Dark patches* are areas where cemented surface crust is found.

less freshwater algal marsh which is essentially a meadow of a single blue-green alga *Scytonema*; (2) a central channeled belt 3 to 4 km wide, which is a complex of channels, bars, levees, and shallow ponds fringed along their shores by *Scytonema* algal marshes; (3) a nearshore marine belt, the inner part of the

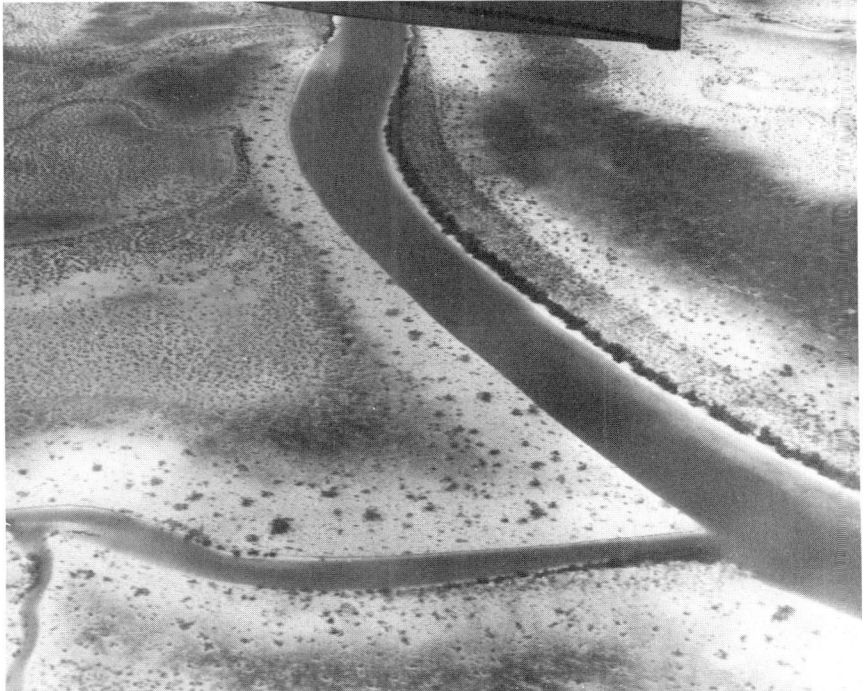

Fig. 7. Two oblique aerial views of parts of Three Creeks tidal flats. *Upper photo:* offshore, beach-ridge, channeled-belt complex. Light areas along channel edges are levees. Dark patches are *Scytonema* algal marshes at edges of ponds, which are seen as the light areas between the channels. *Lower photo:* channel-levee-pond zonation. Very light-colored zones along channel margins are levees patchily covered with black mangroves. Dark areas are *Scytonema* algal marshes. Light areas away from channels are ponds dotted with red mangroves. Note the red mangrove thicket along point bar at inside of channel bend.

shallow Great Bahama Bank. The areal distribution of these physiographic environments and their subdivisions is shown in Figure 6B and can be clearly seen in the aerial views in Figures 6A and 7. More detailed descriptions of these different environments are given below.

It is worth pointing out here that with the exception of the main channels, which may be as deep as 3 m near their mouths, the Three Creeks tidal flat is indeed flat, relief being measured in centimeters. This is illustrated in Figures 9A and 9B. In spite of the small elevation differences across the flats, the sedimentary structures from one subenvironment to another are strikingly different (Shinn et al. 1969; Ginsburg et al. 1970; Ginsburg & Hardie 1975). This sensitivity of the sedimentary record to exceptionally small elevation differences is one of the outstanding features of tidal flat sedimentation and will be discussed later.

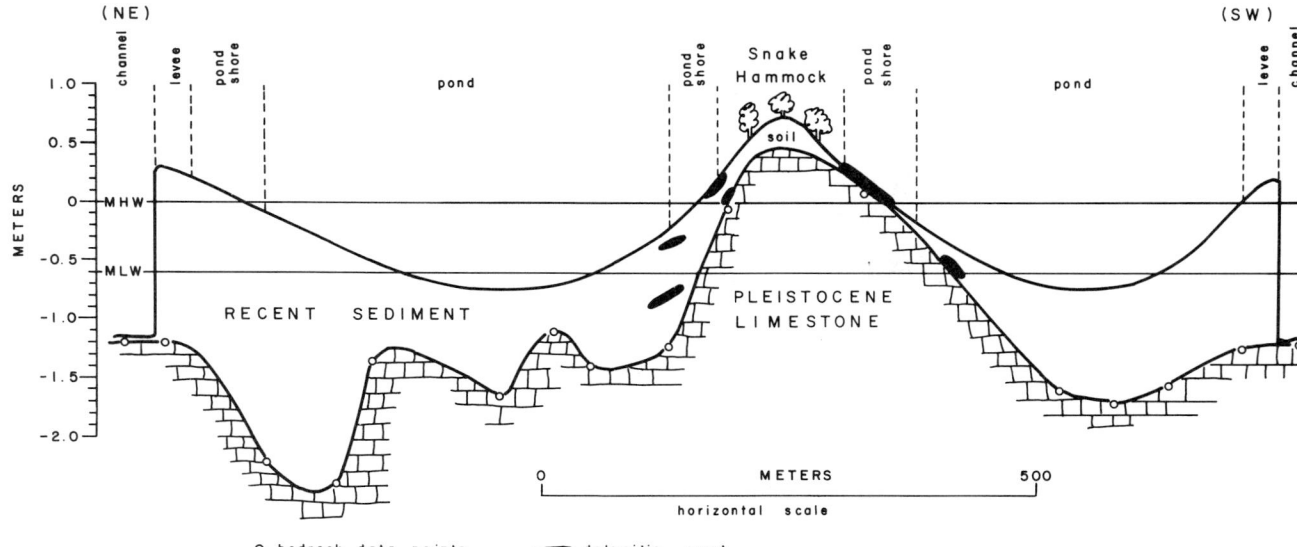

Fig. 8. Cross-section through "palm hammock" at Pumpion Cay just north of Williams Island to show the uneven Pleistocene basement and the rock core to the hammock. See Fig. 88 for precise location of the section.

5. Climate and Weather

Andros Island lies just north of the Tropic of Cancer, is of low relief and completely surrounded by ocean water. The climate is, therefore, typically tropical-maritime with long, warm showery summers and mild, dry winters. Temperatures are very uniform throughout the year (Table 1). *In contrast to the arid and semi-arid conditions of the carbonate tidal flat areas of the Persian Gulf and Western Australia, Andros Island is wet and rainy.* Annual rainfall averages about 129 cm a year and can be as high as 232 cm or as low as

Table 1. Weather data for Nassau, Bahamas (25°05'N, 77°21'W, elev. 4m; data from U.S. Naval Oceanographic Office Publ. H.O. 128)

Weather element	Value	Years of record
Temperature (°C)		
Annual average	25	42
Extreme high	35	25
Extreme low	6	25
Humidity (%)*		
Annual average	77	20
Rainfall (cm)		
Annual average	128.8	59
Extreme high	232.4	70
Extreme low	64.9	70
Wind direction (% time, annual average)*		10
N	9	
NE	23	
E	26	
SE	19	
S	7	
SW	3	
W	2	
NW	5	
Calm	6	
Wind Speed (annual average)		10
m/sec	3	
knots	6	

* Readings made at 1500 hours local standard time.

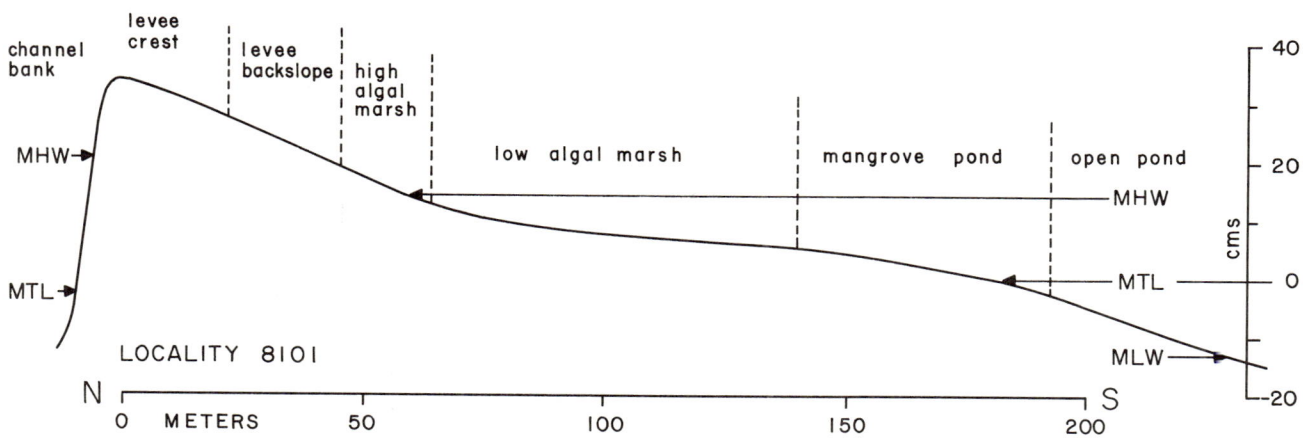

Fig. 9A. Carefully levelled profile across a levee from channel to pond (locality 8101, see Fig. 2), showing how the sequence of subenvironments or zones varies with elevation and horizontal distance. Note the differences between tide levels in channel and pond

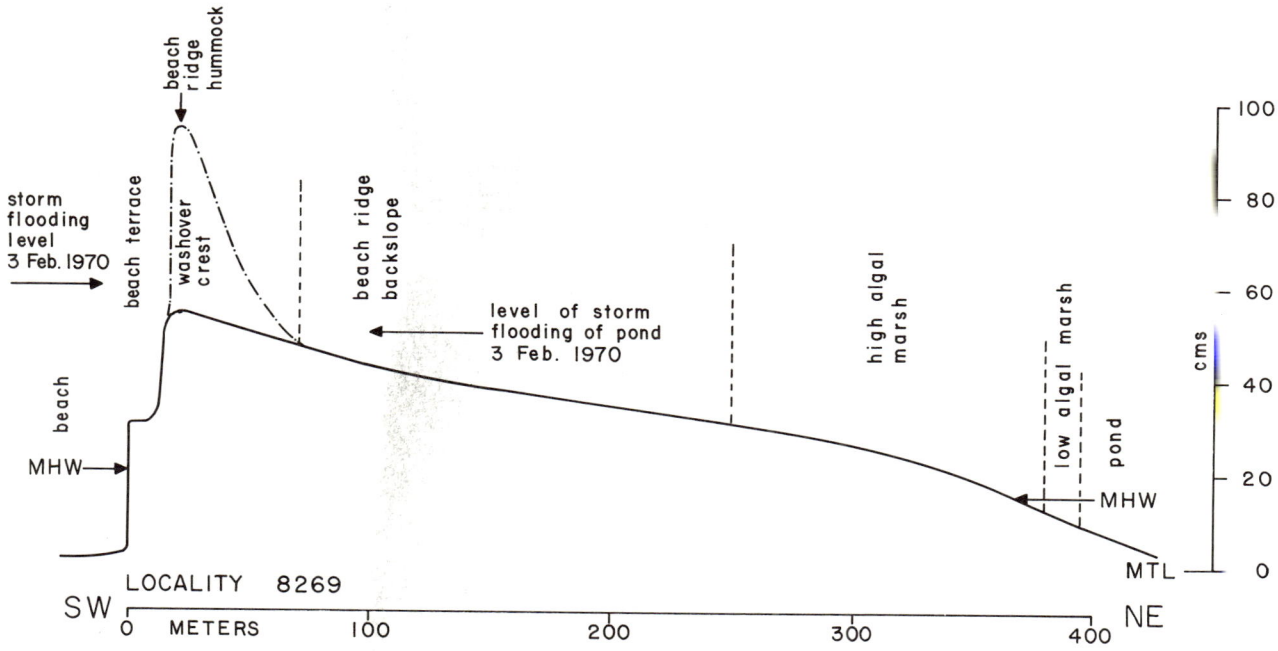

Fig. 9B. Carefully levelled profile across a beach ridge from beach to pond (locality 8269, see Fig. 2) showing the vertical distribution of subenvironments at the shoreline.

65 cm (Table 1). About three-quarters of this rain falls in short, heavy squalls during the summer months of May to October. This rainy period is also the season of low barometric pressures generated by the northward expansion of the Azores–Bermuda High (Dunn & Miller 1960, pp. 16–18). In mid-summer this anticyclone lies at about latitude 35°N, producing, through its clockwise flow of air, persistent southeasterly to northeasterly tradewinds that average between 2 m/sec (4 knots) and 4 m/sec (8 knots). The dry winter months usually bring slightly stronger but more variable tradewinds and higher pressures, as the Azores–Bermuda High moves south to about 30°N. At this time of the year southward-pushing cold fronts from the United States mainland penetrate into the northern Bahamas, perhaps as often as once a week. In a few of these, the pressure gradient is large enough to develop a "norther"; the cold front is then accompanied by squall lines and heavy winds. After the front passes, the weather clears and strong (8 to 10 m/sec or more) westerly,

Table 2. Range and height of tides on Three Creeks tidal flats

	Tide range (cm)			Tidal height (overall MTL = 0) (cm)	
	Mean	Spring	Neap	MHW	MLW
POND					
Overall: 24 June 1969– 11 May 1971	28.7	40.6	16.8	+14.3	−14.3
Seasonal:					
Jun 24–Aug 9, 1969	27.3			+16.0	−11.3
Aug 10–Oct 11, 1969	23.1			+20.6	− 2.5
Oct 12–Dec 31, 1969	27.9			+12.4	−15.5
Jan 1–Mar 3, 1970	30.2			+ 3.3	−26.9
Mar 4–Jun 15, 1970	32.8			+ 2.8	−30.0
Jun 16–Sep 12, 1970	27.4			+17.0	−10.4
CHANNEL (near mouth)					
Overall: 4 Jan 1969– 11 May 1971	46.0				
CHANNEL (about 3 km upstream)					
June 1968	18.0				

northwesterly and finally northerly winds blow steadily for a day or two. As we show in Chapter 5, these winter storms are truly significant depositional events on the tidal flats.

The Bahamas lie in the main track of West Indian tropical cyclones, and hardly a year passes without one northwest-heading cyclonic storm affecting some part of the Bahama area. The hurricane season is June to October, when over 90% of all known tropical cyclones have occurred. September is the peak month. In the years 1900 to 1937, thirty-two tropical cyclones passed within 100 km of Andros Island; eleven of these passed directly over the island (from maps in Tannehill 1938).

6. Tides

The "tides"[3] near the mouth of one of the main channels and in one of the ponds (Fig. 2) have been continuously monitored over a period of 2½ years. We found the tide to be semi-diurnal, each day two high waters and two low waters occur in approximately 12 hr 23 min cycles. Mean tide range is 46 cm at the channel mouth and 29 cm in the pond (Table 2); spring and neap tides change these values by about +10 cm and −10 cm respectively (Table 2). There is a marked dampening of the tide wave as it moves up the channels: 2 to 3 km upstream the mean tide range is only about 18 cm, and this decreases to near zero a further 1 to 2 km up-channel at the seawardmost edge of the inland marsh where the channels terminate.

Tidal height, or water level, determines which subenvironment will be flooded or exposed. Relative heights reached by the water surface depend on both astronomical and meteorological (mainly wind, rain, and air pressure) factors. Normally, spring and Perigean tides will produce both the highest and lowest levels, however, the day-to-day tide pattern on the tidal flats is so strongly influenced by the day-to-day weather patterns that water level (and hence flooding) is not at all predictable from the astronomical tide factors. For example, steady easterly winds blowing against the incoming tide will keep water level,

---

3. By tides we here mean "all fluctuations in water level." This is necessary because on shallow tidal flats water level changes significantly with the weather, mainly wind and rain, as explained in the text.

GENERAL ENVIRONMENTAL SETTING

even at spring tide, well below normal for as long as the wind stress is maintained. Prolonged strong easterlies, especially if coupled with a barometric high, will keep the ponds dry for days at a time. This is most frequent in the dry winter months when the Azores–Bermuda High moves south (see above). Onshore winds (westerlies) and low pressures have the opposite effect, raising water level and keeping the ponds from draining at the turn of the tide. The highest water occurs when storms generate very strong onshore winds. During these windy spells most of the tidal flats may be under water. The longest single period of overall immersion of levees and ponds that we have recorded in the three-year period 1968–71 is 16½ hours; water depth over the highest levee in the channeled belt reached 23 cm.

Rainfall has an enormous effect on the flooding pattern of the shallow tidal flat belt. In summer, when as much as 168 cm of rain may fall in five months (as recorded on Andros in 1969), the water levels are generally very high and the ponds and inland algal marsh are seldom completely drained. Runoff and groundwater from the pinelands to the east ensure that the inland marsh stays wet with freshwater through most of these showery summer months. Prolonged periods of heavy rain, when a low-pressure cell hovers over the island, will cause water level in the channels to rise to brimfull at high tide and a film of water may lap over the levees. During these periods the water across the entire channeled belt will be "brackish," that is, seawater diluted to a greater or lesser extent by rainwater. The rainfall effect is so seasonal that a subenvironment like the ponds will appear to be "subtidal" in the summer, but in the winter must be classed as "intertidal" (see seasonal variation in tidal height in Table 2). This, coupled with the daily influences of wind, makes the tidal-zone classification of subenvironments on the Andros tidal flats misleading. Instead, Ginsburg et al. (1970; and Chapter 3) have suggested using the "Exposure Index" as a measure. We have followed this procedure; the method is outlined in Chapter 3.

Tidal currents in the channels were estimated by timing the movement of floats to be in the order of 25 cm/sec at peak flood. In some of the small, shallow sinuous channels the bottom currents are strong enough to move small (about 1 cm long) gastropod tests into ripples with 10-cm avalanche faces. In the offshore area, surface currents ranging from 7 to 25 cm/sec were measured over several tide cycles. In the ponds the tidal currents were not measured, but were observed to be very sluggish.

Because of the shallow shelving approach to the west Andros shoreline, waves are not normally a significant factor at Three Creeks. During normal day-to-day conditions of weather and tide, waves along the shoreline are so small as to be hardly noticeable. However, during severe onshore storms, waves a meter high can be generated across the 100-km fetch of the Great Bahama Banks, in spite of its shallowness.

7. Salinity and Water Composition

Chemical analyses of the waters of the Three Creeks tidal flat over two seasonal cycles have been carried out by Owen P. Bricker and will be published later. Here we will offer only a very brief outline. Salinities of the tidewaters on the flats are normally about 39–42 ‰: they may reach 47 ‰ during the dry winter months, but may be as low as 5 ‰ during a prolonged rainy period. This freshening of the incoming Bank water by rain is undoubtedly one of the major biologic stress factors that account for the low diversity of marine fauna on the tidal flats (Chapter 6). Porewaters in the sediment of the levees may reach salinities as high as 65 ‰ but this is found only in the uppermost centimeter or two during the dry months.

Water composition expressed in major ion ratios is almost always close to that of normal seawater: Mg/Ca mole ratios stay near 5 whether the tidewaters are diluted by rain or slightly concentrated by evaporation.

8. Surface Features of the Depositional Subenvironments

Each of the three major physiographic belts of the Three Creeks tidal flat—the nearshore marine belt, the channeled belt, and the inland algal marsh—is itself a complex of smaller depositional subenvironments, each made readily recognizable by its distinctive physiography, physical surface features, and biological communities. The particular sedimentological significance of these distinctive surface subenvironments is that each was found to carry a distinctive set of sedimentary structures and textures in the immediately underlying sediments. Further, many of these structures directly record features existing at the sediment surface, as the chapters to follow will amply demonstrate. For this reason we have described below, in some small detail, the surface features (both physical and biological) of the significant subenvironments. We recognize four major subenvironment complexes—the levee-pond subenvironments, the shoreline subenvironments, the channels and the channel bars, and the inland algal marsh—and have dealt with them below in that order.

A. The Levee-Pond Subenvironments

The channeled belt between the shoreline and the inland algal marsh is 3 to 4 km wide and is transected by sinuous, branching tidal channels bordered by low natural levees, which, like the levees of terrestrial flood-plain rivers, are built up by overbank flooding. Between the levees of the major channels lie very shallow "ponds" analogous to the backswamps of deltas; these ponds are dammed on the shore side by the beach ridge and on the land side by the inland algal marsh. Normally the ponds are flooded twice a day by tidewaters which exchange through secondary distributary channels leading into the ponds from the main channels (Figs. 6 & 7). As noted above, mean tide range in the ponds is only 29 cm and in the main channels is 46 cm near the mouth (Table 2). The levees remain dry much of the year (they are known locally as "sidewalks"), except when severe onshore storms cause temporary flooding of almost the entire tidal flats. These levees vary in size from one channel to the next, but in general are highest (up to 25 cm above MHW) and widest (up to 250 m across) near the mouths of the main channels, where they merge into the beach ridge. Also, especially notable is that on channel bends the upstream (east) levees are *wider* than the downstream (west) levees. This is easily seen in Figure 6. It clearly points to an offshore (west) source for the levee sediment (Chapter 5). The levees become progressively lower and narrower upstream, until about 2 to 3 km from the shore the levees are virtually nonexistent (Fig. 6). Here, near the inland marsh, the channels and the ponds blend into one continuous body of water at high tide.

Zonation of Surface Features. Across the large seawardmost levees from channel to pond we recognize an ordered sequence of six different zones or subenvironments, different in terms of their physical surface features and biota. These subenvironments are: (1) channel bank, (2) levee crest, (3) levee backslope, (4) high *Scytonema* algal marsh, (5) low *Scytonema* algal marsh, (6) pond (Fig. 9A). This zonation of surface features across the levee is mainly due to the interplay between frequency of exposure (or flooding) and frequency of sedimentation, which in turn are functions of elevation and proximity to the sediment source offshore. These factors largely determine (a) the distribution of organisms, which in turn profoundly influences the character of the sediment record, and (b) the desiccation gradient, which affects the mechanical processes of mud-cracking and erosion to produce a zonation of disruption features. The superposition of these biological and mechanical phenomena result in a distinctive set of surface subenvironments, *each with a distinctive set or assemblage of sedimentary structures and fabrics in the immediately underlying sediment.* This zonation is mappable (see Fig. 10; Wanless 1969; Ginsburg et al. 1970). Upstream, as the levees decrease in elevation and width, the zonation is less complex: the levee crest and backslope zones rapidly contract and disappear, so that near the landward edge

Fig. 10.
Map of levee-pond complex (at locality 70-35, see Fig. 2), showing the detailed areal distribution of subenvironments.

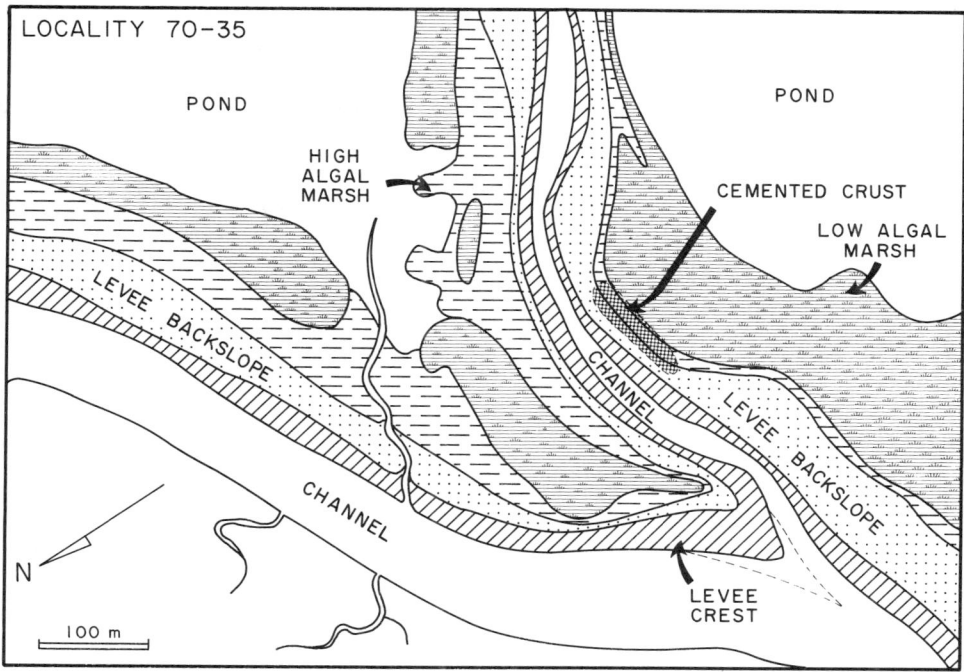

of the channeled belt the levees consist only of the algal marsh zones separating the channels from the ponds. In the lowest areas of this landward part of the channeled belt, the pond zone reaches right to the channel bank without any marsh fringe.

The spatial relations and scale of these levee-pond subenvironments are shown clearly in Figures 7 and 10. The range of annual average percent of time the surface of each subenvironment is exposed—the Exposure Index of Ginsburg et al. (1970; and Chapter 3)—is given in the text below.

Channel Bank. (Exposure Index 40 to 90%) On many of the channel banks, from mean tide level to just above mean high water the sediment surface is a complex of small (2 to 10 cm), rounded unlithified knobs (Fig. 11) covered by a pink-grey, tough filamentous blue-green algal mat, mainly the small, thin-filamented *Schizothrix calcicola*[4] with filaments $< 5 \mu$ across. In cross-section these knobs are seen as beautifully laminated hemispherical proto-stromatolites[5] (see Chapter 5; Fig. 34). The knobs are formed by the mat draping over the mounds between the multitude of fiddler-crab (*Uca* sp.) burrows that riddle the channel banks, or by draping over intraclasts caught between the knobs.

These knobby channel banks are mainly to be found along the straighter reaches and on some of the gentler bends of the channels. On the outside bends of very sinuous channels slow but active undercutting has produced a near-vertical erosion cliff. The inner banks of many channel-bends are narrow, mud-cracked, *Schizothrix* algal-mat-covered, sloping point bar-like features, commonly overgrown by red mangrove thickets (see Fig. 7).

Levee Crest. (Exposure Index 98 to 99.7%) Above the knobby channel bank the levee surface is quite smooth and flat (Fig. 12A). Low amplitude undulations (in the order of 1 cm) occur where sediment has collected preferentially as lee-

---

4. Unless otherwise noted, all of the identifications of the blue-green algae in the present paper were made by Dr. Harold J. Humm of the University of South Florida.
5. The term "stromatolite" has come to be used rigorously for lithified, algal-laminated structures. To avoid confusion we (reluctantly) use here the term "proto-stromatolite" for unlithified, blue-green algal-laminated structures.

Fig. 11. View of complex knobby upper surface of a channel bank at low tide. The bank is riddled with fiddler-crab burrows. *Schizothrix* mats cover the entire surface around and between burrow entrances. Note the polygonal prism-cracks. Grasses and red mangroves on the levee. Scale bar is 30 cm long.

shadows behind vegetation clumps or where a sediment lamina has been rippled or "current lineated" during deposition (Fig. 12B). There is a sparse cover of vegetation (Fig. 12) consisting typically of stunted bushes of black mangrove, red mangrove, and buttonwood, and small succulents and grasses. This is the levee crest and, like the channel bank, it is bound by a smooth, pink-grey, tough, algal-filament complex that makes a firm, resistant, springy surface when dry, but is slick and slippery with mucilage when wet. The algal complex is composed essentially of *Schizothrix calcicola*,[6] but scattered strands of *Scytonema* occur. The surface sediment is so tightly bound by this algal mat that neither flooding by seawater nor pounding by torrential rains were observed to dislodge even mud particles. Despite long periods of exposure (Exposure Index $> 98\%$) the surface is conspicuously *free of mud-cracks*, presumably because the pervasive algal binding imparts considerable elasticity to the sediment. However, disruption of the surface does occur: isolated, small (a few centimeters across), shallow (about 1 cm deep), flat-bottomed depressions are left as scars, where patches of the mat-bound sediment have been torn out (Fig. 13A). Their origin is thought to be as follows. Doming of the mat occurs during saturating rains coupled with very high water, when air in the underlying sediment is forced up through a pinhole by the rising water table and becomes trapped below the mat-bound surface sediment layer. The sediment parts along the bedding and the upper algally

---

6. Dr. Harold Humm suggested that the alga is actually an ecophene (Drouet 1963) of *S. calcicola*; he therefore preferred the identification as *Phormidium crosbynam*.

Fig. 12. Levee crest surface: (A) General view of levee crest with smooth surface and vegetation of grasses, *Rachicallis* (*next to knife*), *Borrichia* (*right foreground*) and *Conocarpus* (*left middle distance*). Note the sediment mounds collected behind the grass tufts. Red mangroves colonize the opposite channel bank at a lower elevation. (B) Close-up view of levee surface, showing "current lineations" oriented in direction of current. Sediment is tightly bound by *Schizothrix* algal mat. Note the bird tracks and irregular depressions produced by spalling off of matbound surface sediment (see Fig. 13). Note the absence of mudcracks in this very dry subenvironment. Pencil for scale.

Fig. 13. Levee crest surface: (A) "Scars" in surface where algal mat-bound sediment has been ripped out. Note smooth edges repaired by overgrowth of algal mat. Pencil for scale. (B) "Air bubble" dome in algal mat-bound sediment. Pencil for scale.

Fig. 14. Close-up view of mud-cracked levee backslope surface, showing the size and extent of cracking of a newly deposited sediment lamina. Note the sediment collection behind mangrove pneumatophores. Levee crest is to the top of the picture. Box-core in foreground is 10 cm wide.

bound layers are pushed up into a small dome (Fig. 13B). We have seen these domes rise in a matter of minutes. The softening and horizontal expansion of the mat as the wetted algal filaments generate mucilaginous sheaths undoubtedly contributes to this doming process. On subsequent drying out of the sediment ceiling within the hollow of the dome, the dome cracks around the base and platy, curled pebbles ("cookie-sized" clasts) are produced. Storm-flooding and heavy rainwash carry the clasts down the levee backslope toward the pond. Slow growth of the mat "repairs" the walls of the depression left in the surface by draping over the edges (Fig. 13A).

The only traces of animal life on the levee crest are a few small cup-shaped burrows of insect larvae and the tracings of immediately subsurface ant tunnels.

The upper 5 to 15 cm of sediment of the levee crest is magnificently laminated, the individual laminae rarely exceeding 1 mm in thickness (Fig. 36A).

Levee Backslope. (Exposure Index 95 to 98%) Down from the levee crest onto the levee backslope, the smooth levee surface shows the first signs of mud-cracking. Very shallow (a millimeter or two), closely spaced ($\sim$ 1 cm), complete and incomplete mud-cracks may break up large patches of the surface sediment into small polygons the size of small ceramic tiles (Fig. 14). Cracking of a sediment layer apparently begins soon after the depositing floodwater has drained from the levee. The cracking process may be quickly arrested, probably by rapid

Fig. 15. Chipped surface of levee backslope: a sediment lamina, deposited during a storm, later becomes desiccated and cracked into small polygons which vary in size with the thickness of the layer. These polygons may become separated and further broken as flat chip intraclasts, as in this scene.

"growth" of *Schizothrix*, giving "incomplete" mud-cracks, or it may go to completion, producing isolated polygons one lamina thick. These polygons, if not welded to the surface by the algae, may be eroded and transported a few meters downslope to accumulate in depressions as pockets of flat, sand- and granule-size chips (Fig. 15).

Vegetation is similar to that on the levee crest, except that grasses are more abundant. The burrow entrances of the fiddler and marsh crab (marked by piles of mastication pellets) are found in the lower part of the zone, especially around the shrub roots.

Like the levee crest, the upper 5 to 10 cm of levee backslope sediment is layered into uniform millimeter-scale lamination, but the laminae are commonly disrupted by shallow mud-cracks (see Chapter 5).

High Algal Marsh. (Exposure Index 85 to 95%) Further down the backslope of the levee, near the pond high water mark, the small-mud-crack zone tapers away and a dark, flat, tufted mat of *Scytonema* covers the surface like a carpet (Fig. 16A). This flat *Scytonema* mat zone we call the *high algal marsh*. *Scytonema* is a large filamented (individual filaments 10 to 30 μ across) blue-green alga, essentially a freshwater lover, that grows in stubbly upright tufts up to several millimeters in length (Chapter 7; Black 1933; Monty 1967). In addition to the algal cover, the high marsh is dotted with stunted red mangrove bushes and halophyte grasses. Near the high water mark, the burrow entrances of fiddler and marsh crabs are common in places, but they are rarely as abundant as on the channel banks.

The surface features in the uppermost part of the high algal marsh vary from levee to levee and from time to time on the same levee, dependent on the relative rates of sedimentation and *Scytonema* growth (Chapter 7). We have found three main kinds of high marsh edge surfaces. On the first kind, small bundles and tufts of *Scytonema* filaments grow along the crests of small ripples or current lineations but not in the troughs, which are covered by *Schizothrix* (Fig. 16B). The slightly elevated position of the *Scytonema* filaments has apparently protected them from burial (and hence extinction) by the most recent sediment laminae deposited here. On the second kind of surface, we find tufts and bundles

Fig. 16. High algal marsh zone of levee-pond complex: (A) close-up surface view of somewhat wrinkled *Scytonema* "carpet." Pencil (for scale) points to small perforated algal tufted "dome." (B) Upper edge of high algal marsh, near boundary with levee backslope, showing sparse growth of *Scytonema* (dark) on the crests of current lineations; troughs (light) are covered by *Schizothrix*. Pen for scale.

Fig. 17. High algal marsh zone of levee-pond complex: (A) Coarse *Scytonema* filaments protruding through the surface sediment like a stubbly beard. The surface layers are here broken into polygons with raised edges by shallow desiccation cracks. These polygons make Black's type C algal heads. Scale in inches. (B) These mounds and shallow depressions result from the patchy growth of *Scytonema* mats. The interior of the mounds are actually thick, cemented crusts which preserve the mat structure of *Scytonema*, while the depressions are floored by paper-thin crusts, colonized by *Schizothrix*. Scale in cms.

of *Scytonema* filaments protruding through a *Schizothrix* mat along the edges of slightly upturned desiccation polygons (Fig. 17A). Again, it is only the *Scytonema* tufts in the slightly elevated positions that have survived the latest sediment influxes. These polygons may become recracked and even detached. Flooding then washes these "cookie clasts" downslope toward the pond. On the third kind of surface, the *Scytonema* is found as isolated, low, rounded knobs or mounds and as patchy, flat, discontinuous mats that stand above a smooth *Schizothrix*[7] surface (Fig. 17B). This kind of patchy growth is characteristically found in areas where sediment has not been deposited for at least several years. Also characteristically, this kind of high algal marsh surface is the zone where contemporary cementation is occurring (Shinn et al. 1965, 1969; Chapter 7): in scattered areas this patchy high marsh mat is cemented by high magnesian calcite and aragonite, producing lithified stromatolitic heads (Chapter 7; Black 1933; Monty 1967; Hardie 1969). In some of these areas, the upper few centimeters of sediment is thoroughly lithified by a void-filling cement of aragonite and/or high-Mg calcite, making very hard pavements. Some of the pavement crusts carry proto-dolomite or very high magnesian calcite (Shinn et al. 1965, 1969; Hardie 1969).

Downslope toward the pond the patchy boundary zone *Scytonema* mat joins into a continuous, thin (1 to 2 mm), flat, stubbly carpet that covers the entire surface (Fig. 16A). Small round tufts (<1 cm across) like tiny pompons dot the mat where growth is more vigorous. During the wet summer months when rains and tidal flooding are frequent, the mat is dark olive green, soft, and spongy. In the dry winter months, however, the mat is dull grey, stiff, and bristly. It wrinkles with drying and may even become polygonally cracked and curled. This latter feature is especially true of the upper part of the zone, and it is this cracked mat when covered by sediment laminae that makes the draped saucer polygons discussed above and shown in Figure 17A.

The near surface sediment below the high algal marsh is beautifully laminated on a millimeter scale, with the sediment laminae separated by dark, wrinkled, *Scytonema* mat laminae (Chapter 5; Fig. 43).

Low Algal Marsh. (Exposure Index 65 to 85%) Fringing the pond, between MHW and MTL, *Scytonema* changes from a continuous thin mat into larger and larger isolated tufts ("pincushions," 1 to 3 cm thick and 5 to 10 cm across) made of bundles of thick filaments (individual filaments 10 to 30 μ in diameter). This *low algal marsh* environment is continually wetted during the summer and the surface becomes a lush meadow of dark green, spongy *Scytonema* pincushions (Fig. 18A). For most of the winter, however, this marsh remains dry, or at least suffers long periods of dryness, and the pincushions shrink into stiff, bristly, dull grey polygons, clearly revealing the soft-pelleted, lime mud substrate to which they are very loosely anchored (Fig. 18B).

Other vegetation in this zone is sparse, consisting mostly of dwarf black mangroves in the upper half and red mangroves in the lower half. Two species of grass are also sparsely present. *Uca* and *Sesarma* burrows are found in places in the upper part of the zone. Cerithid gastropods, so common in the pond, may be found in small numbers in the lower fringes of the low algal marsh, and other pond animals, such as polychaetes and foraminifera, occur in the mud between the *Scytonema* tufts.

The sediment beneath the *Scytonema* meadow is layered in lenticular beds that range from millimeter-laminae to thin beds up to 3 cm thick (Chapter 5).

Pond. (Exposure Index 0 to 65%) At the lower edge of the algal marsh the *Scytonema* pincushions become smaller and widely scattered and then abruptly die

---

7. A sample of this *Schizothrix* mat from one locality was found to carry some *Anacystis aeruginosa* and *A. marina*, as well as a little *Scytonema*.

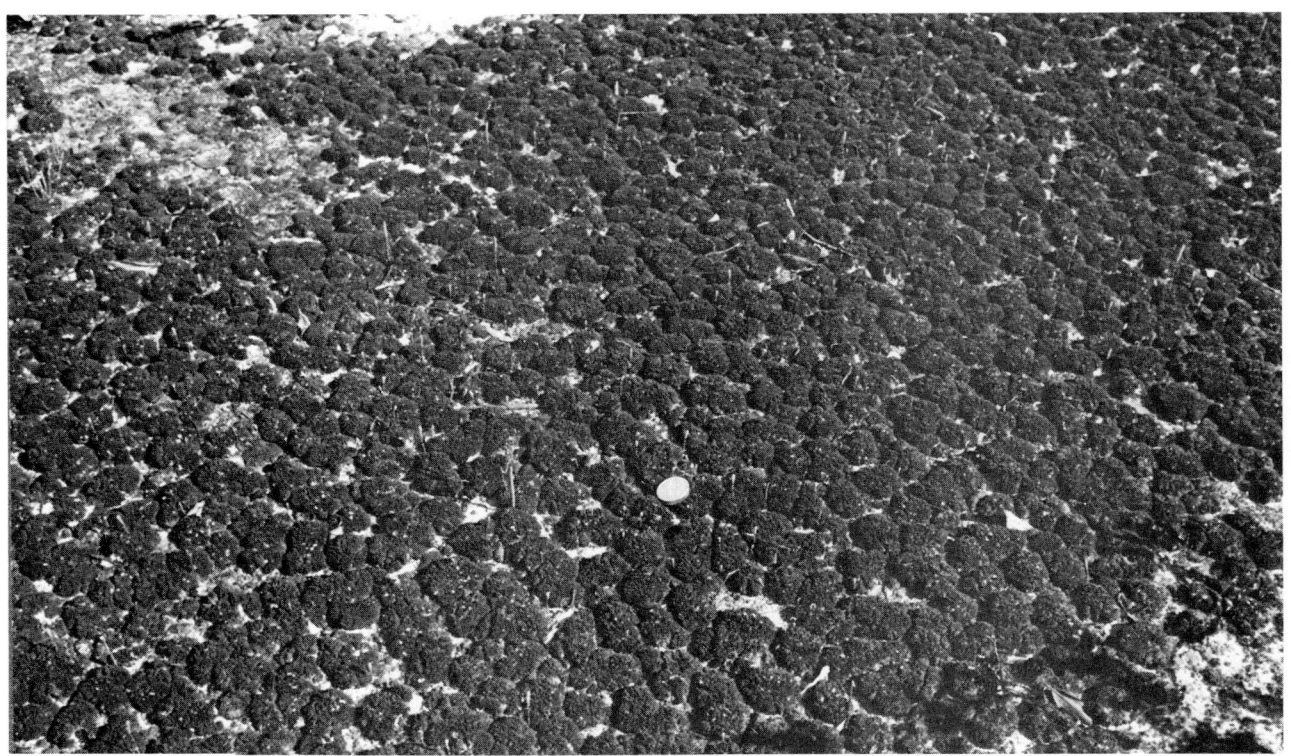

Fig. 18. Low algal marsh at edge of pond: (A) View of surface of luxuriant "pincushions" of *Scytonema* at low tide during summer. Each "pincushion" is about 7 cm in diameter. (B) Desiccated *Scytonema* "pincushion" meadow during dry winter months.

out altogether. This is a sharply defined boundary and marks the start of the *pond* subenvironment.

Ponds occupy about 60% of the area of the seaward channeled belt, and about 90% of the landward channeled belt. Their very irregular shape (Fig. 6) is determined by the configuration of the channels which, together with their levees, divide the ponds into compartments or long arms and embayments.

The depths of the ponds vary with their width and distance from the channel branches which feed them. Generally, the pond surface slopes gently down from neighboring levees: thus wider ponds are deeper. Also the points at which channels enter the ponds are marked by low levees and channel-mouth bars which prevent the ponds from draining completely at the lowest low tides. During winter and spring the general water level is lower than in summer and fall by some 10 to 15 cm (Table 2), and the ponds may remain unflooded for weeks (Fig. 19A), during which time most of the ponded surface water is removed by evaporation as well as slow seepage by subsurface drainage. Conversely, in summer and fall the ponds may remain virtually undrained through several tide cycles.

The pond surface is muddy, though its consistency varies from place to place depending on its exposure index and on the species of infaunal burrowing animals present (Chapter 6), but generally one's foot sinks in between ½ cm and 5 cm. The surface shows no major irregularities or undulations (Fig. 19A), though pellet mounds of one worm—*Marphysa*—give the surface a dimpled appearance (Fig. 20A). On a small scale, the surface can be seen to consist of fecal pellets of cerithid gastropods and polychaetes (Fig. 19B) held loosely together by an incoherent algal mat.[8] This mat is grazed in the intertidal zone by abundant cerithid gastropods whose shells litter the surface, as do the tests of peneroplid forams. The centers of the ponds are frequently bare of any macro-flora, though *Batophora* may densely or sparsely cover large areas, and *Diplanthera*, *Thalassia*, and rarely *Penicillus* may be present, the latter two only subtidally. Around the pond margins, however, are red mangroves, which occur as scrubby plants spaced so that their aerial roots do not intertwine.

Beneath the surface the sediment is unlayered, burrowed, pelleted mud containing cerithid gastropods and peneroplid forams and a lacework of mangrove roots and root hairs. During winter and spring low water levels, much of the pond surface starts to dry out, but while still remaining moist, the sediment shrinks and cracks deeply, producing "prism" polygons 20 to 30 cm in diameter which extend into the mud some 15 to 30 cm (Fig. 20B). During times of high water, these cracks heal and completely disappear from view.

Due to the activities of grazing and pelleting animals, particularly the gastropods, the pond surface is not protected from erosion by the algal mat (in contrast to the situations described in Neumann et al. 1970). Even in fair weather, algal-bound clots of pellets become dislodged and roll slowly along the surface to rest at least several meters away. With the combination of a heavy rainfall and a falling tide, the pond water can turn milky with suspended sediment. And there is some evidence that the central parts of some ponds have been eroded out during storms: red mangrove roots and rootlets are found in abundance in the centers of these ponds, where no live mangroves are now to be found, and dead, exhumed cerithid gastropod shells litter the surface as a lag deposit. Living red mangroves and cerithids are both restricted to the margins of such ponds.

Because of browsing and burrowing (Chapter 6) the pond sediment is unlayered, only the roots of mangroves, the curving branching worm burrows, and scattered tests of gastropods and foraminifera break the homogeneity (Chapter 5).

---

8. This "mat" is a scum of *Anacystis aeruginosa* with some *Schizothrix calcicola* and a little *Scytonema*, *Microcoleus lyngbyaceus* and *M. vaginatus*.

Fig. 19. Pond features: (A) Pond at low water. The wet portion of this pond (*distance*) rarely drains. Nearer the shore, red mangroves (*right*) begin to colonize and cerithid gastropods (*black dots, foreground*) graze the surface. The pond mud is soft (*see footsteps*) chiefly through the activities of burrowing polychaetes. At high tide the water level reaches to about where the mangrove roots meet the leaf crown. A pond may, during periods of prolonged easterly winds, stay dry like this for several days, or, conversely, during westerly winds it may not drain for days. (B) Pond surface with characteristic flora, fauna, and pelleted sediment. The fluffy heads are of the alga *Batophora*; the leaf has fallen from a red mangrove. Blue-green filamentous algae ("*Schizothrix*") colonize surface sediment and are grazed on by cerithid gastropods (*black*, e.g., above 2.5 and 7.3 on scale). Chief pellet-makers are gastropods (*tiny rods on leaf*) and burrowing polychaetes (*ellipsoids, center*). The disc-shaped foraminifer *Peneroplis* is also visible (e.g., at 10.2 on scale).

Fig. 20. Two views of the pond surface: (A) Pellet mounds of the polychaete worm *Marphysa*. Note the "prism"-cracks. Scale about the same as in (B). (B) "Prism"-cracked surface at low tide in winter. Dark spots are cerithid gastropods.

B. The Shoreline Subenvironments

The shallow marine bank offshore of Andros—the Great Bahama Bank described so well by Newell and Rigby (1957), Newell et al. (1959), Cloud (1962), and Purdy (1963)—is indeed shallow: over its entire interior 100,000 km² area the maximum water depth reaches only 7 m. The bottom shallows so gradually toward the western shoreline of Andros that vessels with a draft as little as 1.5 m cannot approach Three Creeks closer than 10 km. The Three Creeks tidal flat is separated from the open bank by a low beach ridge that reaches a maximum elevation of 1.3 m above mean tide level. The actual shoreline is marked by a 40-cm high erosion cliff which zigzags in and out, making a micro headland-and-bay topography (Shinn et al. 1969, Fig. 14).

A traverse from the beach up across the beach ridge and down to the pond behind the ridge (Fig. 9B) yields a basically similar picture to that outlined above for the levee-pond complex. The beach ridge is really nothing more than a shoreline levee, the low barrier responsible for the development of the protected channel-levee-pond-marsh complex behind it. The major subenvironments we recognize at the shoreline are the offshore, the beach, and the beach ridge. The beach we arbitrarily define as the foreshore between the beach cliff and lowest LW: seaward the beach merges almost imperceptibly with the shallow offshore marine Bank. The beach ridge is a complex feature. It is an undulating ridge parallel to the shoreline, consisting of low, flat, smooth washover fans separated by higher pine- and shrub-covered sandy hummocks. Like the channel levees, the beach ridges are zoned: above the cliff is a smooth, flat beach terrace, which merges landward into either a pine hummock or a washover marking the crest of the beach ridge. Over the crest the beach ridge passes downslope through the beach-ridge backslope into the algal marsh zone that fringes the ponds of the channeled belt. The spatial relations and scale of all these shoreline subenvironments are shown in profile in Figure 9B and in plan in Figure 21.

Offshore. (Exposure Index 0%) The offshore surface is a soft, muddy sand very loosely bound by a slimy organic scum of micro-organisms (not identified) and sparsely covered by small plants of the seagrasses *Thalassia* and *Cymodocea*, and the green algae *Penicillus*, *Rhipocephalus*, and *Batophora* (also *Halimeda* and *Acetabularia* further out on the bank). These plants are covered by a slimy film to which mud readily adheres, giving the plants a green and white mottled look. The entire sediment surface is pitted with conical depressions, 10 to 20 cm across and 10 cm deep, that mark the entrances to the burrows of the shrimp *Callianassa*. The shallow resting burrows of *Callinectes*, a portunid crab, are also responsible for some of the surface irregularities. On two transects made in 1968 (March and June) across the Bank from Gun Cay to Point Simon, a distance of about 100 km, plant and shrimp burrow abundances were measured. Distribution of vegetation was very patchy, the grasses ranging from 0 to 70 plants/m² and the green algae from 0 to 20 plants/m². On the other hand, the *Callianassa* burrow entrances were ubiquitous, closely averaging about 4 holes/m² across the width of the Bank. Other animals recognized on the Bank were echinoids, starfish, holothurians, sponges, bryozoa, rose coral, and a variety of molluscs and worms. Burrowing is so intense that the offshore sediment is completely unlayered.

Beach. (Exposure Index 0 to 80%) The beach, as here described, is taken to include the graded shore between mean high and the lowest low tide, as well as the channel-mouth bars which lie in the general shore zone.

The width and grade of the beach zone vary with the degree of protection of the shore. Thus, at the exposed headlands the beach is narrow (25 m) and relatively steep, whereas in embayments and around the mouths of the channels the beach zone grades off gently into the offshore marine environment, and a

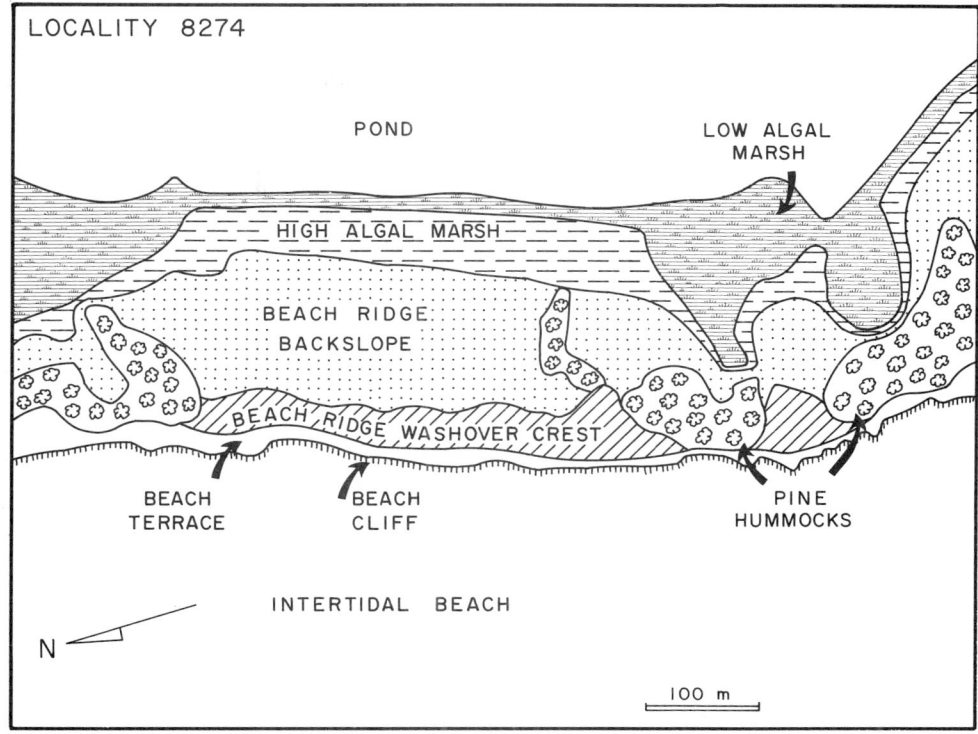

Fig. 21. Map showing the typical distribution of surface subenvironments in the shoreline area of Three Creeks. Locality 8274 (see Fig. 2).

zone several hundred meters wide may be exposed at the lowest tides. Several quite different types of "beaches" occur.

Beach mounds and hollows (Fig. 22) are typical of most embayments between headlands. They extend through the lower intertidal and grade off gently into the subtidal. The local topography is one of small mounds and hollows of irregular shape, size, and elongation. The mounds are generally 2 or 3 m in diameter and are built up to ½ m above the troughs of the hollows. The tops of the mounds are covered with a slightly bumpy, pink *Schizothrix* mat (Fig. 22C) through which *Marphysa* (worm) and *Callianassa* (shrimp) burrow mounds protrude. *Alpheus* (shrimp) burrow entrances are common at the bases of the mounds and are partly responsible for mound erosion. There is usually no epifauna on the mounds, but in the more protected embayments many *Batillaria minima* (gastropods) graze over the surface in gregarious herds. Beneath the surface the upper 2 mm may be layered, but below all is homogenized by burrowing.

The intermound hollows are always filled with water, even at low tides. Their surface mud is softer than that forming the mounds, lacks a cohesive algal mat, and is pock-marked by many pellet mounds of burrowing polychaetes, mounds and pits of *Callianassa*, and the large excavations of the crab (*Callinectes*). There is also abundant evidence of the activities of bonefish, which leave small, roughly circular feeding pits, and small, mound-shaped fecal piles (Fig. 22B). Vegetation in the hollows consists of diminuitive plants of *Thalassia* and *Penicillus*.

Planar beaches occur as three different and unrelated types:

1) A wave-cut bench of exhumed and eroded pond sediments is present where there is little or no sedimentation to cover it. Such a bench occurs in front of all the exposed headlands and also at the mouths of some blocked creeks. Its surface is rough-textured due to the irregular erosion of the cohesive pond sediment. Out of this sediment large red mangrove root systems, root hairs, and many pond shells such as *Cerithidea*, *Batillaria*, *Geloina*, and *Peneroplis* are actively being eroded (Fig. 23A). A few *Batophora* plants recolonize the eroding sediment.

Fig. 22.
Beach mounds and hollows: (A) General view of beach mounds near to low tide. This is a major depositional locale in the beach zone. (B) Surface view of mound and hollow development. Mounds are colonized by algal mats, while hollows remain subtidal with a subtidal flora and fauna (*Thalassia*, *Penicillus*, and *Callianassa*). Small mounds in hollow are the fecal piles of bonefish which graze extensively in this zone. (C) Bumpy pink "*Schizothrix*"-type algal mat on crest of beach mound. This mat traps sediment within its filaments during times of flooding with sediment-laden water, but the sediment is not laminated below the top few millimeters due to the activities of burrowing polychaetes, such as *Marphysa*, whose pellets are strewn over the surface. Scale in cms.

2) A smooth, flat surface occurs on the landward side of most mound and hollow developments. It has a strongly cohesive algal-mat surface, bears no macroflora and no epifauna, except occasional herds of *Batillaria*.

3) Rippled mud surfaces (Fig. 24) occur directly in front of the cliff which bounds the beach ridge. The surface is soft mud, the rippled grains being mostly sand-sized muddy fecal pellets rather than hard grains of sand. Because the grains are frequently in motion, no algal mats, other vegetation or desiccation cracks are found, though *Callianassa* and some few polychaetes, which live in the deeper undisturbed sediments, push out small fecal pellet mounds onto the surface.

Channel-mouth bars are broad, poorly defined shoals which completely block the channel mouths at the lowest low tides. The seaward margins of these bars grade gently into the typical offshore environment, and the side margins grade into beach subenvironments—typically the mound and hollow beach type.

The sediment surface on the bars is almost level, the only irregularities being small current ripples and *Callianassa* burrow openings (the large mounds that *Callianassa* usually develops over its burrows are washed away here). The vegetation is similar to that in the offshore environment and consists of sea-grasses, *Penicillus* (Fig. 23B), and a few sponges. The sediment is similar to offshore sediment.

As noted above, the beach ends abruptly against the shoreline at a low cliff some 40 cm high that represents the seaward edge of the beach ridge (Fig. 24A). The cliff is burrowed by crabs and eroded by storm waves, producing rounded pebbles where the cliff sediment is massive (Fig. 24B), and flat pebbles where the cliff is of laminated sediment or where cemented crusts outcrop (see Chapter 7). These intraclasts accumulate along with coarse shell debris (mainly gastropod tests) in pockets at the base of the cliff as "round-stone" or "flat-pebble" gravels (Fig. 24B).

Beach Terrace. (Exposure Index approximately 85 to 95%) Above the beach cliff is a smooth, gently upsloping terrace 10 to 30 m wide that either leads abruptly into a pine hummock or merges smoothly with a low washover lobe (Figs. 9B & 21). This beach terrace is covered by a pink-grey, smooth, tough, rubbery *Schizothrix* algal mat precisely like that of the channel levees (Fig. 25A). In places near the seaward edge of the terrace, *Scytonema* and *Rivularia* (identified by L. A. H.; cf. Monty 1967, p. 85) filament tufts protrude through the *Schizothrix*-bound surface, making dark grey-black blotches in the pink mat. At the ragged edge of the cliff face, storm waves have ripped up the algally bound surface layer of sediment, breaking it into flat, plate-like pebbles up to 15 cm across and 1 cm thick (Fig. 25B). The tough, rubbery mat makes these pebbles cohesive enough to be carried intact by storm waves up the terrace and onto the washover crest. A few of these intraclast pebbles are left stranded on the smooth terrace where they become tightly welded to the surface by the exposed algal mat of the pebble joining with the terrace mat (Fig. 25A). The sediment of the beach terrace down to a depth of 30 cm is beautifully laminated (Chapter 5).

Beach-Ridge Washover. (Exposure Index approximately 95 to 99.7%). The beach terrace passes smoothly into the washover lobes that make wide, low, flat saddles between the sandy *Casuarina*-covered hummocks at the crest of the beach ridge (Figs. 9B & 21). These washover plains are much like the channel levee crests: the surface is smooth and white, covered by a tough, rubbery *Schizothrix* mat, and sparsely dotted with low halophyte shrubs, grasses, and black mangroves. Ants and earthworms are the only burrowers in this zone.

Two surface features are particularly outstanding on the washovers: the patches of flat-pebble gravels and the thin sheets of rippled sand. In places at the crest of the washovers there are narrow (a few meters wide) trains of flat pebbles,

Fig. 23A.
Exhumed and planed pond sediment being eroded immediately seaward of the beach ridge. Clearly visible is the root system of a red mangrove and dead cerithid gastropods (*black dots*)—both are characteristic of pond sediments.

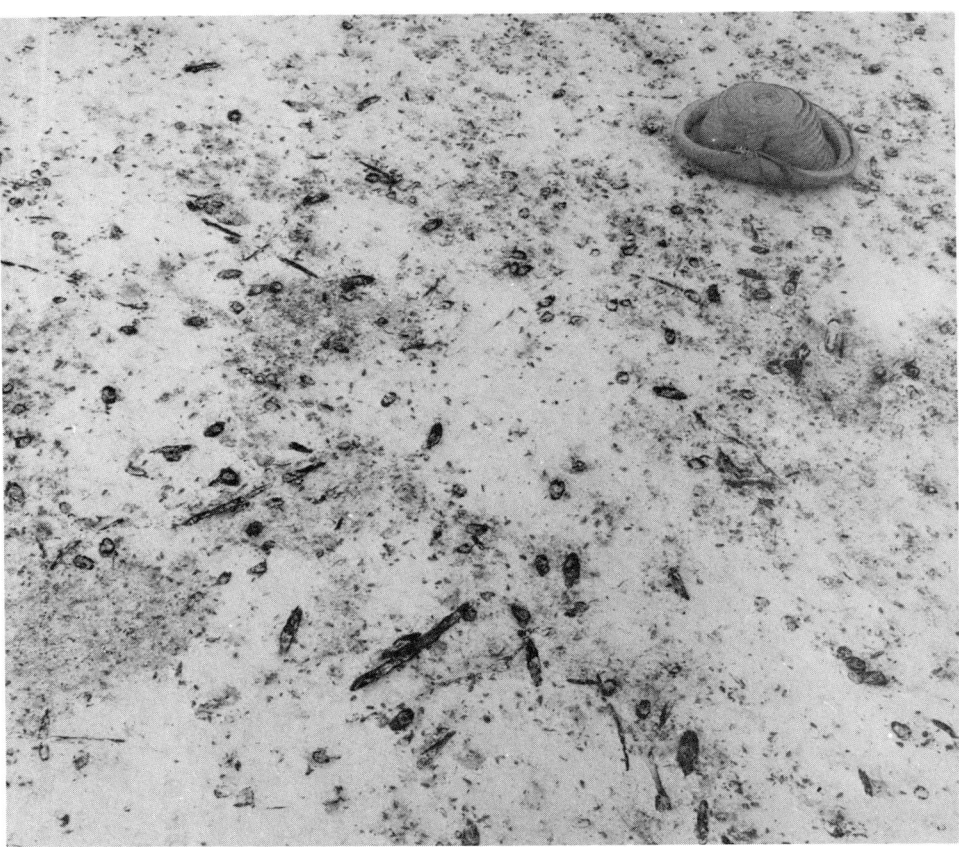

Fig. 23B.
A crop of *Penicillus* sp. exposed (fatally) on a channel mouth bar during an exceptionally low tide. The occurrence of this codiacean alga is significant because it is believed to produce much of the aragonite mud of the sediment. Footprint for scale.

which are mainly plates and slabs up to 15 cm across and 2 cm thick, and consist of hard fragments of cemented crust eroded out of the beach cliff face and soft fragments of algally bound, laminated sediment ripped up from the beach-terrace surface (Fig. 26A). The soft, laminated pebbles are completely unlithified and still flexible enough to bend; they have remained cohesive enough to with-

Fig. 24.
Beach cliff: (A) View looking shoreward toward pine-covered beach ridge, showing eroded beach cliff and rippled beach surface at low tide. Note the indented footprints—the sand is actually composed of soft peloids. (B) Close-up of eroded beach cliff, showing pebble mass at the base, crab burrows in the cliff face, and beach-ridge vegetation in the background.

A

B

stand transport, owing to the dryness of the terrace sediment and to the top surface binding by the still attached *Schizothrix* mat. Some of these soft, flat pebbles show rounding, even though they have been moved no more than a few tens of meters. The crust pebbles have undergone significant dissolution (probably during subaerial weathering) so that they have deeply pitted but smooth surfaces. Often associated with the gravel patches, but by no means exclusively so, are very thin discontinuous sheets of coarse skeletal and peloid sand and granules. These sheets of sand are commonly spread into flat cuspate ripples, generally only a few millimeters high (although we have found ripples with

Fig. 25.
Beach terrace surface: (A) Smooth *Schizothrix*-bound pelleted mud surface strewn with angular flat algal-bound pebbles derived from the cliff edge. Note especially that the algal-surfaced pebbles are now welded to the terrace by the mat; these produce LLH-S proto-stromatolites. Those pebbles with rough surfaces were deposited with their algal surfaces downward and so no welding occurred. Note the faint impression of a footprint in the upper right-hand corner of the picture: this clearly demonstrates the cohesiveness of the mat. (B) View of the eroding algal-bound laminated sediment of the terrace at the cliff edge. The polygonal cracking of the upper centimeter or so of laminated sediment produces the flat pebbles shown in (A).

A

B

amplitudes up to 7 mm) and up to 30 cm apart. The orientation of these current ripples clearly points to landward transport. Atop the cuspate ripples and parallel to the transport direction are conspicuous ridges of sediment, a pronounced current lineation (Fig. 26B). The distribution of the sand is patchy, small surface depressions and the leeside of bushes being favored accumulation sites. The sand particles consist mainly of broken fragments of gastropod and pelecypod shells, the whole tests of foraminifera, and dense ovoid peloids. The granule particles may also be of skeletal fragments but a significant percentage consists of flat mud chips, eroded from the beach cliff and terrace.

Fig. 26A.
Beach-ridge washover crest —imbricate flat pebble trains on a smooth algal-bound sandy surface. Note typical halophyte vegetation. The beach cliff, the source of the pebbles (Fig. 25), is to the left of the picture.

Fig. 26B.
Close-up view of beach-ridge washover crest surface, showing patchy cover of loose sand smeared landward in flat current ripples with current lineations on their backs. Sea is to the left, land to the right. The dark patches around the stake are exposed parts of a red pigment poured on the surface two months earlier as a rate-of-deposition experiment-marker (Chapter 5). Note the raindrop impressions. Scale is 20 cm long.

Lamination is the essential structure of the sediment underlying the beach-ridge washovers (Chapter 5).

Beach-Ridge Backslope and Marsh. (Exposure Indices as for levee backslope and marsh) From the washover crest to the pond shore behind the beach ridge, the washover plain slopes smoothly and very gently downward and landward (Fig. 9B). Precisely the same surface zonation as occurs on the levee is found on these washover plains, that is, between the washover crest and the open pond the sequence of zones is beach-ridge backslope, high algal marsh, low algal marsh, and pond (Fig. 21). These zones are much wider and more extensive than those of the levees, but their essential surface features are the same: the descriptions, therefore, will not be repeated here. Perhaps the only noticeable difference is that on the beach ridge the *Scytonema-Schizothrix* of the upper part of the high

Fig. 27. Beach-ridge backslope surface showing curled, chipped, desiccated algal mat and typical higher plant cover. Scale is about 30 cm long.

algal marsh shows much more evidence of very intensive desiccation. Typically, the thin mat is rumpled, bubbled, and cracked in irregular patches. The cracked parts of the mat commonly are rolled up, peeled off, and even fragmented into granule-sized chips that litter the surface (Fig. 27).

Beach-Ridge Hummocks. (Exposure Index > 99.9%) The highest features on the tidal flats are the low, narrow, discontinuous, sandy, pine-covered hummocks of the beach ridge (Figs. 9B & 21). These sand mounds, elongated parallel to the shoreline, although reaching a maximum elevation of only 1.3 m above MTL, are only overtopped by the sea during hurricane surges. They are essentially a string of small, isolated "islands" that cause funneling of the winter storm floodtides into the channel openings, and, during the larger storms, over the beach-ridge washover plains. These hummocks are essentially terrestrial environments, as is testified by the flourishing stands of Casuarina pine trees and smaller hardwood shrubs and bushes, the drinkable groundwater, the burrows of the large land crab Cardisoma (Shinn 1968), and the pine-needle litter over a sandy soil-like surface. Tree-rings of a large Casuarina trunk growing near the mouth of the Three Creeks main channel indicate an age of at least forty years for the trees, so that these high, pine ridges are quite stable features.

The sediment making up these hummocks consists of a porous, very coarse sand- and granule-sized skeletal hash. Trenches cut into this sediment reveal a strong landward-dipping (ca. 30°) inclined bedding (Shinn et al. 1969, Fig. 15). We did not do any extensive coring into these hummocks, but the few cores we took and the trenches we dug showed that these high ridges are just coarse sand caps (up to 50 cm thick) over homogeneous pelleted mud (containing gastropod tests and red mangrove roots that point to old pond sediment).

Both the beach terrace seaward and the beach-ridge backslope landward of these hummocks are somewhat different from those of the washover lobes. In

Fig. 28. Channel bar in the axis of a small distributary channel. This is an "intertidal" bar, showing deep prism-cracks and a *Schizothrix* algal cover.

places on the narrow ledge between the pine hummock and the beach, wide, flat, skeletal sand bars (several meters across and 10 to 20 cm high) have accumulated where the hummocks have blocked landward transport. These bars are of very loose sand easily moved by flood currents. Landward, the well-vegetated hummocks pass quickly into a dry, cracked and chipped high algal marsh zone and no mud chip-strewn backslope surface typical of the washovers or the levees occurs.

C. The Channels and Channel Bars

Shinn et al. (1969, pp. 1205–07) have described and illustrated the morphology of the channels of the Three Creeks area. A good idea of the shape and scale of the tidal channel systems can be gained from Figure 6. Careful measurements of aerial photographs by Tom Tourek (1968, unpublished manuscript) showed that active and abandoned channels cover 14% of the surface of the Three Creeks channeled belt. The channels vary from mere ditches no more than a meter wide and 20 to 30 cm deep at their "headwaters" in the ponds and inland marsh, to

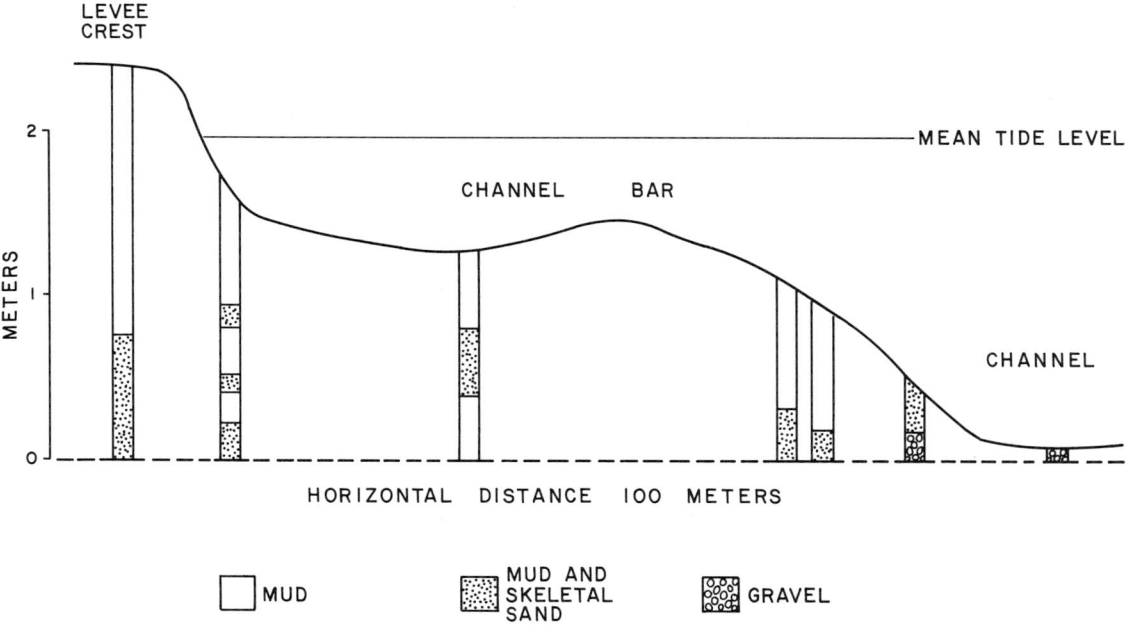

Fig. 29. Sediment core profile across a "subtidal" channel bar which is attached to channel bank. Note that the lag gravel in the main channel cannot be traced across the bar.

wide creeks up to 100 m across and 3 m deep near their mouths. These deep seaward reaches of the channels generally cut down to the Pleistocene bedrock.

Every major and secondary channel must be navigated with caution, because the thalweg winds from side to side between long (up to 300 m or more) bars (maximum relief 1½ m; usually submerged at high tide) that hug the inside bends either as directly attached point bars or separated from the bank by a very shallow tidal "chute." In some of the straigher parts of the smaller channels, bars have developed in the axis of the channel, making a small central island looking like a whaleback (Fig. 28). Tourek (1968, unpublished manuscript) estimated that bars cover 1.6% of the surface area of the whole channeled belt at Three Creeks.

Most of the channel bars are largely permanently submerged, their crests exposed only during abnormally low tides (Exposure Index < 2%). These submerged bars, as well as the channel floors, are covered by a "soupy," flocculent, organic, slime-rich, muddy surface sediment which supports a crop of *Thalassia* and *Diplanthera* grasses and the algae *Batophora* and *Laurencia*. Seasonally (usually in winter when salinities are highest because rainfall dilution is minimal), blooms of small *Penicillus* plants cover these bars. *Callianassa* burrow entrances commonly riddle the bar surface.

Although most channels appear to be floored with a somewhat sandy gravel of skeletal debris, intraclast lumps, and Pleistocene bedrock fragments, this "lag" generally cannot be traced laterally into the bars (Fig. 29). Perhaps, as Tourek (1968, unpublished manuscript) suggests, the bars are not really point bars, but start in the middle of the channel between the ebb-tide thalweg and the flood-tide thalweg (Ahnert 1960) and grow by vertical rather than lateral accretion, aided by vegetation fixation and sediment trapping. In this case the bars would not migrate, but instead when the channel widens by outer bank erosion a new bar forms in the middle of the new channel and the old bar simply gets "silted in." This would certainly explain both the unusual morphology of the bars and the lack of continuity of the channel lags. Channel migration is certainly going on at present, as is demonstrated by the steep, caved-in banks and large, slumped blocks on outside bends and along some reaches (Fig. 30), but the process is

Fig. 30. Slumped blocks along a small channel. Levee to the left, small point bar to the right.

exceedingly slow: aerial photographs taken twenty-five years apart show little significant migration.

In some of the small, tightly meandering distributary channels leading into ponds, the channel floor is blanketed at the bends with a well-sorted shell gravel. *Peneroplis* and gastropod shells up to 1 cm long are moved back and forth as current ripples (amplitudes up to 10 cm) with each tide. There is no organic binding here, and the shells are free to move with the current.

There are some bars that stand rather higher than average and are exposed as much as 90% of the time. These "intertidal" bars usually are coated by a fairly cohesive *Schizothrix* mat and may even support a growth of red mangroves. Commonly, one observes the multiple burrow entrances of *Alpheus* and the pellet piles of *Marphysa*. At low tide the most impressive aspect of these bars is the polygonal cracking; the sediment is cut into large polygonal columns (20 to 30 cm wide) by "prism"-cracks that are many tens of centimeters deep (Fig. 28).

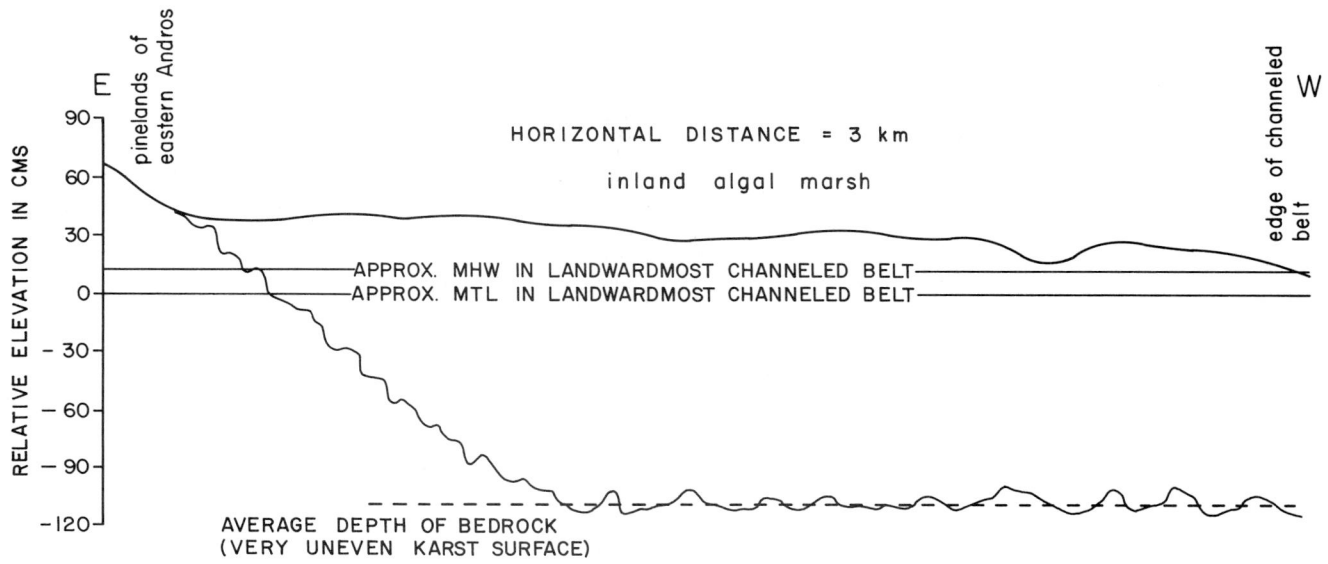

Fig. 31. Levelled profile of inland marsh surface, showing the essential flatness of the surface (elevation variation of only 30 cm over 3 km distance). Note, however, the small undulations that make ponding of water a common feature of the inland marsh.

Usually at least the upper few centimeters of these prism-cracked bars are beautifully laminated on a millimeter scale (Chapter 5).

D. The Inland Algal Marsh

At the landward edge of the channeled belt the channels are very shallow and sinuous and the bordering levees are so small that the entire area is virtually a single, broad, shallow, channel-dissected pond. This wide pond complex shoals gently landward and ultimately passes into a bordering *Scytonema* algal marsh zone (Figs. 6 & 31), precisely as we find at the shores of the inter-levee ponds further downstream. However, here, some 3 to 4 km from the sea, the marsh is not simply a narrow fringe zone, but is a vast meadow of *Scytonema* that stretches 4 to 8 km eastward to the pine-covered mainland and runs for at least 50 km along the length of the island. This extensive *inland algal marsh* makes up the largest single depositional subenvironment of the entire northwest Andros Island tidal flats (Fig. 6), covering an area of between 200 and 400 km².

The boundary between the inland marsh and the channeled belt is sharply marked, as aerial views such as Figure 6A show so well. It is a scalloped boundary, indented where the shallow ($< \frac{1}{2}$ m deep) creeks—the "headwaters" of the major tidal channel systems—drain the rainfall runoff from the marsh. These creeks cut back into the marsh less than 1 km before completely losing their identity. This narrow channeled fringe of the inland marsh shows the same *Scytonema* zonation as does any pond shore in the channeled belt. A lush meadow of spongy *Scytonema* "pincushions" (a low algal marsh zone) rims the pond complex and borders the channels that extend into the inland marsh (Fig. 32A). Landward, this "pincushion" zone merges upslope into a continuous carpet of flat, tufted *Scytonema* (Fig. 32B), a high algal marsh zone. In several places along this seawardmost edge of the inland marsh, the surface is gently raised into low, flat crowns that carry only a patchy growth of flat *Scytonema* lightly cemented with high Mg-calcite to make lithified stromatolitic heads, just as occurs at the upper edge of the levee high marshes; between these *Scytonema* mounds the sediment surface is covered by a white, paper-thin, brittle, aragonitic crust that cracks like thin ice when walked on (Fig. 33; see also Chapter 7). Such "algal crust" areas can be seen on the aerial photograph in Figure 6A as the light colored patches between the channels along the scalloped edge of the inland marsh.

Fig. 32.
Two views of the inland algal marsh surface: (A) *Scytonema* "pincushions" that cover the low depressions of the marsh. Note the mass of dwarf red mangroves. Photo taken in wet summer period. Each "pincushion" is 5 to 10 cm across. (B) Dried and cracked, fleshy, flat, continuous *Scytonema* mat that covers the "high" ground of the inland marsh. Photo taken in dry winter period.

A

B

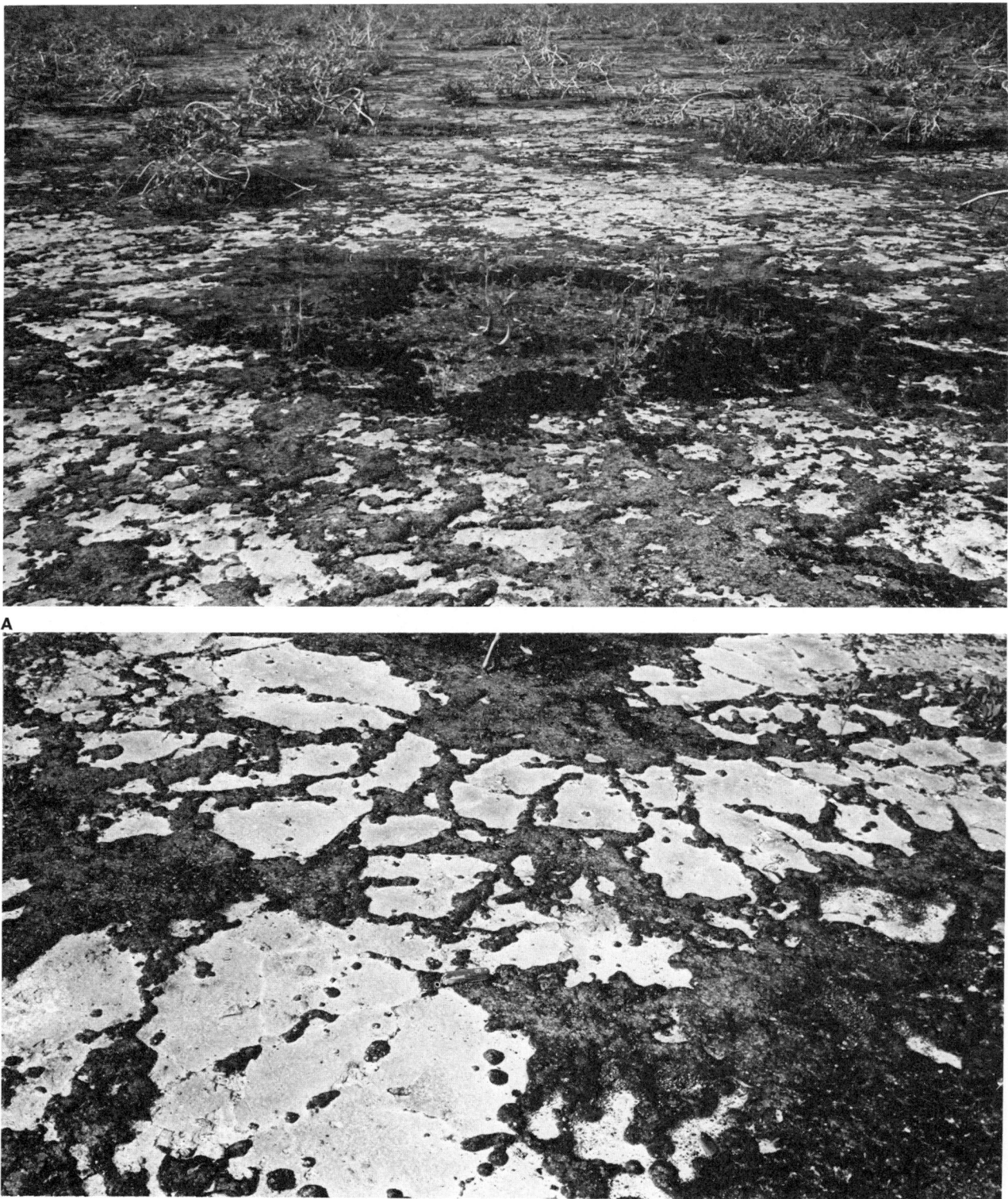

Fig. 33. Two views of a high algal marsh "crown" along the seawardmost edge of the inland algal marsh: (A) General view of the surface which consists of a very lightly cemented paper-crust (*light areas*) and patchy growths of flat *Scytonema* mats (*dark grey areas*). In the low depressions where water remains ponded the *Scytonema* develops into small "pincushions" (*very dark areas*). Small red mangroves (30 to 50 cm high) give the scale. (B) Close-up of the surface showing how the *Scytonema* develops over cracks in the paper-crust where moisture can seep through from the shallow groundwater table. Winter aspect: this area may be under 10 to 20 cm of brackish water in the summer. Small circular "heads" in foreground are ½ to 2 cm across.

GENERAL ENVIRONMENTAL SETTING

The main body of the inland marsh stands at least 15 to 20 cm above mean high water of the channeled belt (Fig. 31). The surface undulates gently and somewhat irregularly, but is essentially a rather flat, featureless plain that rises gradually to meet the Pleistocene bedrock of the mainland to the east. Figure 31 is a carefully leveled profile typical of the inland marsh topography. The entire surface of the marsh proper is covered by a continuous carpet of *Scytonema*, a thick (½ to 3 cm) tufted mat of coarse (up to 40 μ wide), heavy-sheathed filaments and filament bundles, with a widespread but scrubby growth of dwarfed (?) red mangroves (< 1 m high) and clumps of grasses. In the broad topographic lows, where water can become ponded for weeks at a time, the *Scytonema* cover consists of "pincushions" or large, fleshy "lily-pads"; over the higher somewhat drier areas the *Scytonema* takes the flat mat form (Fig. 32B). Nowhere in the interior of the marsh did we find living invertebrate communities, either epifaunal or infaunal, although Shinn et al. (1969, p. 1214) report the land crab *Cardisoma* and the land snail *Cerion* (living on stems and leaves of plants).

Except at the shoreward edge, the marsh is not influenced by the day-to-day tides, because mean elevation is above normal tide levels of the channels. Also, the tide range is damped to near zero by the time the tide wave reaches into the marsh. It is essentially a *freshwater* marsh; in the summer the heavy rainfall keeps the marsh under freshwater for long periods of time, and it is this that promotes the lush growth of *Scytonema*, which, as Black (1933) pointed out, is primarily a freshwater alga (see also Monty 1967). In the winter months when little or no rain falls, the marsh is completely drained and the *Scytonema* cover becomes dry, leathery, and cracked. In places, winter drying has been severe enough that the mat is fragmented into isolated "polygons," which have become draped by successive algal growths to produce Black's Type C algal heads (Black 1933, p. 25, and Figs. 4 & 8).

The sediment below the *Scytonema* carpet of the inland algal marsh is beautifully layered into thick laminae and thin beds (1 mm to 10 cm thick) seen as alternations of white and brownish sediment layers (Chapter 5). The layering extends all the way down to bedrock, a thickness in some places of as much as 1.7 m.

5 LAYERING: THE ORIGIN AND ENVIRONMENTAL SIGNIFICANCE OF LAMINATION AND THIN BEDDING

Lawrence A. Hardie and Robert N. Ginsburg

1. Introduction

In Chapters 1 and 4 of this report on sedimentation on the modern carbonate tidal flats of northwestern Andros Island, Bahamas, we outlined the main objectives of the study and described the general environmental setting of the study area (Fig. 1). The present chapter takes up the problem of the description, origin, and environmental significance of layering in these carbonate tidal deposits of Andros Island.

Of all the sedimentary structures in modern carbonate tidal sediments, layering is the most obvious and the most easily studied, yet is perhaps one of the least understood features. The existing reports (see below) provide valuable data on the macroscopic nature of layering in the tidal zone of South Florida and the Bahamas, the Qatar Peninsula and Trucial Coast of the Persian Gulf, and Shark Bay in Western Australia. These observations, and especially the illustrations of the layering, are the basis for our recognition of analogous layering in ancient carbonates. Valuable as these first-order descriptions are (and many more are needed), they are insufficient to answer more detailed questions about specific aspects of the depositional environment, the sedimentary processes, and the periodicity recorded in the layering.

It is our purpose to present a step toward fulfilling this need, with the results of a comprehensive study of the types of layering in a single modern carbonate tidal flat, their characteristic morphologies, their rates and modes of accumulation, their precise position in the tidal zone and in the overall sequence, and the relative roles of physical sedimentation and mats of blue-green algae. We first describe and illustrate the macro- and micro-features of the different types of layering and relate them to the surface features and specific setting of the different subenvironments in which they are found. Next we present the results of rates of deposition experiments we carried out in the field over a $2\frac{1}{2}$-year period and their implications for the time value and mechanism of deposition of the layering. Particular emphasis is given to the origin of the uniform millimeter-lamination that characterizes the most elevated subenvironments of the tidal flats. Then we focus on the criteria for recognizing "Andros-type" layering in the geologic record. Finally, the question of the distribution of layering, its preservation and destruction, and its implications in environmental and stratigraphic reconstruction are discussed.

2. Previous Work on Layering in Modern Carbonate Sediments and the Need for More Intensive Study

Pioneering work on tidal flat stratification has been carried out on the modern silici-clastic deposits of the North Sea (see, for example, the review of early work in Bucher 1938; and the more recent work of van Straaten 1954, 1961; Evans 1965; Reineck 1967, 1972; Reineck & Singh 1972; Reineck & Wunderlich 1969). In the North Sea area the tidal range is high (3 to 8 m), the climate temperate, and the source of sediment distant, so, unfortunately, the results have only limited application to carbonate environments of low tidal range, tropical climates, and more local sediment supply.

There are several descriptions of layering in modern carbonate tidal flat environments. Black (1933) first pointed out the major role played by mats of filamentous blue-green algae in producing distinctive lamination on the tidal flats and nearby freshwater swamps of Andros Island, Bahamas. Black's careful observations remain unsurpassed. Ginsburg et al. (1954) (see also Ginsburg 1955) found similar algal-laminated sediments in the South Florida area. They showed how algal mats could actively trap sediment locally on wave-swept shores. By marking the sediment surface they demonstrated that in a restricted environment like Florida Bay sediment laminae were deposited by winter storms. They recognized two major types of layering: (1) laminae of sorted, fine sand alternating with thinner, algal-rich laminae; this type of lamination is characteristic of the seaward edges of mudflats and mangrove islands; (2) thick laminae and thin beds of muddy sediment separated by thick masses of algal debris; this type of organic-rich layering is found in the ponded interiors of mudflats and islands. Shinn et al. (1965, 1969) illustrated the macro- and micro-features of layering on the northwest coast of Andros Island, the study area of the present report. They found that layering is confined to sediments in the upper intertidal and supratidal zones. From their work we can distinguish three major kinds of layering: (1) flat millimeter lamination on natural levees of tidal creeks and on beach ridges; (2) millimeter lamination with "palisade" structure in cemented crusts that occur just above the normal high water mark; (3) thick lamination and centimeter-scale thin beds in the fresh water algal marsh. They pointed out the abundance of algal filament molds and the graded internal fabric of the lamination. Illing, Wells, and Taylor (1965) showed the exclusive development of lamination on intertidal algal flats of the Qatar area of the Persian Gulf. They noted that these "stromatolitic algal laminations" can be used to recognize the buried algal flat environment in cores put down through the supratidal sabkha. Kendall and Skipwith (1968) made the same general observation for the Trucial coast of the Persian Gulf but they were able to recognize within the intertidal algal flat four algal mat zones, each with a different surface mat morphology. In both these studies the layering is illustrated in core-slabs, but details are not shown nor described. A most careful recent description of modern carbonate sediment stratification is that of Davies (1970a) on the "intertidal" algal-laminated sediments of the Gladstone embayment in Shark Bay, Western Australia. Davies described and illustrated two types of layering, both intimately associated with algal mats: (1) "high sediment volume, low organic content" laminae, found in the outer intertidal zone; these laminae are essentially flat and show little "deformation," only "micro-unconformities"; (2) "moderate sediment, high organic content" layers found in partly blocked tidal channels; deformation of the laminae in the form of fragmentation and overfolding, shrinkage cracks, and gas bubbles is characteristic of these sediments, as is the presence of thick, coarse, graded "storm deposits." Davies emphasized two internal features of the laminae—well-preserved molds of algal filaments and graded bedding. The mechanism of deposition of the layering was not discussed in detail, but Davies stated that the "algal-laminated sediments are composed of clastic grains that have been trapped and bound by an algal film." Coarse graded beds he ascribed to deposition by "storms or cyclones"; "cyclic" alternations of thin silt beds and algal laminae he believed to be a winter-summer seasonal cycle, and the flat laminae of the outer intertidal zone he implies are due to semidiurnal tidal flooding. Most recently Logan et al. (1974) have described the relation between sediment fabric and the successive zones from subtidal to supratidal of distinctive algal mats in Hamelin Pool in Shark Bay, where magnificent Holocene stromatolitic structures occur. Logan et al. aimed their study at these stromatolitic structures and did not specifically describe the detailed features of the lamination. However, they emphasized the importance of mat types in determining the fabric of the sediment and showed from dye-marker experiments that accretion rates are in the order of 1 to 10 mm/yr, and in the middle and upper intertidal zones, "sediment influx is mainly limited to transport during storms and abnormal tides."

These descriptions of layering in the tidal zone of modern carbonate environments give an association of flat and undulating lamination that closely resembles that of many ancient carbonates. Indeed, the general similarity between recent and ancient lamination has been used repeatedly as one of the diagnostic attributes of deposition in the intertidal zone; for example, by Fischer (1964), Laporte (1967), Roehl (1967), Matter (1967), among many others. Yet there remain many unanswered questions about the origin and environmental significance of layering in carbonate tidal deposits. Even when the "intertidal" position of a section of laminated carbonates is well established, through the use of sequence and association of sedimentary structures, fauna and paleogeography, one is immediately led to consider the bewildering variety of changes in the attitude, continuity, and thickness of the layering. What factors produce the changes from flat lamination to undulating, crinkled or vertically stacked, convex-upward lamination within a few centimeters vertically or laterally? What is the significance of the change from lamination to thin-bedding? Why is some lamination characterized by fenestral pores, the loferites of Fischer (1964), but others formed in apparently the same zone have none? Are the laminae in ancient carbonates produced like those in modern sediments, by the alternation of physical sedimentation with the growth of algal mats? If so, which laminae were the algal mats? Are the sediment-rich laminae settle-outs from turbid water or are they traction deposits? Are millimeter laminae varves? or semi-diurnal tidal layers? or storm layers? Questions such as these can only be approached by studying modern examples where it is possible to calibrate the relationship between lamination architecture, processes, and environmental factors. Answers to these kinds of questions would provide us with a depth of understanding of the environments and depositional processes of ancient carbonates that would be orders of magnitude more refined than that which we presently possess.

An equally important reason for further study of the fine structure and mode of origin of tidal flat stratification is to search for features that might distinguish "intertidal" lamination from similar lamination formed in quite different environments. The inherent danger of placing too much reliance on the generalized macroscopic features of lamination is already evident. Laminated carbonates in the Triassic of the Dolomites have many features of the intertidal lamination in modern examples (Bosellini 1967), and they overlie a section of unstratified deposits with normal marine fauna. However, further study has disclosed that the laminae are more likely to have been formed in the subaerial environment as caliche (Bernoulli & Wagner 1970). The need for diagnostic features within laminated carbonates is critical in the subsurface, where only core borings are available. Again, we can best begin to explore for distinctive characteristics of lamination by studying modern examples in detail.

Finally, the need for understanding the time value of different types of lamina and thin bed is underscored by the attempts several workers (McGugan 1967) have made to use stromatolitic laminae as time-markers in the manner of coral growth rings. Without established calibrations the dangers are obvious.

3. Layering in the Andros Island Tidal Flat Sediments: A Description

Hardie and Garrett in Chapter 4 made the point that many of the sedimentary structures in the Andros Island sediments directly record features existing at the sediment surface. This is especially true for stratification, and so we would urge the reader to consult their descriptions and illustrations of the environments in conjunction with the present chapter.

Table 3. Types of layering in the Three Creeks tidal flat sediments

Layer type	Subenvironment (s)	Exposure Index
I. Thin lamination		
A. smooth domal lamination	channel bank	40–90%
B. smooth flat lamination	levee crest, beach terrace, beach ridge washover crest, intertidal channel bar crest	98–99.7%
C. disrupted flat lamination	levee backslope, beach ridge washover backslope	91–98%
D. crinkled fenestral lamination	high algal marsh, inland algal marsh	80–91% (up to 99 in inland marsh)
E. wavy fenestral lamination	high algal marsh-levee backslope boundary, beach terrace	85–95%
F. lamination with "palisade" structure	high algal marsh-levee backslope boundary, inland algal marsh	85–95% (up to 99 in inland marsh)
II. Thin bedding and thick lamination		
A. algal tufa-peloidal mud interbeds	inland algal marsh	approx. 70–99%
B. disrupted fenestral bedding	low algal marsh	60–80%
C. flat-pebble gravel	beach ridge washover crest, low algal marsh, inland algal marsh, channel floor, beach	approx. 70–99% (0–5% in channels)
D. round-pebble gravel	beach, channel floor	0–approx. 45%
III. Thin to thick cross-bedding		
A. rippled skeletal sands	channel floor	0–5%
B. festooned skeletal sands	beach ridge hummocks	98–99.9%
IV. Thick to very thick bedding		
A. bioturbated peloidal mud	pond, subtidal channel bars, beach, offshore	0–60%

Our techniques for collecting samples and impregnating them in plastic have been outlined in Chapter 2. To facilitate comparison with ancient carbonate rocks, the stratification features are described and illustrated in vertical polished slabs and in thin sections of the plastic-impregnated cores. Bedding thickness nomenclature follows that of McKee and Weir (1953).

The sediments below mean tide level in the offshore, beach, and ponds are essentially *unlayered* (for example, Figs. 58A & 59A), homogenized by burrowing crustaceans and polychaete worms (in some channels cross-bedded skeletal sands do survive bioturbation). However, above mean tide level layering is beautifully preserved as (1) thin millimeter lamination ("laminites," typical of the levees, beach ridges, and channel bar crests); (2) thin beds and thick lamination (restricted to the algal marshes); (3) cross-bedded skeletal sands (found only on beach ridges). Within these major subdivisions we recognized several distinctive layer types, each of which is characteristic of a distinctive subenvironment.

The distinctive types of layering we found in the Three Creeks area and the subenvironments in which each occurs are given in Table 3.

I. Thin Lamination **A. Smooth domal lamination.** The surface of many of the banks of the tidal channels above mean tide level is a complex of small, rounded, *unlithified* knobs (up to 10 cm across; see Figs. 11 & 34A) covered by a tough coating of blue-green algae, mainly the tiny filamented ($< 5\ \mu$) *Schizothrix calcicola*, with rare, scattered, large filamented *Scytonema* strands.[1] This algal coating on the knobs is not an easily detachable mat, but is an ingrained pervasive sediment-binding filament complex. When these knobs are broken open they show beautiful subconcentric hemispherical or domal internal lamination with laminae 0.1 to 0.7

---

1. These large filamented *Scytonema* (perhaps also *Rivularia*?) are particularly common at the base of the knobs beneath the overhanging edges, where very little sediment accretes.

Fig. 34. Smooth domal lamination: (A) Close-up of channel bank surface showing the *Schizothrix*-draped knobs between fiddler-crab burrow entrances. Note the oblique ridges across the knobs and the coarse dark *Scytonema* filament stubble at the lower sides of each knob. Scale in inches. (B) Vertical section through plastic impregnated knob, showing the beautifully laminated internal structure. Note the disconformities and fenestral pores. Sample oriented with channel to the right and levee crest to left. Scale in mm. (C) Vertical section through a knob that has accreted over a vertically oriented clast of laminated sediment. Scale in mm.

mm thick (Figs. 34B, C). The laminae are thickest at the crest and pinch out down the sides. For the most part the laminae mimic the surface topography of the knobs, indicating that the morphology of the knobs is due to sediment accretion and not to erosion. An especially notable feature of the knobs are curving ridges that give the surface a fluted aspect (Fig. 34A). The ridges run oblique to the channel bank and appear like low profile current ripple crests. Slabs cut at right angles to these ridges show that the surface lamina ends at the ridge and is absent on the landward (levee) side of the knob. This indicates that at times only the surface facing toward the channel will accrete sediment, producing an asymmetrical dome that can be used as a depositional current indicator (cf. Hoffman 1967). This explains the "unconformities" commonly seen within the knobs (Fig. 34B): selective deposition on the surface facing the current leaves an incomplete lamina (a micro-ripple?) which is draped by succeeding more extensive laminae. In this way the landward-facing side of the knob becomes steeper than the current-facing side (Fig. 34B).

Other significant macro-features of the domes are large tubular perforations that cut cleanly through the lamination, smaller irregular "fenestral pores," and horizontal hair-line "sheet-cracks" that separate some of the laminae (see Fig. 34B). The tubes are burrows made by worms, ants, and insect larvae and have been discussed by Garrett in Chapter 6 of this book. The fenestrae are ovoid to amoeboidal in outline, up to a millimeter long, their longest axes vertical or subvertical, and are usually strung out in zones along bedding. Some of these fenestrae are clearly associated with isolated bundles of large *Scytonema* filaments and so may represent air-pockets trapped among the filaments (primary "holes"). On the other hand, the irregular fenestrae that carry no filaments nor resemble filament-bundle molds are probably shrinkage pores (secondary "holes") produced on drying of muddy laminae. Such drying must also have produced the horizontal sheet-cracks.[2]

In thin section, Figure 35, the internal domal lamination of the knobs is seen to consist of poorly sorted laminae of brownish clotted mud with dark "suspended" peloids and skeletal grains alternating with laminae of quite well-sorted peloids and skeletal grains. The sorted peloid laminae are an open framework of dark, dense, ovoid peloids, lighter colored peloids with somewhat ill-defined borders, and minor amounts of broken gastropod debris and whole tests of tiny foraminifera. Peloids range in length from 30 to 150 μ, but very fine sand of 70 to 100 μ is the most common size by far. Skeletal grains are about the same size as the peloids, but rare curved fragments of gastropod shells may reach 250 μ. Garrett, in Chapter 6, has discussed the origin of the dense peloids. These peloids are hard, indurated, porcelaneous grains that survive wet-sieving of the sediment and are most likely micritized skeletal grains and fragments of *Peneroplis* tests. The light-colored more irregular shaped peloids are probably soft fecal pellets and/or intraclasts slightly deformed by compaction and drying.[3]

The poorly sorted muddy laminae carry dense peloids and skeletal grains of the same size as the sorted peloid laminae but in smaller amounts. The bulk of each mud lamina is a brownish micrite. Close examination under medium power reveals a vague but undeniable clotted texture. In places this clotted texture bears a strong resemblance to that made by the compacted soft peloids in the sandy laminae. This suggests that the micrite was deposited originally as fine sand and silt made up mainly of soft peloids (and mud flocs ?) which have now

---

2. Some of these desiccation features may have been formed when the samples were oven-dried, but many are undoubtedly natural, as examination of samples in the field at the time of collection showed.

3. These knobs are wetted and dried frequently in the natural chain of several events, but because the cores were oven-dried before impregnation with plastic, it is conceivable that some deformation occurred in the laboratory processing.

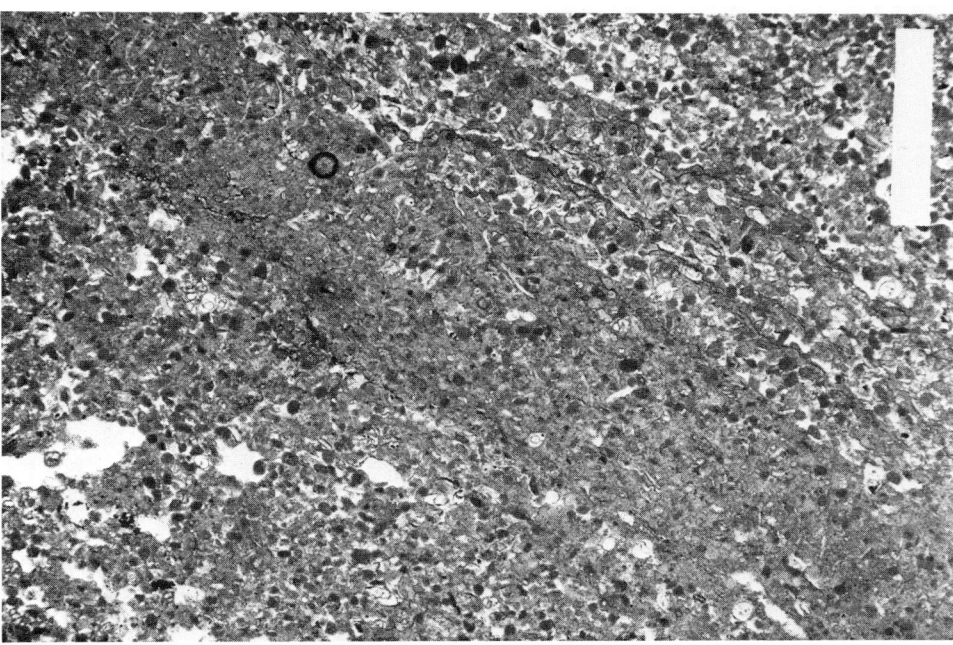

Fig. 35. Thin section photomicrograph of part of a channel bank laminated knob. Scale bar is 1 mm. Note muddy layers alternating with peloid-rich layers. Note wavy, dark stringers which mark remnants of *Schizothrix* surface mats. Thread-like voids at right angles to layering probably represent *Scytonema* algal filament molds. Layering obscure in lower left below thick mud lamina in center; above this mud layer the peloid laminae are well defined and notably discontinuous.

become almost completely deformed and welded together (by drying and compaction).

An especially interesting feature of many of the muddy laminae are little thread-like voids (10 to 30 μ wide and up to 300 μ long) that pierce upward through the clotted micrite, bending around dense peloids (Fig. 35). Careful examination shows that many of these voids do not extend through the entire thickness of the thin section and so are really tubes. It is highly probable that they are algal filament molds. They are too large to be the molds of the dominant alga, *Schizothrix*, and so they probably record the sparse presence of single strands of *Scytonema* (and perhaps *Rivularia*). Such well-defined tubular voids are not obvious in the sorted peloid layers, but many pores have a distinctly vertical trend suggestive of filament molds. In fact, under the low power microscope one gets a distinct impression that even the long axes of the ovoid peloids are roughly oriented in a vertical or subvertical direction.

Boundaries separating peloid and muddy laminae are quite sharp (Fig. 35), indicating that separate events produced the two laminae. However, it is with great difficulty that one can see boundaries between one peloid lamina and another or between one muddy layer and another. Slight changes in peloid to micrite proportions are all that reveal that an apparently single layer 1 or 2 mm thick is actually several superposed thin laminae.

Some boundaries are especially sharply marked by thin (5 to 10 μ), dark, wavy lines (Fig. 35). These dark lines are wispy, discontinuous, and bifurcate around peloids, but they clearly follow the trace of the bedding. Under the highest power in convergent light they look like fine, wispy, hairline fractures, but they go around mud clots and peloids and not through them. This, together with the fact that they invariably have a yellow stained fringe, suggests that these wispy lines are oxidized remnants of algal films rather than micro-cracks produced by drying. If this be so, then these dark lines mark the former positions of the sediment surface and represent a period of horizontal growth of algal mats during a distinct pause in sedimentation.

Individual muddy laminae in many cases can be traced across the entire length of a knob, but the peloid layers are notably discontinuous, rarely can they be followed over more than half the knob. Wedging out of the peloid laminae is sharply outlined by draped-over muddy laminae (Fig. 35). The abrupt terminations of these discontinuous peloid laminae produce the surface ridges discussed above.

It should be pointed out that the knobby structure of the channel bank is a direct result of the burrowing by hordes of fiddler crabs (*Uca*) near the high water mark. The knobs are formed by serial accretion of sediment laminae on the mounds between the crab burrow entrance holes. The steep "gravity-defying" laminae of the knobs (Fig. 34) must be due to agglutination of sediment onto the algal surface coating (which is sticky when wet), so that these knobs are properly termed *algal proto-stromatolites* (see footnote 5, Chapter 4). Now, the channel banks are ephemeral features in geologic terms and so these domal proto-stromatolites are unlikely to be preserved as distinct beds. Their significance, then, lies in the insight they provide into the *processes* that produce stromatolitic lamination, rather than in the environmental information they provide. In this regard it is clear that on Andros Island, *Schizothrix*-type blue-green algae will coat any surface—flat or bumpy—so long as the surface provides mechanical stability and there is sufficient sunlight. In the absence of browsers, this algal coating will thrive (Garrett 1970) and stromatolite formation is possible by sediment agglutination. With these conditions met the external morphology of the stromatolite will then depend mainly on the substrate morphology. If the substrate is "discontinuous," such as the highly burrowed channel banks of Andros Island or the indurated flat pebbles scattered in mobile sand off the headlands of Hamelin Pool, Shark Bay, then laminated stromatolitic *heads* will develop. If the substrate is continuous and flat, then flat lamination will result, there will not be any heads (i.e., no stromatolites in the commonly used sense). One final additional feature to note is the size of grains that are "trapped" by a *Schizothrix*-type mat. The channel bank domes were found to carry *vertical* laminae with peloids as large as 150 μ, well into the fine-sand range. The tiny filamented algae clearly can attach particles many times their own size! The role of blue-green algae in forming lamination will be discussed again later in this chapter.

B. Smooth flat lamination. The surfaces of the levee crests, beach-ridge washover crests, beach terraces, and "intertidal" channel-bar crests are all quite smooth and flat and bound by a tough, pervasive *Schizothrix* filament complex that tinges the white sediment a pink-gray color (see descriptions and illustrations of these subenvironments in Chapter 4). The sediment underlying each of these subenvironments (minimum exposure index is 50%) down to depths of 5 to 60 cm (most commonly 10 to 30 cm) is beautifully laminated on a millimeter-scale; individual laminae range from 0.1 to 2.0 mm in thickness. Characteristically, the lamination appears in core-slabs as alternations of dark and light very thin, smooth, flat to very gently undulating, parallel to subparallel laminae (Fig. 36A). Close examination shows that the dark laminae are extremely fine-grained, uniformly thin, tabular, and continuous, at least on a scale of many centimeters. The light-colored laminae are silty to sandy and typically are discontinuous and lenticular; in places, the upper boundaries of these light laminae have evenly spaced undulations (arrow, Fig. 36A) that almost certainly are the ripples and lineations that are a common feature of the surface of these subenvironments (see, for example, Fig. 12B).

Notable macro-features associated with this flat lamination are open tubular burrows (up to 2 mm across) of ants and insect larvae, horizontal pellet-filled earthworm burrows, roots and brown-red stained root-molds of grasses, shrubs, and mangroves, rows of small (< 1 mm) irregular to ovoid fenestral pores strung out along bedding, and horizontal hair-line sheet-cracks. The earthworm burrows are made by the oligochaete *Pontodrilus* (Garrett, Chapter 6) and are found most abundantly in the beach-ridge washover sediment, where their spectacular development may destroy almost all trace of lamination (Fig. 36B). Some

Fig. 36. Smooth flat lamination: (A) Slab of levee crest sediment, showing typical uniformly thin lamination. Note flat "starved" sand ripples or current lineations near middle of slab (*arrow*) and small sand-filled depression 5 scale units below arrow (see 12B for surface view of these types of features). Two dark horizons at the top are red pigment-markers put down as part of a rate of deposition experiment; lack of strong color contrast between laminae as seen in bulk of the slab is the norm. Scale in mm. (B) Slab of beach-ridge washover sediment, showing destruction of lamination by oligochaete (*Pontodrilus* sp.) burrowing. The pellet-filled burrows have a tendency to run parallel to bedding. Round and tubular fenestrae are unfilled burrows. Scale in mm.

LAYERING

of the rows of vertically oriented fenestral pores may be trapped air-pockets associated with algal filaments (as in the domal lamination of the channel banks) or shrinkage pores on drying. Horizontal, tubular fenestral pores such as those shown in Figure 36B are open, horizontal oligochaete burrows but some may be grass-root molds. It is most interesting that there are *no mud-cracks* in the "supratidal" laminated sediments which are exposed for more than 85% of the time. Only horizontal sheet-cracks and perhaps irregular fenestral pores point to the extreme desiccation these sediments actually experience. Quite contrary to expectations, it is the sediment below mean high water that shows the most spectacular mud-cracks, such as the deep, widely spaced prism-cracks typical of the ponds (Fig. 20B) and of the "intertidal" channel bars we are concerned with here (Fig. 28).

In thin section, Figures 37A and B, the smooth flat lamination is basically composed of laminae of mud, alternating with laminae of peloidal sand, similar to the fine structure of the domal lamination of the channel bank proto-stromatolites. The peloid laminae are typically an open framework of well-sorted silt and fine sand grains, which consist mainly of dark dense sharply ovoid peloids and light-brown irregular to ovoid peloids (30 to 200 μ, with most about 70 to 100 μ) with scattered abraded skeletal fragments (up to 750 μ), whole tests of small foraminifera (up to 500 μ) and large (up to 750 μ) irregular aggregates of pelleted mud and skeletal fragments (probably *Callianassa* fecal pellets). The notable feature of the sandy laminae is their lateral discontinuity and variability in thickness. When these laminae are traced over distances of only a few centimeters they usually either wedge-out (Fig. 37A) like lenticular sand-filled depressions, or they pinch-and-swell (Fig. 38A) like starved micro-ripples.

The muddy laminae appear as brownish micrite in which are "suspended" silt and fine sand grains of dark, dense, ovoid peloids and skeletal fragments. The proportion of "suspended" grains varies from lamina to lamina; it may be as high as 50% and low as 5%. Some of the sandier micritic laminae have an unmistakable clotted internal fabric (Fig. 38B) that strongly suggests that the mud was originally deposited as a sand composed of a mixture of soft and firm peloids, and the soft peloids have been deformed and merged by drying. On the other hand, other muddy laminae appear markedly homogeneous, a tightly packed mix of dense brown micrite and dark peloids, with the peloids restricted to silt size, suggesting that these laminae were deposited as mud and not as sand. The tops of many of the mud layers are marked by a thin (50 to 100 μ) denser zone, sharply and smoothly bordered at the top, but more irregularly connected to the parent layer below (like those shown in Fig. 42A). This dark cap or rind differs from the parent mud lamina in opacity only, which strongly suggests that it represents a cemented crust like the paper-thin *Schizothrix* crusts that are a particular feature of the inland marsh surfaces and some of the levee backslopes (Chapter 7).

The boundaries between the muddy and sandy laminae may be either sharp or gradational. Most commonly, but not invariably, the bases of the sandy laminae and the tops of the muddy laminae are sharp, so that the combination makes a graded bed. A good illustration of this is shown in Figure 37B. However, careful examination shows that if these graded laminae are traced laterally it is usually found that the sandy basal part pinches out, while the muddy top continues quite unchanged! This is clearly seen in Figure 38A (details are given in the figure caption). This indicates that most of the muddy laminae and sandy laminae were deposited by *separate* processes or events. This inference is strongly supported by the common occurrence of homogeneous muddy laminae with sharp tops and bases sandwiched between two sandy laminae (Fig. 37A). However, it must be noted that some individual sandy laminae are clearly internally graded, where the base is a firm peloid sand and the top is mainly clotted soft peloids. Finally, some boundaries, particularly between two muddy laminae, are

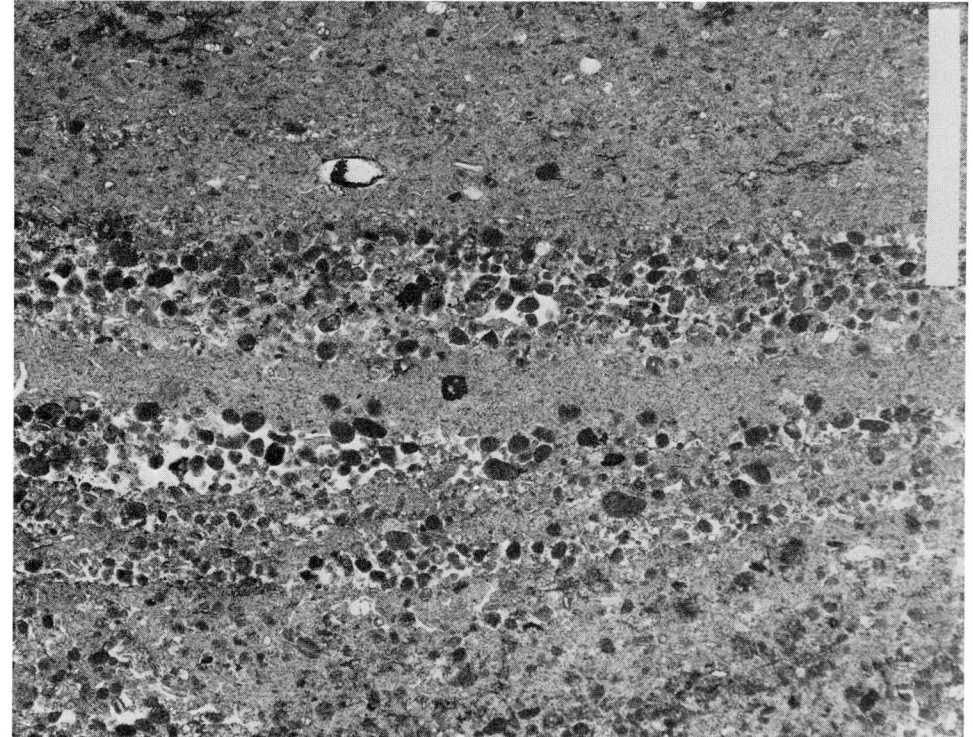

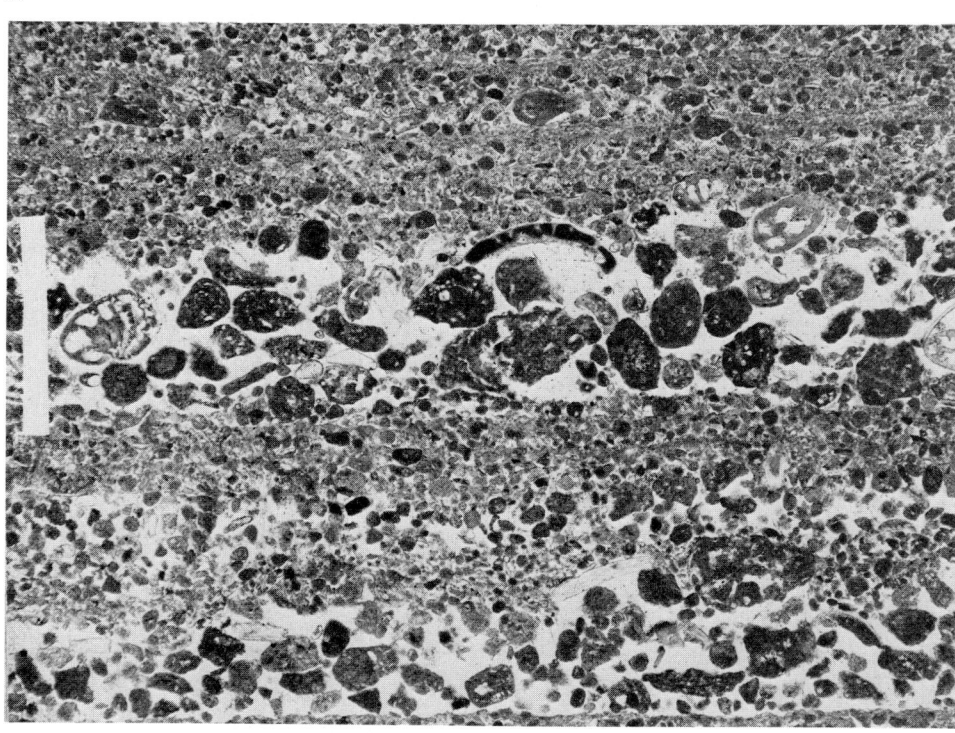

Fig. 37. Thin section photomicrographs of smooth flat lamination: (A) Levee crest sediment. The notable features here are: (i) the basic alternation of relatively well-sorted peloid sand laminae (the dark peloids are hard grains, the light peloids are of soft deformable mud) and clotted mud laminae; (ii) the sharp boundaries between the peloid laminae and mud laminae; (iii) the lenticular structure of the peloid sand laminae which pinch out against the mud laminae, as is beautifully shown in the center of the section. Note that the mud layers at the top and bottom of the section are actually compound layers made up of several muddy laminae that have no peloid sand partings; dark wrinkled tracings (algal mat remnants) mark the boundaries of laminae within these layers (best seen in the top muddy compound layer). Scale bar is 1 mm. (B) Beach-ridge washover crest sediment. Basically the same as in (A) but the peloids are larger, there is more sand, there are more skeletal grains, the sand laminae are thicker. An *apparent* graded bedding is typical here, but in fact if the peloid laminae are traced laterally they pinch out, while the muddy "caps" do not (exactly as shown in (A) above and in Fig. 38A). Scale bar is 1 mm.

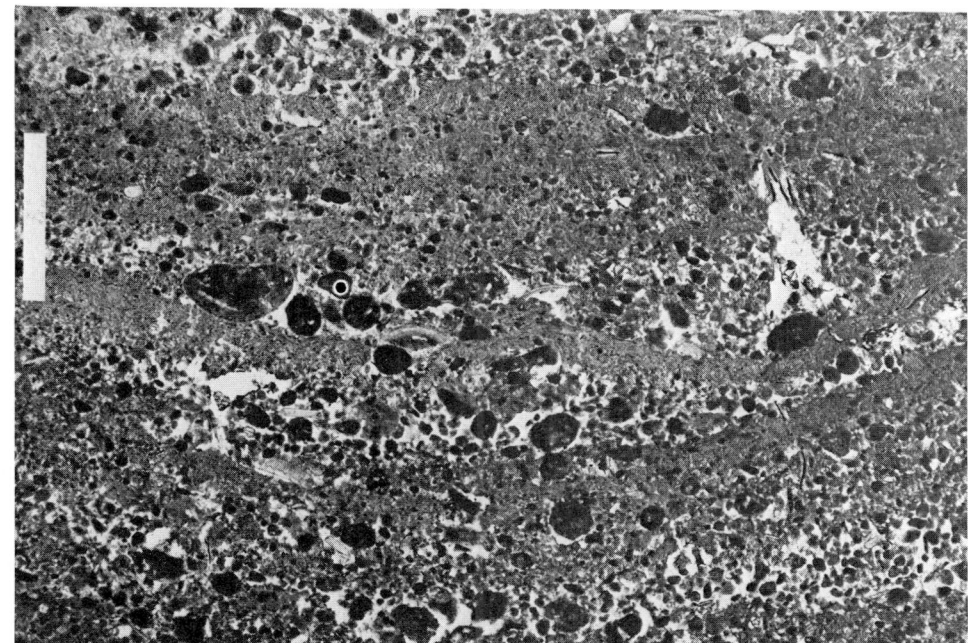

Fig. 38. Thin section photomicrographs of smooth flat lamination: (A) Beach-terrace sediment. This thin section clearly demonstrates the discontinuous "pinch-and-swell" nature of peloid sand laminae and the continuous, even thickness, and draping nature of the thin mud laminae of the smooth flat lamination. Note especially that the peloid lamina just below the scale bar to the left grades upward into the overlying mud laminae, but if this "graded bed" is traced to the right the peloid "base" pinches out while the mud "cap" continues unchanged. These are clearly two separate laminae and not a single sedimentation unit. Note the large vertical fenestra at center right; this is a tuft of *Scytonema* (the dark threads at the top of the fenestra are entombed filaments). Scale bar is 1 mm. (B) Close-up view of a thick compound muddy layer, showing the vague to well-defined clotted (peloid) fabric so typical of the Andros tidal flat mud layers of all types. This clotted fabric is thought to be due to merging (on drying and wetting) of soft, deformable muddy peloids originally deposited as reasonably well-sorted silt. The skeletons are whole empty tests of peneroplid forams (hydraulic equivalents of silt). Scale bar is 500 μ.

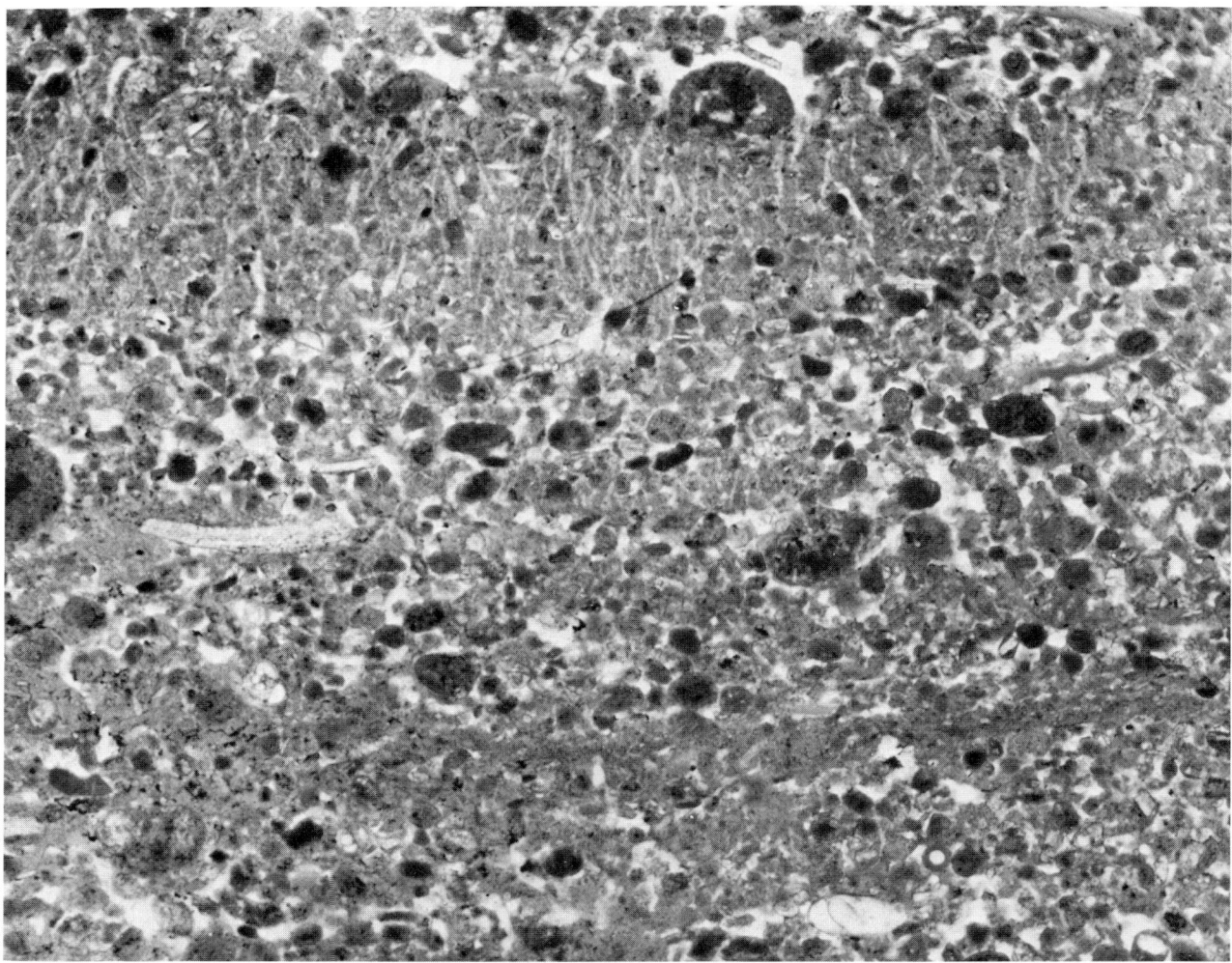

Fig. 39A. Thin section photomicrograph of smooth flat lamination, showing vertical algal filament molds in a muddy lamina. The molds are about 10–15 μ thick. In the peloid laminae, a vertical trend to the voids, undoubtedly also due to algal filaments, can be made out on careful examination. These filament molds are too large to be *Schizothrix* molds, but probably are *Scytonema* or *Rivularia* molds; ironically, then, these molds do not record the presence of the essential mat-forming alga of the smooth flat laminated sediment environments!

marked by wavy, dark lines exactly like those described for the smooth domal lamination; their description will not be repeated here.

Throughout the flat laminated sediment one can find thin, thread-like voids (10 to 30 μ wide, up to 300 μ long) that are very likely filament molds (Fig. 39A), such as are common in the channel-bank proto-stromatolites. These tubular voids are particularly conspicuous in the muddy laminae, but they can also be discerned in the peloid laminae where wispy pores separate nearly vertically oriented peloids.

Filament molds are also associated with some peculiar ovoid to cup-shaped "grains," 1 to 1.5 mm across, found very sparsely scattered in some flat laminated sediment. These large smooth-surfaced "grains" have a well-sorted peloid sand core and a well-defined micrite rind which has a serrated inside edge and is pierced by radially oriented filament molds (Fig. 39B). Their origin is not entirely certain, but they look like cross-sections through sand-filled, pelleted, mud-lined oligochaete or ant burrows.

In summary, the fabric of the smooth flat lamination is characterized by alter-

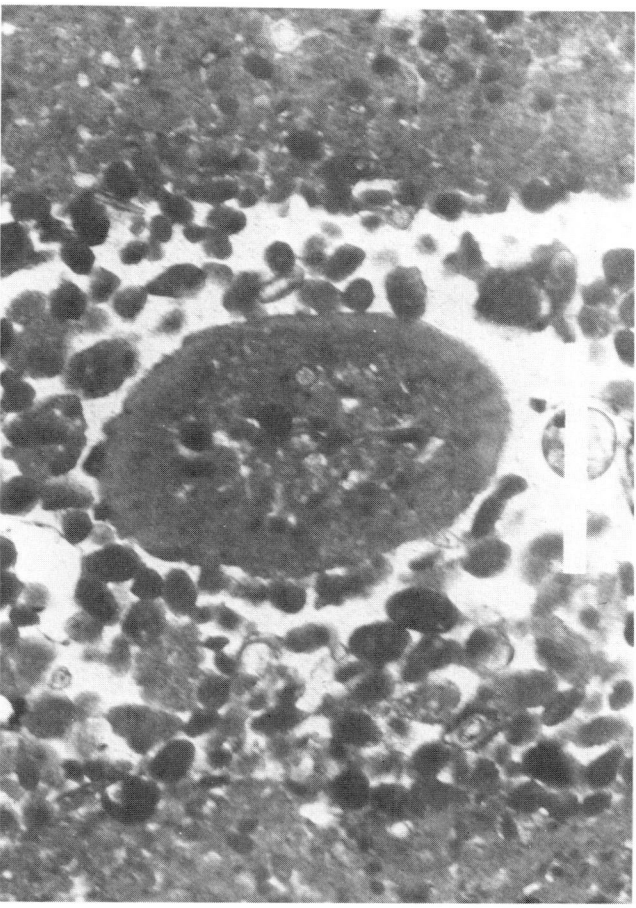

Fig. 39B. Thin section photomicrograph of large ovoid grain with peloid core and muddy rim. Vague traces of radiating filament molds can just be made out. These peculiar grains are found in peloid layers of both the smooth and disrupted flat lamination and appear to be eroded fragments of mud-lined, peloid-filled oligochaete or ant burrows. Scale bar is 500 μ.

nations of persistent sandy micrite laminae alternating with discontinuous lenticular laminae of well-sorted peloidal sand. The variation in thickness from one lamina to another is small, ranging from 0.1 to 2 mm. Many boundaries between laminae are sharp, but diffuse contacts that result in apparent graded bedding are common. Filament molds in the form of tubular voids are typically present, especially in the muddy laminae. These features are common to the smooth flat lamination in all the subenvironments in which such layering occurs, but there are characteristic differences in the lamination from one subenvironment to another. Briefly, it is possible to distinguish the beach-ridge washover and beach-terrace lamination from the levee-crest and channel-bar lamination on the basis of the mean maximum size of sand particles, the composition of the sand, the mean maximum lamina thickness, and the ratio of the number of sandy laminae to muddy laminae. These differences are briefly summarized in Table 4.

C. Disrupted flat lamination. The backslopes of the levees and the beach-ridge washovers above high water (exposure index 91 to 98%) are normally characterized by a *Schizothrix*-bound sediment surface that is mud-cracked into small centimeter-size polygons (Fig. 14). During long periods of nonsedimentation (many months) these backslope surfaces may become colonized by a stubbly *Scytonema* mat as the high algal marsh encroaches up the backslopes. However, new sediment influx will smother the *Scytonema* and a mud-cracked, *Schizothrix*-bound surface will once again prevail (see Chapter 7 for details of the growth habits of *Scytonema* and *Schizothrix*). In places the mud-crack polygons spall off and are washed into depressions by the next influx of water (either rainwash or seawater from overbank flooding) to produce small pockets of intraclast sand and

Table 4. Comparison of features of smooth flat lamination of different subenvironments.

	Beach-ridge washover	Beach terrace	Levee crest	Channel bar
Approx. mean max. lamina thickness	2 mm	1 mm	1 mm	1 mm
Ratio mud laminae to sand laminae	low	low	high	high
Approx. mean max. grain size	750 μ	300 μ	200 μ	200 μ
Sand composition	relatively high skeletal and mud-aggregate proportions	relatively high skeletal proportions	peloids dominant, no mud-aggregates	peloids dominant, no mud-aggregates

granules (Fig. 15). All these processes leave a legible record in the backslope sediment as a distinctive layer type, the disrupted flat lamination.

Figure 40 shows a surface and a cross-sectional view of levee backslope sediment collected a few months after a 1 to 2 mm mud layer had completely covered a thin *Scytonema* mat. The top view shows the mud-cracked surface with the dark, dried-out, buried *Scytonema* mat showing through where polygon chips have flaked off. The sectional view shows the beautifully laminated structure of the sediment. The laminae vary from very thin (0.1 mm) through thin (2 mm) flat to crinkled alternating layers of peloidal sand and mud. Dark, crinkled partings are particularly conspicuous in the upper few centimeters and mark where thin, rumpled *Scytonema* mats are buried and the organic matter is still preserved. The sediment laminae are characteristically *disrupted* by mud-cracks, sheet-cracks, and oligochaete burrows.

The mud-cracks are found only in the thicker laminae; they occur as nearly straight-walled, wedge-shaped openings (ca. 1 mm wide) that reach down no further than the underlying lamina, as the fragmented surface layer in Figure 40 shows. In slabs these mud-cracks are only obvious when filled with sand from an overlying lamina or with dark *Scytonema* growth: they are far more easily recognized in thin sections (see below).

The sheet-cracks are narrow, horizontal, open hair-line fractures that part two laminae along their bedding plane contact. They may be up to 5 to 10 cm long. *Vertical* fractures may connect two parallel horizontal fractures to make a stepped sheet-crack, or they may be small, isolated hair-line openings that quite randomly disrupt the lamination. These horizontal and vertical sheet-cracks, easily seen in Figure 40, are most likely *internal* desiccation cracks produced not at the surface like mud-cracks, but within the sediment (hence under a vertical confining pressure).

The earthworm *Pontodrilus* disrupts and destroys the lamination by building horizontal feeding-burrow galleries that are commonly partly filled with its own fecal pellets and with washed-in peloids, making a spongy porous lense (Fig. 40). The open *Pontodrilus* burrows are seen in slabs as sharp-edged, tubular to nearly circular fenestrae 1 to 2 mm in diameter (Fig. 40). The tubular fenestrae are mostly horizontal, but a few scattered vertical tubes are found. In places the sediment-churning by *Pontodrilus* produces a pocket of fecal pellets, peloids, and flat fragments of uningested laminae that looks like a small lense of a chaotic sandy mud-chip gravel. These bioturbated pockets can be easily confused with the scattered lenses of *mechanically deposited* flat-lying granule- and sand-sized intraclast mud-chips derived by erosion of a mud-cracked surface lamina (Fig. 15). In slabs these clastic lenses are surprisingly uncommon, considering their abundance at the surface: where encountered in cores they are seldom more than

Fig. 40. Disrupted flat lamination. Levee backslope sediment. *Top* is surface view, *bottom* is vertical slab showing layered structure. On the surface where mud polygons have spalled off, the underlying dried and curled *Scytonema* algal mat is clearly exposed. In cross-sectional view the mud-cracked surface layer covering the dark *Scytonema* mat is well shown. Below this surface zone, alternations of thin muddy layers and dark *wrinkled* thick-filamented algal mats emphasize the laminated structure of the levee backslope sediment. With depth the bacterial decay of the mats removes the dark color of the algal partings. About half way down the slab can be seen a back-filled horizontal oligochaete burrow. Scale in mm.

a few centimeters long and a few millimeters thick (Fig. 41). The relatively good sorting helps to distinguish these flat-chip, intraclast "grit" pockets from the earthworm-produced lamina-fragment and pellet lenses.

In thin section (Fig. 42), the disrupted lamination consists basically of alter-

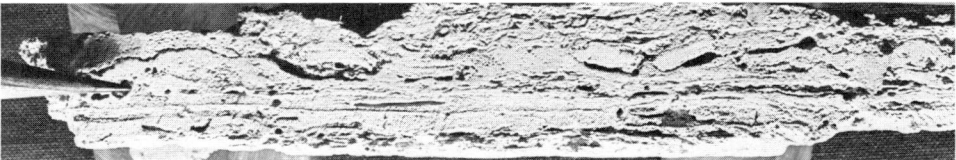
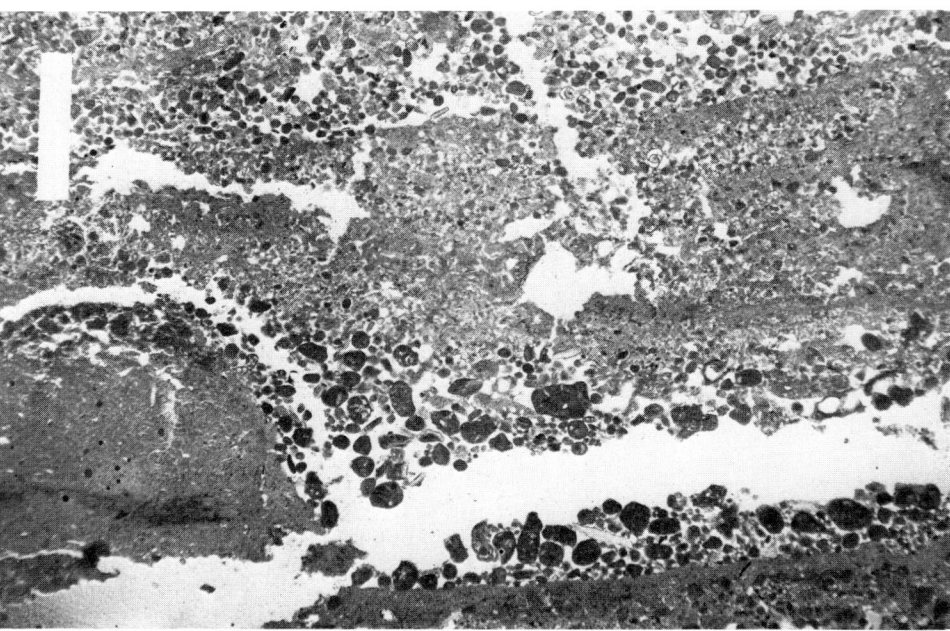

Fig. 41. Intraclast chips in disrupted flat lamination: (A) Slab showing large laminated flat chips in peloidal matrix of levee backslope sediment (*lower edge*). The horizontal fenestrae are oligochaete burrows. Pencil point for scale. (B) Thin section photomicrograph of levee backslope sediment, showing erosional scar (where a mud-crack polygon has been removed by sheetwash) in thick mud lamina filled with large and small peloids. These peloids are very likely intraclasts derived locally by erosion of mud-crack polygons. Compare this with Figs. 14 and 15. Large crack in this intraclast layer is probably a sample-preparation accident. Scale bar is 1 mm.

nations of dense, muddy laminae and porous, peloid-sand laminae. Contacts between these two basic kinds of laminae generally appear quite sharp, much more so than in the smooth lamination that this disrupted lamination resembles under the microscope.

The sandy laminae are characteristically quite well-sorted. Particle size varies from lamina to lamina and encompasses the relatively large range of 30 μ to 3 mm, although most laminae consist of grains 30 to 200 μ, with the average near 70 to 100μ. The grains are mainly dense, dark, stubby, ovoid peloids and light-brownish, irregular to elongate ovoid peloids with minor amounts of skeletal debris (gastropod fragments and whole tests of small foraminifera). The coarsest sand and the granules are flat, rounded, intraclast chips of pelleted mud. Internal grading is not a normal feature of these sandy laminae, but a few layers do show a preponderance of dense ovoid peloids at the base and light irregular peloids at the top. On the other hand, rare laminae were seen with *reverse* grading, having a base of fine dense peloids which passes rather sharply upward to a coarser dense peloid top (Fig. 42B). The geometry of the sandy laminae is typically and notably *discontinuous* laterally on a scale of centimeters; most of these laminae are highly lenticular and appear to represent depression-fills and starved ripples.

The muddy laminae consist of a light-brown dense micrite in which are "suspended" darker dense peloids (up to 100 μ long, but averaging 50 to 70 μ) and skeletal grains (mainly, whole tests of tiny foraminifera up to 150 μ across). These micrite laminae have a vague to distinct clotted texture when examined carefully under medium power, but this is by no means an obvious feature and is decidedly debatable in some cases. What is obvious in most of the thicker muddy laminae is a thin (100 to 150 μ), very dark, dense rind or capping at the top of

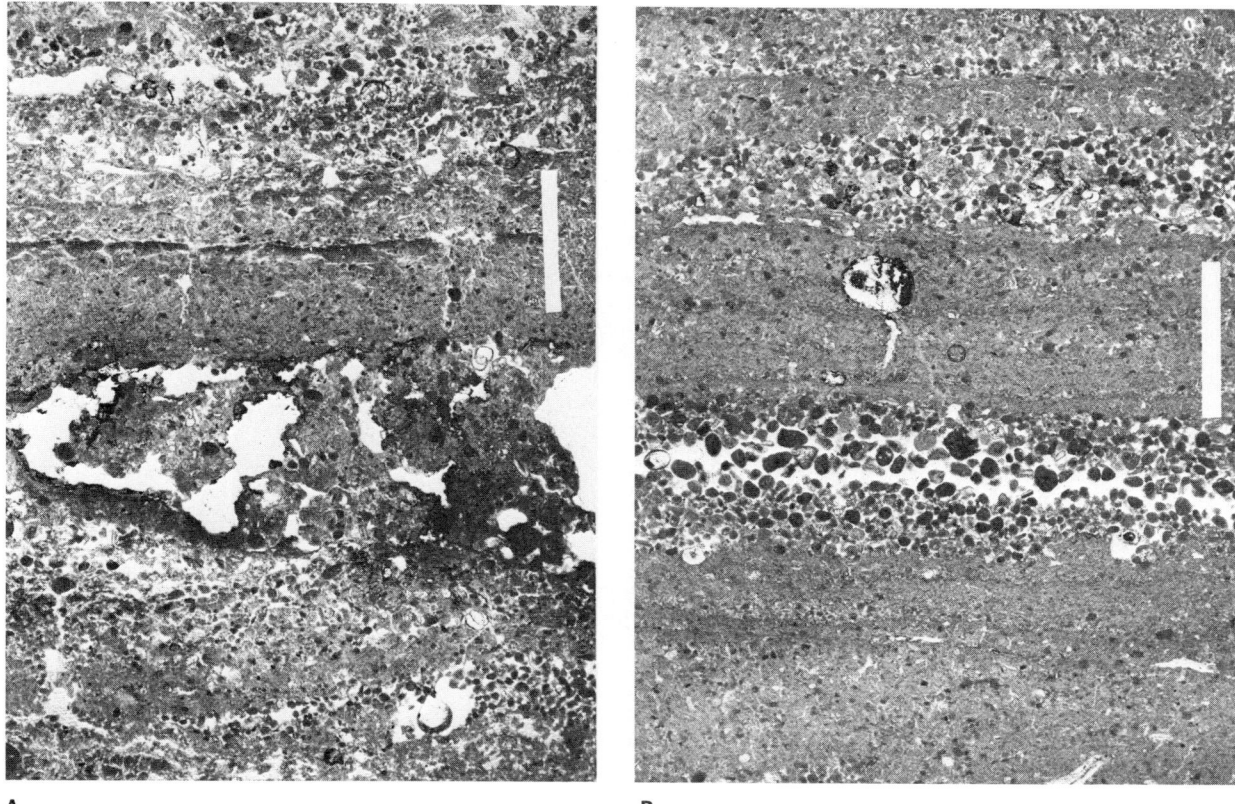

Fig. 42. Thin section photomicrographs of disrupted flat lamination of levee backslope: (A) The significant features illustrated here are: (i) mud-cracked muddy laminae; note sediment-filled, V-shaped mud-cracks to left of scale bar; (ii) dense, dark (cemented?) caps to muddy laminae (see especially layer to left of scale bar); (iii) mass of filament molds and tiny horizontal sheet-cracks concentrated near tops of laminae (see upper part of photo); (iv) disruption of layering by oligochaete burrows; the center of the photo is dominated by a horizontal partly pellet-filled burrow; (v) discontinuous lenticular peloid laminae; a peloid sand lense (starved ripple) draped by a thin mud laminae is seen in lower right hand corner. Scale bar is 1 mm. (B) This thin section shows the sharp separation of peloid sand laminae and mud laminae of the disrupted flat lamination. Note how the mud layers are actually compound layers of as many as six individual laminae, with dark wavy (algal mat) partings. Of special interest here is the thick peloid lamina in the middle of the photo: it shows a well-defined "inverse or reverse grading" with fine peloids at the base grading up to coarser peloids on top. The large circular feature to the left of the scale bar is a mangrove root mold. Note the masses of filament molds in the muddy laminae near the top and bottom of the photo. Scale bar is 1 mm.

the lamina (Fig. 42A); this cap probably is a cemented surface paper-crust, as suggested above for the same features in the smooth lamination. The tops of many mud laminae are also characterized by a highly disrupted breccia-like cap zone made by a complex of tiny, wavy, horizontal sheet-cracks and filament molds (Fig. 42A). Where not churned by oligochaete burrowing the muddy laminae normally can be traced across the width of a large thin section (several centimeters), although a few laminae do wedge-out or come to an abrupt end at an erosion scour (Fig. 41B). The very thinnest muddy laminae (about 100 to 200 μ) are commonly very uniform in thickness and will drape over any irregularity, regardless of its topography; these thin seams sharply outline the lenticular structure of any sandy layers they overlie (Fig. 42A). Finally, two of the most outstanding features of the muddy laminae are mud-cracks and filament molds. The mud-cracks, which are the distinguishing feature of the disrupted lamination, are restricted almost entirely to the thicker mud laminae and appear in thin section as peloid-filled, downward-pointing wedges or straight-walled "dikes"; Figure

42A describes these tiny mud-cracks more adequately than words. The filament molds are small wispy to tubular voids, 10 to 30 μ wide and up to several hundred microns long, that wind horizontally and vertically through the mud, particularly near the top of a lamina (Fig. 42A). They are especially abundant in the crinkled laminae, where they make the clotted-mud fenestral-pore complex so typical of the *Scytonema*-entrapped sediment layers described in detail below under "crinkled fenestral lamination": the description of these crinkled laminae will not be repeated here.

The oligochaete burrows in thin section appear either as a very porous zone of large (250 to 750 μ), irregular mud pellets (Fig. 42A) or as simple tubular openings. Both the partly filled and open burrows cut randomly through the lamination but tend to be horizontal and follow bedding like mine galleries. The pores between the pellets in the partly filled burrows make a complex of randomly linked amoeboidal fenestrae (Fig. 42A). The sheet-cracks appear as the tiny (up to 500 μ long), wavy, horizontal and vertical openings at the tops of muddy laminae noted above (Fig. 42A) and as larger (several centimeters long) irregular but essentially parallel-sided, mainly horizontal cracks.

D. Crinkled fenestral lamination, flat and domal. Much of the inland algal marsh as well as the high algal marshes surrounding the ponds of the channeled belt are covered by a continuous flat stubbly carpet of the large-filamented, tufted blue-green alga *Scytonema* (see Hardie and Garrett, Chapter 4; Black 1933; Monty 1967). The sediment below this flat *Scytonema* mat is beautifully laminated, with white sediment laminae 0.5 to 3 mm thick alternating with dark brown organic partings (Fig. 43A). Although the overall aspect of the lamination may be either flat or domal (see below), the individual laminae are characteristically *crinkled* or wavy on a millimeter-scale (Fig. 43A). Some laminae are even and continuous across an entire slab, others pinch and swell markedly over distances of a few centimeters. The dark organic partings are quite continuous but very uneven and represent the degraded remains of buried *Scytonema* mats. To a large extent, the morphology of an individual sediment lamina is determined by the form of the algal mat the sediment has smothered; for example: (1) where the mat was relatively flat and smooth the sediment lamina above is quite flat and even (see uppermost white sediment lamina in Figure 43A); (2) where the mat was dried, cracked, and rumpled, the overlying sediment lamina is highly crinkled and uneven in thickness (see sediment lamina 4 to 5 mm below the surface in Fig. 43A). Whether this crinkled lamination is essentially flat or raised into domes depends upon the extent of disruption by desiccation cracks. In the absence of widespread, deep, polygonal cracking the crinkled lamination is *flat* over large areas of the marshes. However, where the surface mat and underlying sediment are broken into shrinkage polygons (Fig. 17A) then subsequent deposition of sediment will lead to draping of the upturned edges of the polygons and evenly spaced (5 to 15 cm), "raised disc," *domal*, laminated structures will result. These domal features are the Type C algal heads of Black (1933). Black has carefully described and pictured these algal heads and the reader is referred to this classic work for details.

In addition to the crinkled form of the laminae, this type of lamination, whether flat or domal, is outstanding for its *fenestral* fabric. The fenestrae, seen as dark patches in Figure 43A, are predominantly horizontal (much like bulbous, lenticular sheet-cracks) and vertical (irregular, amoeboidal or tubular, less than a millimeter long). In the upper few centimeters of sediment, these fenestrae are partly filled with *Scytonema* filaments, filament bundles and degraded organic matter; there seems little doubt that the algae and the fenestrae are genetically related. The fenestrae appear to represent (1) primary voids (air-pockets) protected from sedimentation by filament clusters and (2) secondary voids, both individual tubular filament molds and irregular "molds" of the entire mat fila-

Fig. 43. Crinkled fenestral lamination: (A) Slab of sediment from high algal marsh, showing the characteristic crinkled white sediment laminae alternating with dark organic partings, the remnants of *Scytonema* mats. Note how some sediment laminae pinch out laterally. Most significant here are the fenestrae: in the upper part of the slab the fenestrae are partly filled with dark *Scytonema*, but in the lower part where much of the algal material has been degraded by bacterial action the fenestrae are mainly empty. Scale in mm. (B) Vertical slab of high algal marsh sediment treated with acid to dissolve carbonate and reveal the complex meshwork of entombed *Scytonema* algal filaments. Scale in mm.

ment complex. The intense network of *Scytonema* filaments in this fenestral lamination is clearly seen in Figure 43B, which is a vertical slab of high marsh sediment from which the sediment has been dissolved to reveal the incredible filament strand complex that permeates the sediment. These buried filaments are soon broken down by bacteria, so that all that ultimately remains are the fenestrae (see the lower part of the slab in Fig. 43A).

In thin section, Figure 44A, the sediment laminae appear wavy and notably discontinuous; they are separated by thin (30 to 150 μ), red-brown to yellowish, wispy partings of compressed, poorly preserved algal filaments and oxidized, degraded organic matter. As Black (1933, p. 174) noted, it is impossible to recognize individual filaments in thin section, except in the very surface layers where the algae are still alive or newly buried. The dark organic partings commonly bifurcate and strongly outline the lenticular nature of many of the sediment laminae (Fig. 44A). Small twisting branches and wisps of these dark organic laminae pass upward into the overlying sediment and further emphasize the small-scale discontinuities of the lamination. These vertical branches are easily seen in Figure 44A, and the etched slab shown in Figure 43B also illustrates the point well. The sediment laminae between the algal layers are

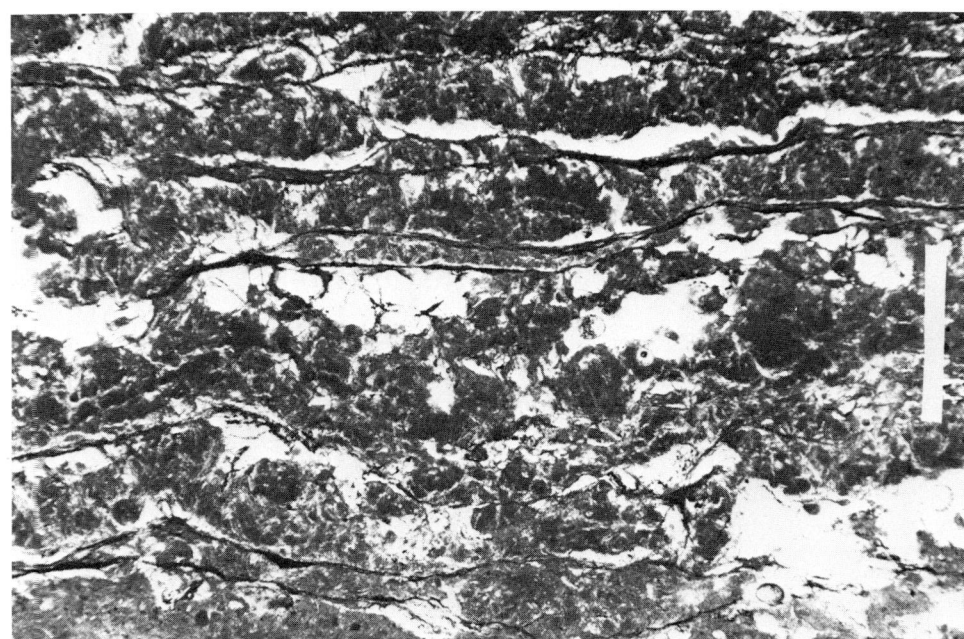

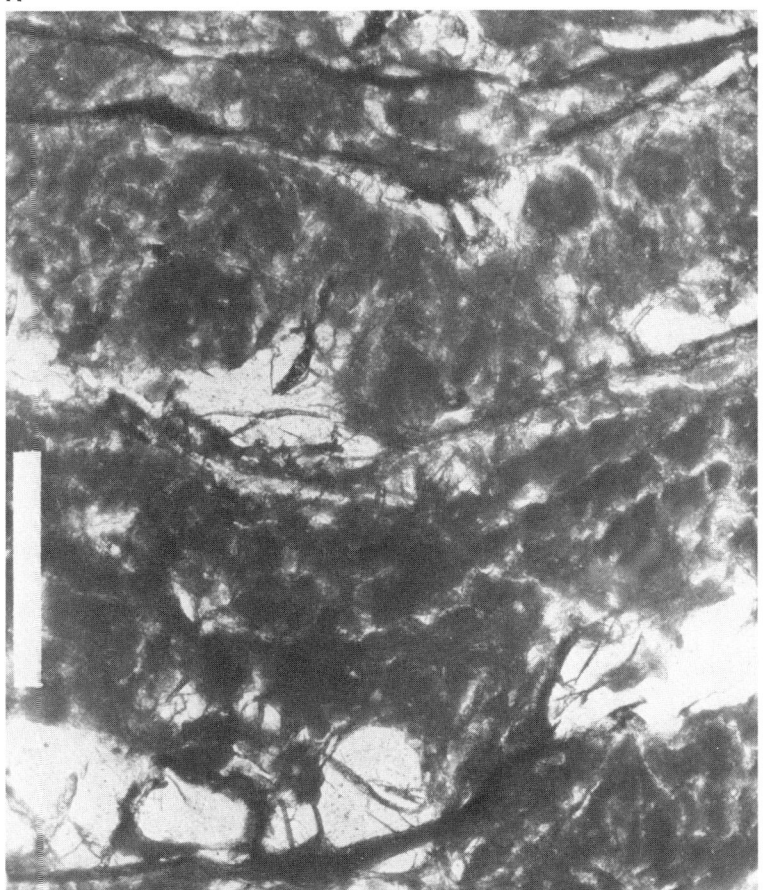

Fig. 44. Crinkled fenestral lamination: (A) Thin section photomicrograph, showing typical clotted sediment laminae separated by thin, dark, wavy organic partings. Note the lenticular form of the laminae emphasized by the bifurcating organic partings. The sheet-like, tubular, and irregular open fenestrae (*white*) are a spectacular feature of this lamination type. Scale bar is 1 mm. (B) Close-up photomicrograph view of crinkled laminae, showing the clotted (soft peloid?) fabric of the sediment and the twisting mass of vertical filaments and filament molds smothered by the sediment clots. The dark bifurcating horizontal layers and vertical wisps are degraded remnants of *Scytonema* mats and tufts. Note especially the open fenestrae. Scale bar is 500 μ.

mainly clotted micrite (soft peloids ?) with scattered, dark, dense peloids (30 to 70 μ) and skeletal fragments (also 30 to 70 μ). The clotted appearance of the mud is very obvious in thin sections and is enhanced by (or perhaps produced by ?) the network of upward-anastomosing tubular fenestrae (filament molds up to 40 μ wide) that permeate all the sediment laminae (Fig. 44B). Although a few laminae do carry abundant firm, dark peloids, the alternations of peloid sand layers and mud layers, so typical of lamination types A, B, and C, are essentially absent in the crinkled fenestral lamination; mud is the overwhelming constituent here. Most striking under the microscope are the large fenestral pores that occupy from 5 to 50% of the volume of the sediment. These fenestrae are irregular to amoeboidal voids up to 1 millimeter across, and sheet-like horizontal openings up to several centimeters long; Figure 44A shows these large fenestrae very clearly. It is interesting to note that if this sediment were to become cemented and the fenestrae filled with sparry calcite, this crinkled lamination would look very much like the loferites of Fischer (1964).

The clotted mud fabric, the lenticular form of the laminae, the "raised disc" structures, and the thorough permeation by a network of filaments strongly suggests that much of the deposition of sediment has occurred within the *Scytonema* filament complex which has acted as a widespread and pervasive current baffle and sediment trap or riffle. The sediment has been mechanically "sieved" out rather than captured by agglutination of particles onto the sticky algal sheaths.

The crinkled fenestral lamination of the inland algal marsh is intimately interbedded with thin sediment beds and algal tufa beds (see below); the lamination zones mark those periods when that part of the inland marsh was covered by a flat *Scytonema* mat, whereas the tufas record periods of more lush growth of cushion *Scytonema*.

E. Wavy fenestral lamination. Lamination of this type is not common and is not found except in association with either smooth flat lamination or crinkled fenestral lamination, but it is a quite distinctive feature. At the upper edge of the high algal marshes, where they encroach on the levee and beach-ridge backslopes, tufts and bundles of *Scytonema* filaments grow along the crests of small ripples and current lineations, but not in the troughs, which are covered by *Schizothrix* (Fig. 16B). These little isolated tufts of large upright filaments preferentially collect sediment by entrainment, or sieving, making porous fenestral muddy clots. These high-standing clots may become draped by smooth lamination or by crinkled lamination, but in either case a distinct wavy lamination over structureless clots is produced, such as shown in Figure 45. Similar draped mud clots are found at the lower edge of the high algal marsh, where tiny pompon-like *Scytonema* growths signal the incoming of lush cushions in the adjacent wetter low algal marsh. Wavy lamination also occurs in the sediment of the beach terrace, where isolated tufts of *Scytonema* and *Rivularia* are common along the seaward edge; the clots in this instance are far more sandy than muddy (cf. Monty 1967, Plate 16-1).

In thin section, the clots consist of an unsorted mass of large and small peloids (both firm and soft) and fine mud; tubular filament molds (living filaments near the surface) up to 30 μ across, and larger irregular to amoeboidal fenestral pores (airpockets between filaments and irregular filament bundle molds) pierce upward throughout the clots (cf. Monty 1967, plate 16-1). The draping laminae have the characteristics of either the smooth flat lamination (beach terrace), or crinkled fenestral lamination (high algal marsh–low algal marsh boundary), or both (high algal marsh-levee backslope boundary).

F. Lamination with "palisade" structure. In slabbed cores this kind of lamination is characterized by thin vertical or near vertical carbonate columns that cut

Fig. 45. Wavy fenestral lamination: (A) Slab of sediment of upper edge of high algal marsh, showing two sediment "clots" where *Scytonema* tufts have entrained sediment. The draping of the clots by thin laminae produces the wavy lamination. Note the new growths of *Scytonema* on the crests of the clots. Scale in mm. (B) Draping of sediment laminae over *Scytonema* entrained clots (*center of picture*) in sediment from high algal marsh—beach-ridge washover backslope boundary. Note the irregular but vertical fenestrae in the structureless clots. Scale in mm.

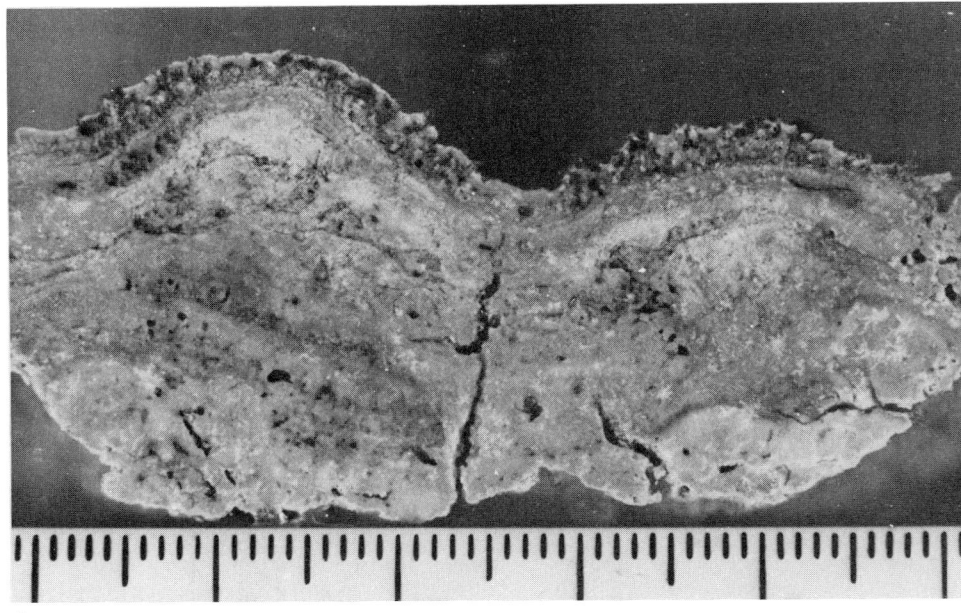
A

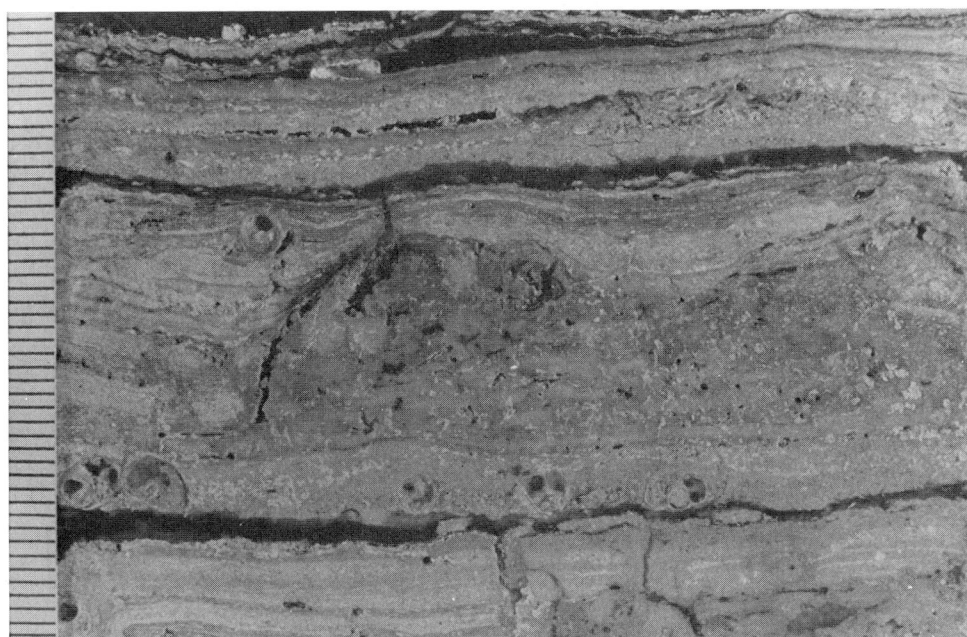
B

upward across the lamination like a fence of pales, a palisade (Fig. 46A). It is the essential structure of most of the thicker, lithified surface crusts (Shinn et al. 1965; Hardie 1969) that form on the upper edges of the high algal marshes of the channeled belt and on high crowns in the inland algal marsh (see Chapter 7 for a description of the distribution of crusts).

Lamination with "palisade" structure can be divided into two subtypes: (i) lamination in which the "fence pales" consist of small individual fibers a few tens of microns across (Fig. 46B), and (ii) lamination in which the "fence pales" are stout columns up to 2 mm wide; in three dimensions these columns are actually *walls* joined in a polygonal honeycombed pattern (Fig. 47). Both subtypes can be either domal or flat laminated (Fig. 46A).

LAYERING

Fig. 46. Lamination with palisade structure: (A) Cemented surface crust from levee backslope—high algal marsh boundary showing both flat (*lower part of slab*) and domal (*upper part of slab*) lamination with columnar "palisade" structure. White layered material is peloidal sediment; dark gray vertically oriented patches are aragonite-cemented sediment "palisades"; vertical and horizontal black patches are open fenestral pores. Note how some adjoining columns in the upper domal part of the crust coalesce upward. Pencil point for scale. (B) Cross-sectional view of lithified mushroom-shaped "head" with fibrous "palisade" structure from inland algal marsh. Black vertical patches are open fenestral pores. Note the rough domal banding. This "head," 3 cm across, is a pure algal tufa with no detrital sediment present.

A

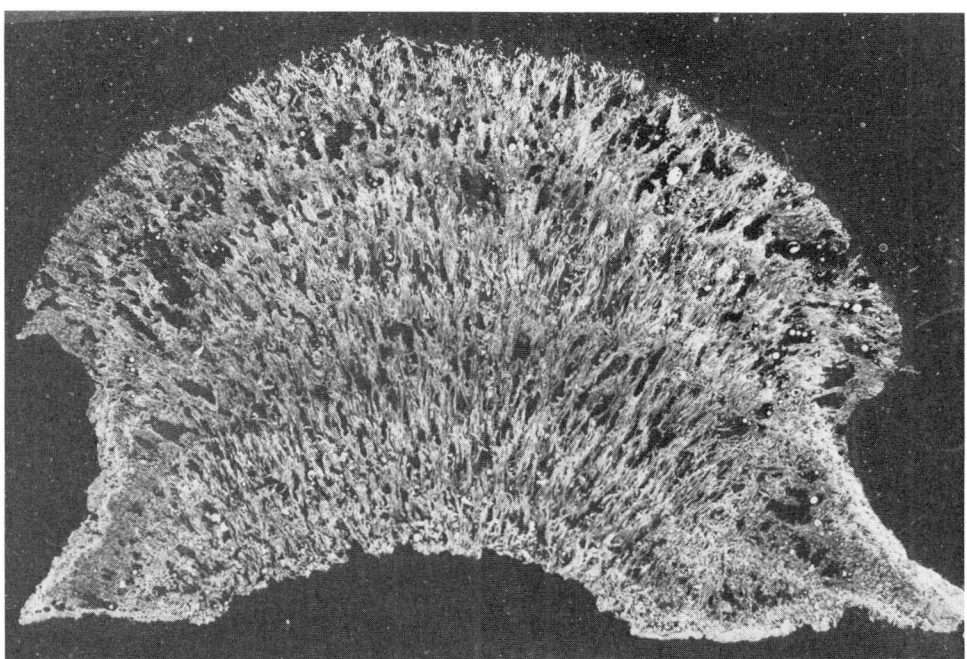

B

The fibrous "palisade" lamination is most spectacularly developed in the inland freshwater algal marsh, but a variation of this subtype occurs in the crusts of the channeled belt. This fibrous "palisade" lamination is mainly domal; it makes mushroom-like knobby structures 1 to 5 cm across (Fig. 17B) and low, flat mounds and sheets with rounded edges up to almost 1 m across (Fig. 33). The internal structure is one of superposed laminae (1 to 3 mm thick) of radiating or fanning very thin fibers (30 to 100 μ wide, up to 3 mm long), a structure powerfully enhanced by elongate, interfiber fenestral pores (Fig. 46B). In thin section, Figure 48, the fibers are seen to be empty crystalline tubes, 10 to 30 μ in wall thickness, made of pale brown-gray micrite; yellow-brown organic matter lines the inner walls of many tubes. The micrite was identified by X-rays as magnesian calcite with 12 to 13 mole % $MgCO_3$. Near the surface of some of these structures *Scytonema* filaments can be seen partly encased in magnesian

Fig. 47A.
Bedding-plane view of "palisade"-structured crust, showing head- and sheet-like mounds with honeycombed surface. Cross-sectional view looks exactly like that shown in Fig. 46 (A). Compare this picture with patchy *Scytonema* mat structure shown in Fig. 17 (B). Pencil for scale.

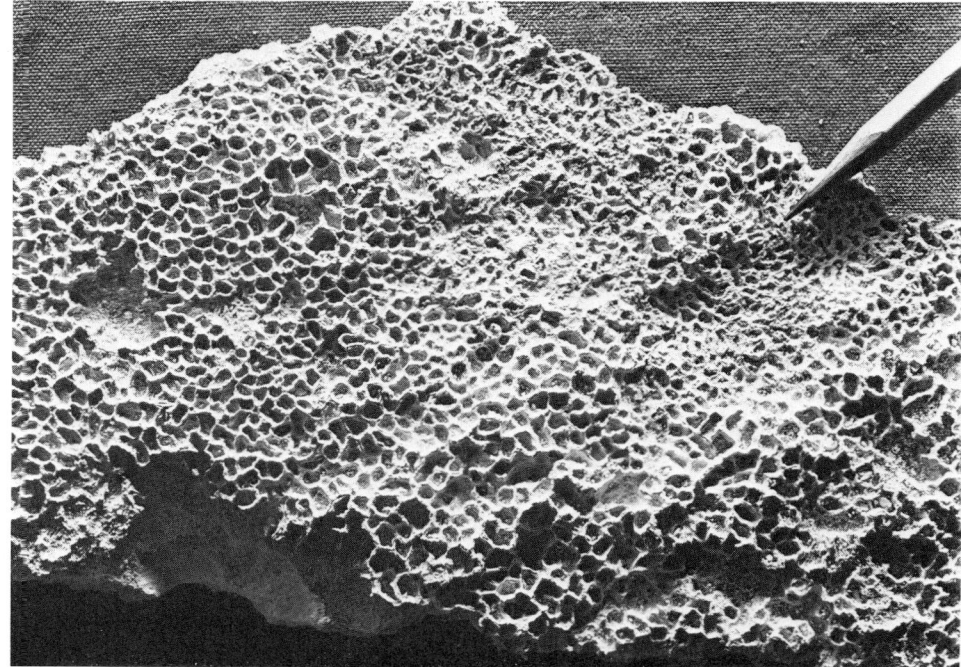

Fig. 47B.
Bedding-plane view of "palisade"-structured crust after the uncemented sediment had been washed from between the aragonite cemented "palisade" network. This honeycomb structure is typical of the stout-columned "palisade"-structured crusts. Single and double hair-line cracks along the crests of these honeycomb walls can just be made out in this picture. Pencil for scale.

calcite, so that there seems little doubt that the crystalline tubes are *Scytonema* filament molds like those described by Black (1933) and Monty (1967) from freshwater lakes further inland on Andros Island. Also, the external morphologies of the knobs and mounds are precisely like those of living *Scytonema* patch mats at the upper edge of the high algal marshes and high crowns of the inland marsh (see Chapter 7 for a more detailed description of the mat morphologies). These magnesian calcite filament molds under the highest power of the microscope appear as a fine-grained mass of tiny ($< 4\,\mu$), clear, blocky (?) crystals that meld together where the tubes touch; they contrast strongly with detrital silt- and sand-sized aragonite peloids that are caught between the tubular filament molds. The sheath calcite appears to be an *in situ* precipitate around *Scytonema* filaments and so is properly termed *algal tufa* (see Chapter 7 and Monty 1967,

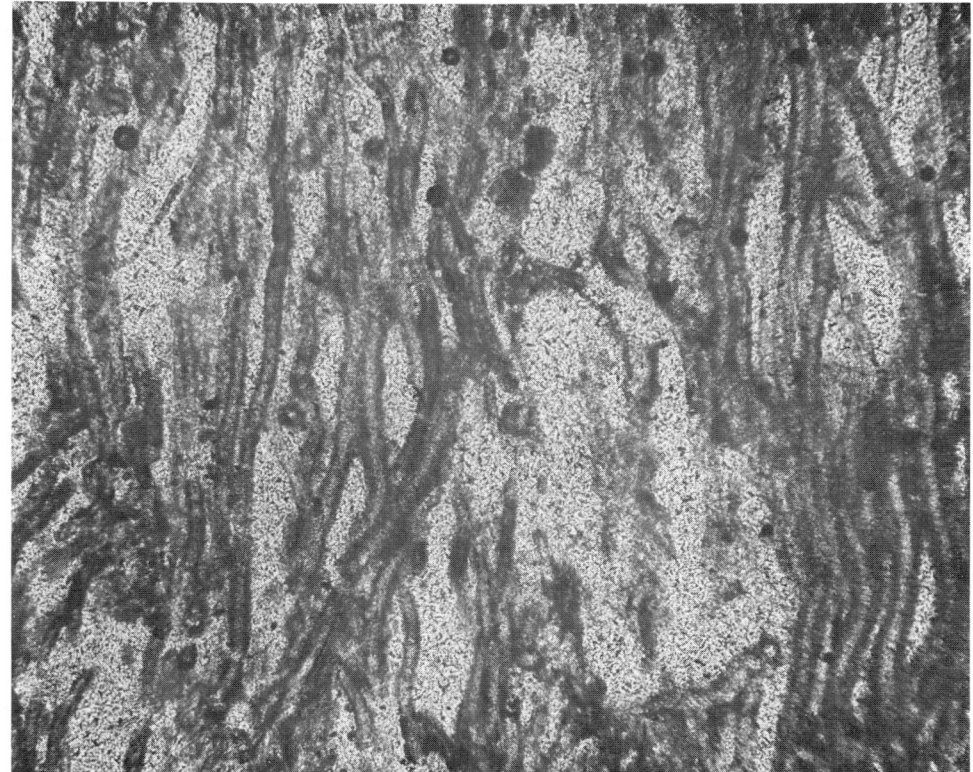

Fig. 48. Thin section photomicrograph of fibrous "palisade" lamination. Close view showing dark Mg-calcite-encrusted tubes (*Scytonema* filament molds) about 50–70 μ wide. Stippled gray areas between tubes are voids (filled here with plastic).

for fuller discussion). These lithified *Scytonema* domes vary from small mushroom "heads" made almost entirely of magnesian calcite filament sheaths (pure algal tufa) to flat mounds and sheets in which detrital sediment predominates. In the pure algal tufa "heads," the domal lamination is marked by color-banding or by irregular "sutures" where one filamentous lamina overtops another (Fig. 46B); the lamination is essentially an algal growth banding. In the sediment-rich structures, *unlithified* aragonite peloid sand laminae and muddy laminae (like those described for the "disrupted flat lamination" above) alternate with laminae made of calcite filament molds smothered by aragonite peloids and muddy clots (Fig. 49). Most of the muddy sediment laminae have dense caps (see "disrupted flat lamination"); these are brittle, paper-thin, *Schizothrix*-mat-bound crusts easily broken into flat sandy intraclasts that get caught between the large filaments of adjacent and upgrowing *Scytonema* mats. Whereas the sediment laminae are well packed, the algal laminae are extremely porous (up to 60% porosity); large irregular to columnar fenestral pores are a spectacular and characteristic feature even in those algal laminae carrying much trapped (entrained or "sieved") sediment. The sediment-rich "heads" could be termed SH stromatolites, although most are really flat-topped, irregular, low mounds rather than hemispheres; Logan et al. (1964) did not provide a niche for these structures in their stromatolite classification. An interesting variation of the tubular "palisade" lamination is found in the hard crusts of the high marsh edges in the channeled belt. The crusts are sediment-rich and usually a compound of subtype (i) and (ii) "palisade" lamination, with encrusting, domal, fibrous-palisade laminae overlying flat, columnar-type palisade laminae (Fig. 50A). The fibrous crusts have a strong *Scytonema* filament tubular structure in hand-specimen, but when viewed under the microscope, rather than sheaths around filaments the

Fig. 49. Sediment-rich fibrous "palisade"-structured domal lamination: (A) Slab showing unlithified domal sediment laminae pierced by vertical lithified tubular fibers. Note how the "head" ends abruptly against intraclastic sand (unlithified) to the left and overlies an unlithified peloid sand lamina (*white*). Note the prominent vertical fenestrae. Scale in mm. (B) Thin section photomicrograph of sediment shown in (A). Mg-calcite micrite-encased empty tubes (*Scytonema* filament molds), 50 to 70 μ wide, smothered by dark irregular shaped peloids. Gray stippled areas are voids (filled with plastic).

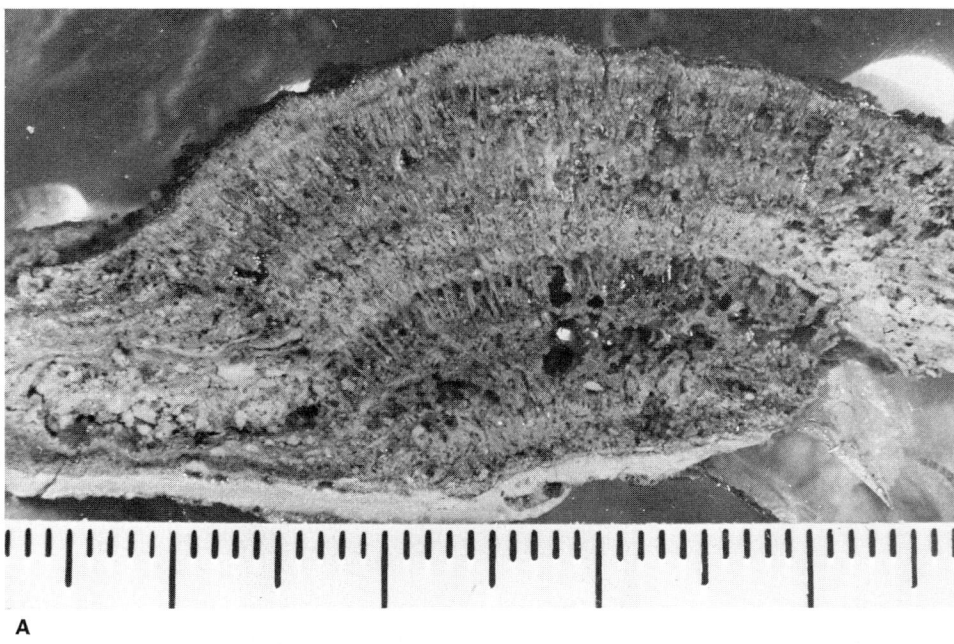

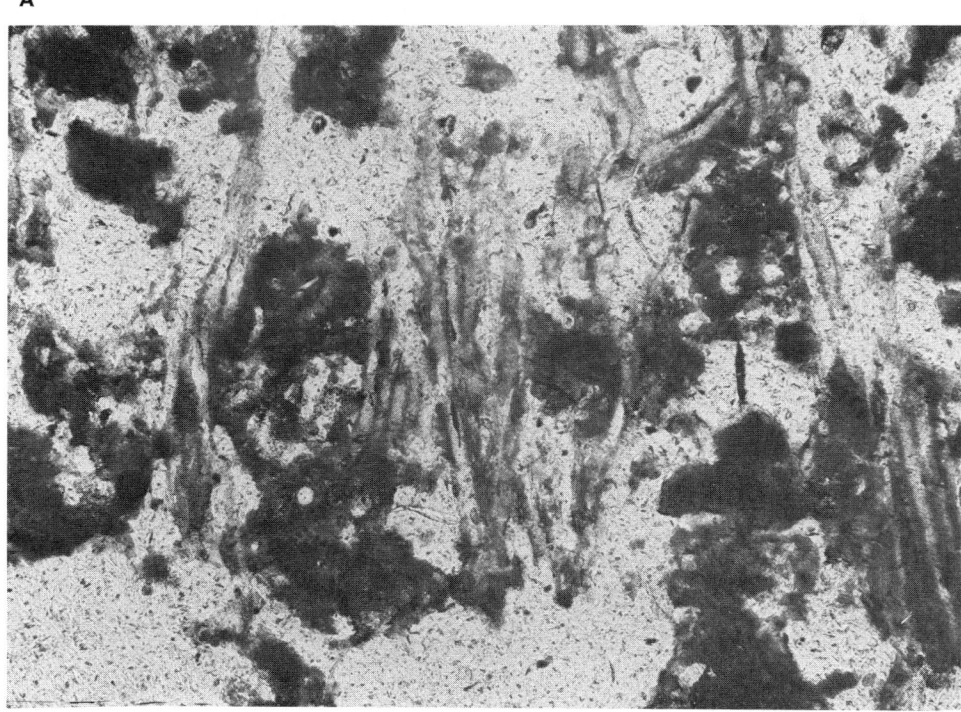

fibrous structure is seen to be filament molds in clotted peloidal mud (Fig. 50B). The clotted mud and peloidal matrix is dense and indurated, but the only obvious sign of cement is a fibrous aragonite that coats the walls of filament molds and fenestral pores. A typical X-ray analysis of this indurated matrix shows approximately 45% aragonite, 35% high magnesian calcite, with a composition of 39 mole % $MgCO_3$, and 20% magnesian calcite with a composition of 12 mole % $MgCO_3$. Many of these fibrous crusts have large volumes highly bioturbated by earthworms; these churned patches are of dense peloidal mud pierced by large (up to 1.5 mm wide) irregular to tubular fenestral pores (typi-

LAYERING

Fig. 50. Lamination with "palisade" structure: (A) Slab of a cemented crust from the high algal marsh-levee backslope boundary, showing fibrous domal "palisade" structure overlying flat lamination with stout columnar "palisade" structure. Note (especially at the left) how the fibrous "palisade" structure has been partly destroyed by oligochaete burrowing. Note also how the laminae in the flat laminated bottom part bend up slightly where they meet the vertical columns. Pencil for scale. (B) Thin section photomicrograph of domal fibrous upper crust shown in (A) above. The radiating *Scytonema* filament molds are a spectacular feature of these crusts. The clotted peloidal mud matrix of this sample carries about 25% very high magnesian calcite (43 mole % $MgCO_3$) cement. Note the large irregular oligochaete burrows at top of photo. Scale bar is 500 μ.

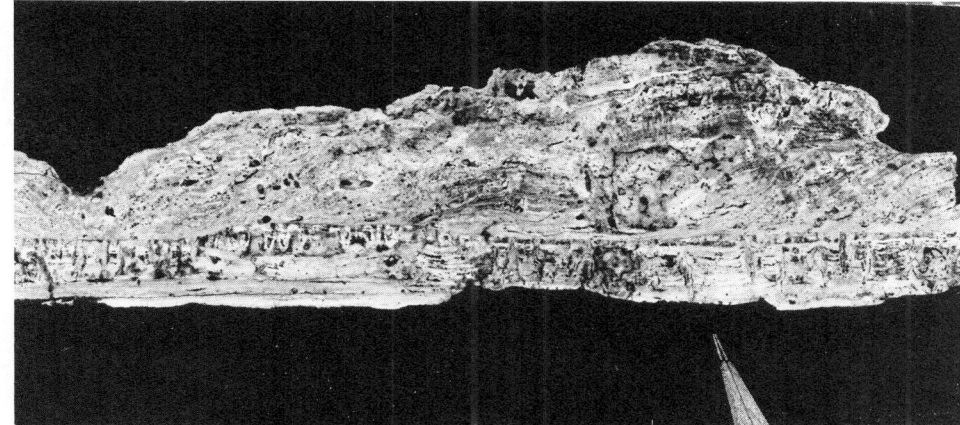

A

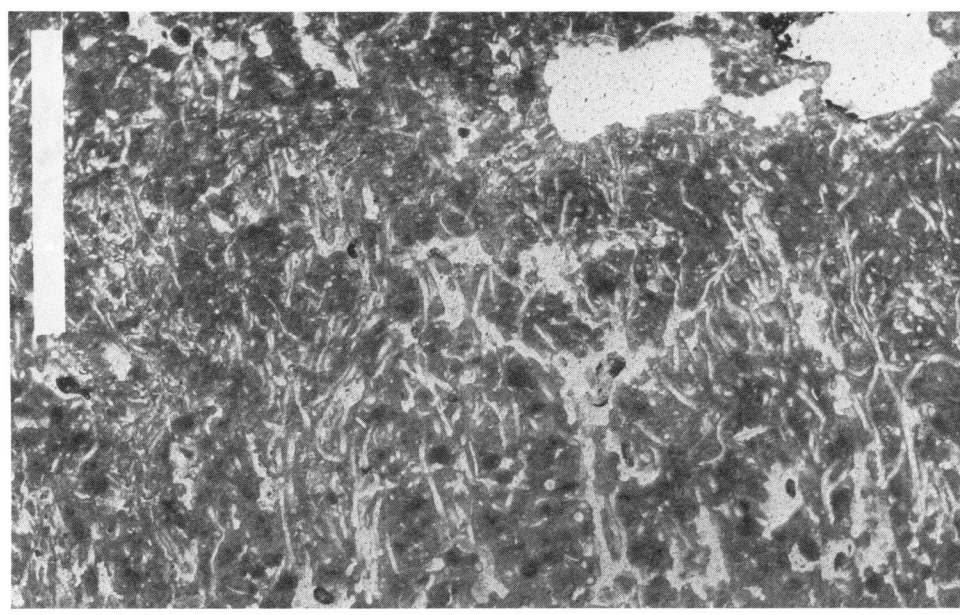

B

cal of oligochaete bioturbation, see "disrupted flat lamination"). Filament molds have survived in these patches, but are not nearly as abundant or as radially disposed as in the unchurned areas.

The second subtype of lamination with "palisade" structure, that with stout, polygonal walls, is found only in the sediments and crusts at the upper edge of the high algal marshes of the channeled belt. This subtype is most commonly flat-laminated, but where domal examples do occur they make very distinctive structures, both in slabs (Fig. 46A) and in three dimensions (Fig. 47A). The domes vary from small, rounded knobs (up to 10 cm across) to more extensive flat, sheet-like mounds (up to 60 cm across). In slabs the blue-gray color of the columns makes them stand out strongly against the pale tan host sediment. The columns of both the flat and domal forms range from 1 mm to 1 cm in length, are up to 2 mm wide, and usually (but not invariably) are spaced no more than 5 mm apart (Fig. 46A). The columns are not straight-sided or V-shaped, but are quite irregular in shape, they may bend, bulge, taper, and fan out in a variety of ways, and in some cases adjoining columns may even coalesce upward (Fig. 46A). The sediment between the columns is laminated on a millimeter-scale and commonly the laminae turn upward where they abut against the columns. Large (up to several millimeters across), irregular fenestral pores, mainly *horizontally* oriented (like wide, discontinuous sheet-cracks), are typical of the intercolumn laminated sediment (Fig. 46A). Flat and domal lamination with this type of columnar palisade structure may occur together in the same crust (Fig. 46A). In the flat laminated structures the columns are vertical, but in the domal structures they fan with the curvature of the lamination, remaining roughly perpendicular to the lamination. In many outcrops of these stout, "palisade"-structured crusts the intercolumn sediment is not lithified and can be easily washed out with a jet of water to reveal the truly striking honeycomb wall structure of the selectively cemented "palisades" (Fig. 47B). In thin section, the columns are seen to be made of detrital sediment which consists of ovoid to irregular dense peloids (30 to 200 μ) and clotted mud; some columns are entirely an open framework of well-sorted, fine sand peloids. The sediment between the columns in the flat laminated structures is similar to that described above for the "disrupted flat lamination," while in the domal structures the intercolumn sediment is like that of the "crinkled fenestral lamination." Characteristically, the columns are segmented by vertical twisting crack-like openings and bifurcating tubular fenestrae (filament molds?) that usually stretch the length of the columns. By contrast the host laminated sediment has mostly horizontally oriented sheet fenestrae (sheet-cracks?) and amoeboidal fenestrae (horizontal algal mat molds or primary air pockets?). The thin sections leave little doubt that the columns are sediment-filled openings, casts of a polygonally patterned set of vertical "cracks." In samples such as that shown in figure 47B, the honeycombed walls look very much like mud-crack fillings, but a mud-crack origin is not supported by the very close spacing (usually < 5 mm), the irregular wall cross-section shapes, coalescence of adjoining columns, and the twisting fenestral pores. We believe that this honeycomb structure is a record of the mat structure of *Scytonema*, for the following reasons. First, careful examination of *Scytonema* mats free of sediment shows that the large vertical filament tufts are not random, but are organized into a decided honeycomb pattern (Fig. 51A), similar in both size and shape to the palisade structure. Second, the mound morphology of the palisade crusts is exactly that of *Scytonema* mats. It is worth noting here that in the knobby mounds the palisade structure does not extend into the enclosing laminated sediment; if the columns are mud-cracks then the mud-cracks formed only on the mounds and not in the sediment between the mounds. Third, thin sections of unlithified, sediment-inundated *Scytonema* mats that are now overgrowing palisade-structured knobby crusts reveal an obvious columnar structure to the *Scytonema*. Figure 51B shows living *Scytonema* filament tufts, separated by sediment (note the upturned edges of the laminae), that bear a striking similarity in size, shape, and spacing to the columns of the palisade-structured crusts. If these filament tufts were to dry out or die and decay they would leave vertical honeycombed openings that would become filled with peloids and mud by the next major sediment influx. Further discussion of the distribution, origin, preservation by cementation, and significance of this striking palisade structure is given in Chapter 7.

II. Thin Bedding and Thick Lamination

A. Algal tufa-peloidal mud interbeds. This kind of thin bedding and thick lamination is found exclusively in the sediment of the inland algal marsh and volumetrically is by far the most abundant type of layering on the Three Creeks tidal flats. This inland marsh is a vast meadow of the freshwater alga *Scytonema*, covering between 200 and 400 km^2 (see Chapter 4 for description).

The sediment below the *Scytonema* carpet of the inland algal marsh is characteristically beautifully layered into vivid alternations of white and brown-gray to tan beds that may be as thin as 0.5 mm or as thick as 10 cm (Fig. 52). The layering extends all the way down to bedrock, a thickness in some places of as much as 1.7 m. Wide and random vertical variation in layer-to-layer thickness is the main feature of the marsh bedding. Individual beds thicken and thin

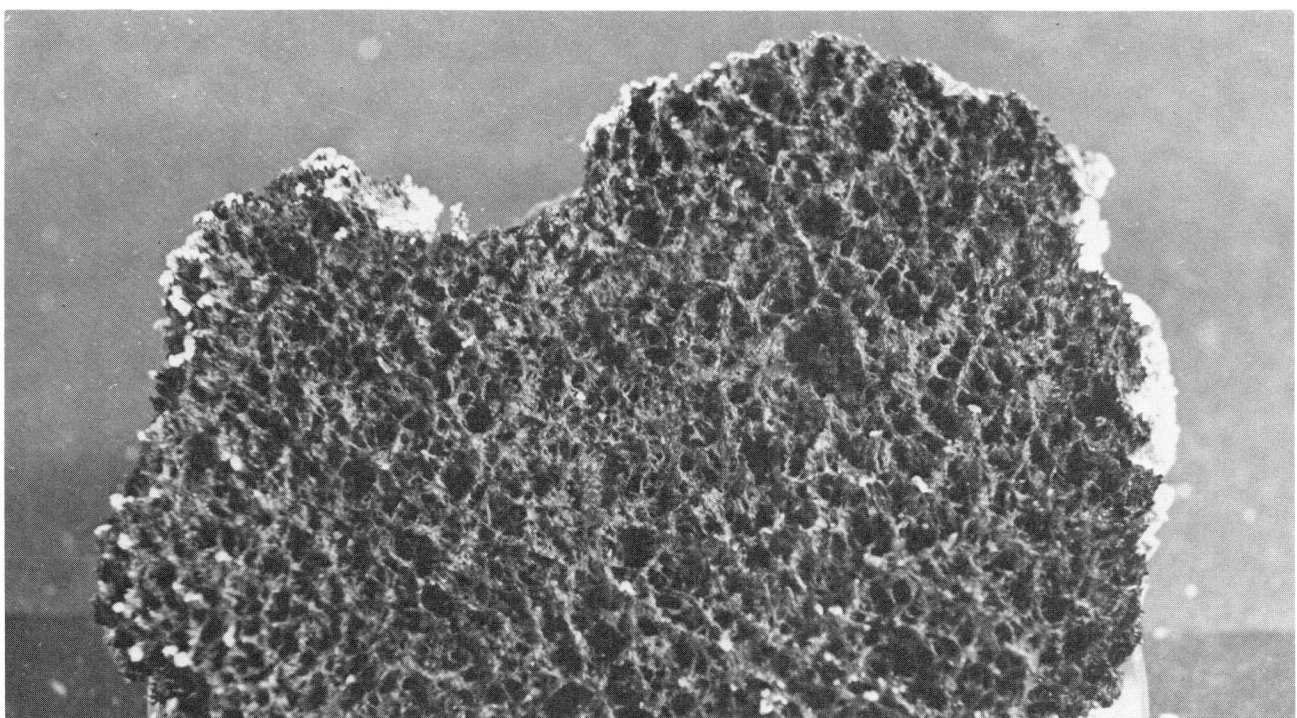

Fig. 51A. View looking directly down onto the surface of a living *Scytonema* mat to show the honeycomb-like structure of the filament tufts. Compare with Fig. 47B. Sample is 9 cm across.

Fig. 51B. Thin section photomicrograph of sediment-smothered living tufted *Scytonema* mat, showing the vertical columnar structure of the algal tufts (actually bundles of filaments). The morphologic similarity to the columns in the "palisade" crusts, such as shown in Figs. 46 (A), 50 (A), is striking. Scale bar is 1 mm.

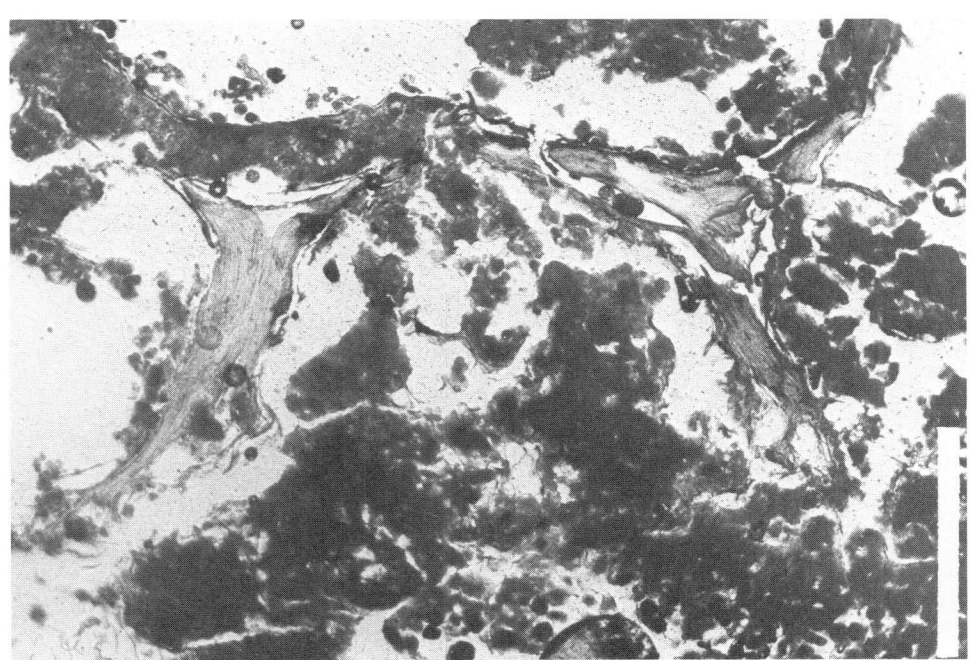

laterally over short distances (see the excellent illustrations in Shinn et al. 1969, e.g., Fig. 22), but in general they can be traced in trenches for many tens of meters and correlated in cores 1 to 3 km apart. Also, typically, the entire sediment is cleanly pierced by a network of red mangrove and grass roots and rootholes. These root structures vary from a mass of the finest hair-roots, that are seen in slabs as tiny threads and dark spots (a few tenths of a millimeter wide), to

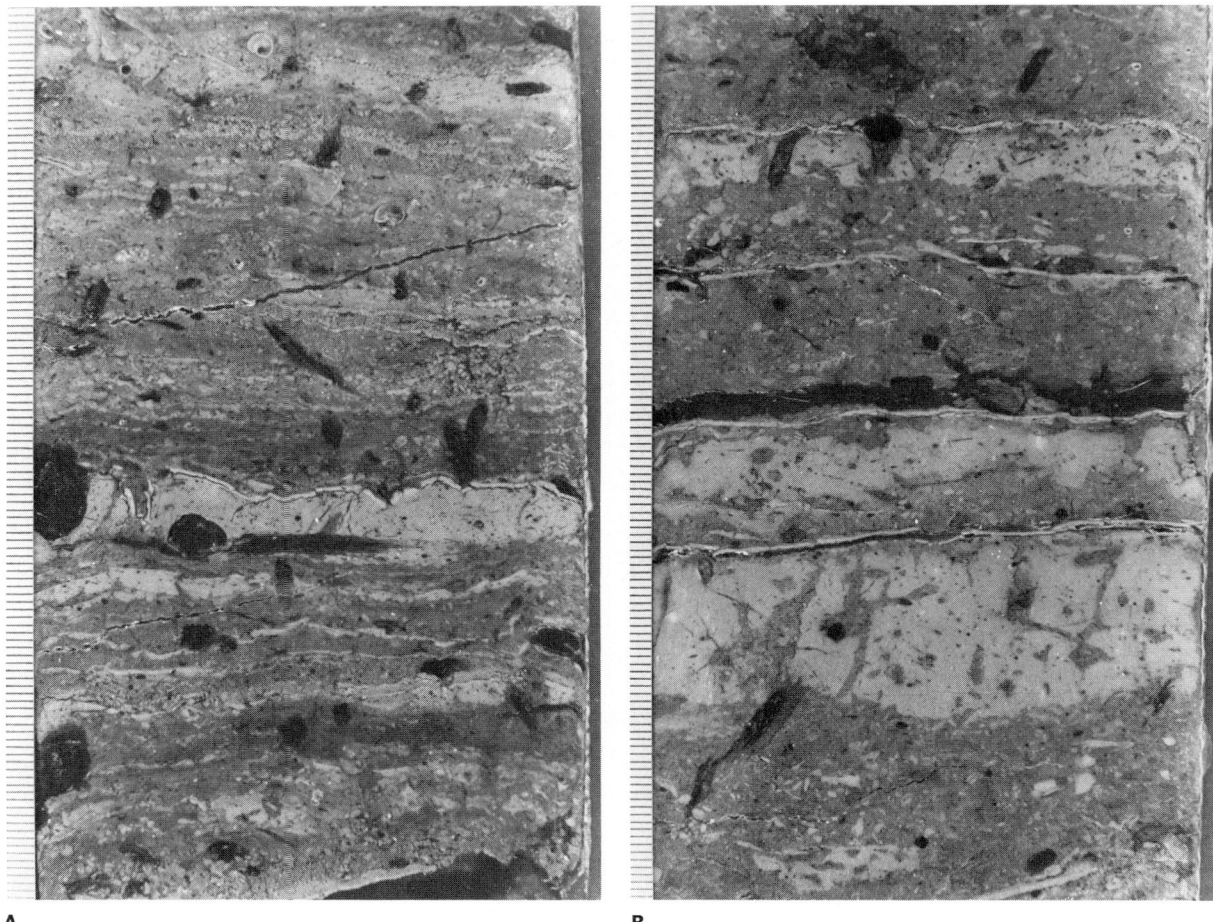

Fig. 52. Algal tufa-peloidal mud interbeds (inland algal marsh sediment): (A) Slab showing typical bedding of inland marsh—alternations of white pelleted mud layers and dark algal tufa layers. Note the characteristic random variation in thickness from one layer to another. About one-third down the right side of the core can be seen a porous "head" of cemented algal filaments. Note also the roots and rootholes. Scale in mm. (B) This slab, from the same core as (A), shows thin white paper-crusts typical of the inland marsh overlying thin beds of sediment which are pierced by tufa-filled rootholes. Note the intraclasts from erosion of the crust and sediment laminae. A strong cyclicity is seen in this core: tufa (with intraclasts) → peloidal sediment → paper-crust.

isolated, heavy prop-roots a centimeter or two in diameter. Few of the layers in 2 m of sediment have escaped penetration by these roots. Many of the white layers are capped by a millimeter-thin dense rind, commonly cracked and askew (Fig. 52B); these are the paper-crusts developed directly on the sediment between the patchy *Scytonema* growths.

The white layers are sediment layers made of lime mud and mud-peloids. Many of these sediment layers are fragmented by desiccation cracks and by roots and root-holes, and in some cases the fragments are strung out along bedding to make discontinuous layers or pockets of flat-pebble conglomerate (Fig. 52B). In thin section, Figure 53A, these white sediment layers are of two kinds: (a) well-sorted, open framework sand layers of dense ovoid peloids and lighter colored irregular peloids 50 to 150 μ across, with scattered skeletal fragments and whole tests of foraminifera and gastropods; and (b) dense, clotted mud layers through which small (up to 75 μ) dense peloids and skeletal fragments are scattered. These sediment layers typically are not cemented; they can easily be washed out with a gentle jet of water and they smear to mud when squeezed between finger and thumb. However, under the microscope many of the sediment layers have a

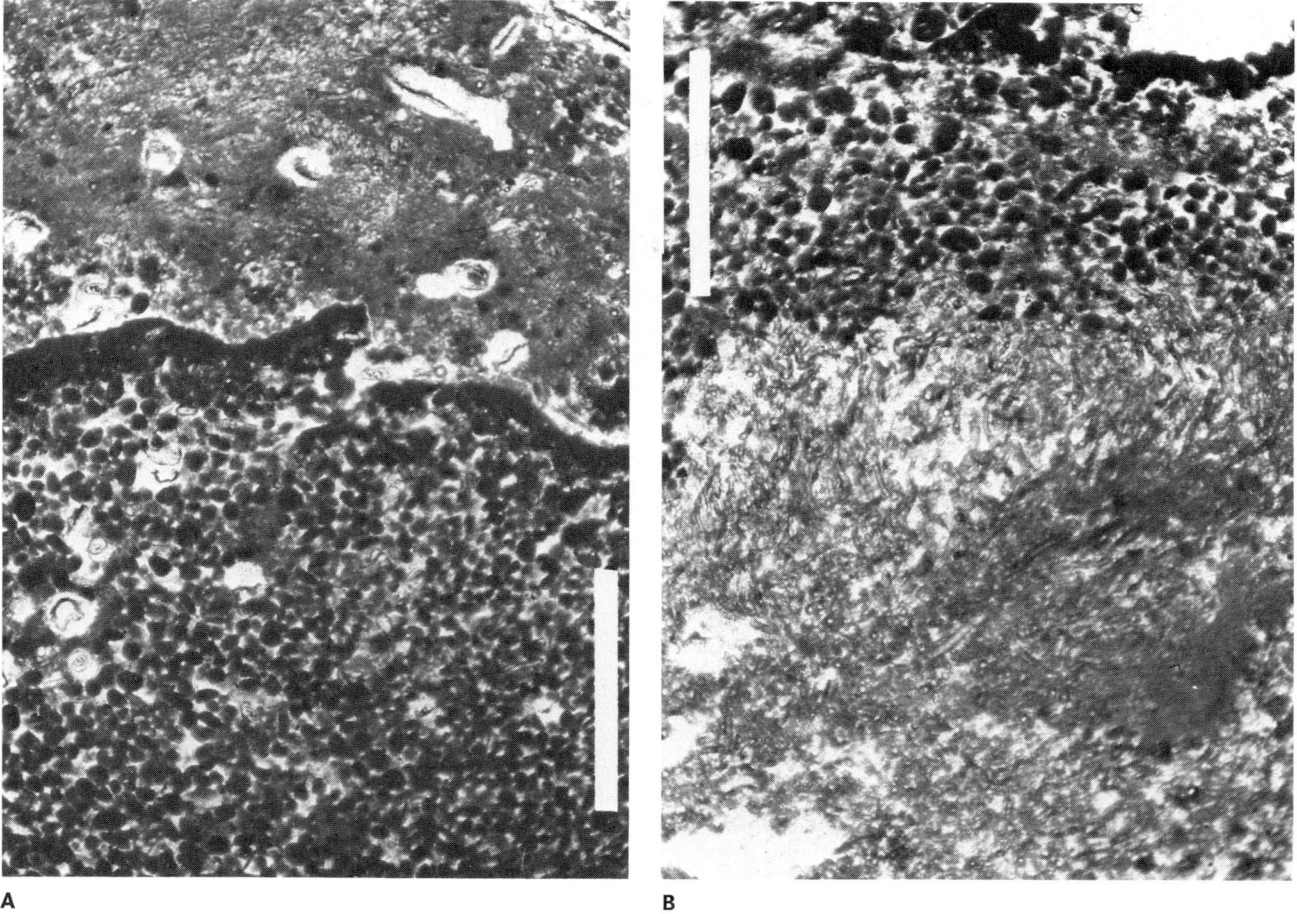

Fig. 53. Thin section photomicrographs of algal tufa-peloidal mud interbeds: (A) Well-sorted peloidal sediment capped by a thin, dark, fragmented rind (paper-crust) and overlain by algal tufa. The circular and tubular fenestrae are mangrove root molds. Scale bar is 1 mm. (B) *Scytonema* algal tufa overlain by a peloidal sediment lamina, which in turn is capped by a dark, wavy paper-crust. While many of the filament molds appear upright near the top of the tufa, the bulk of these Mg-calcite tubes are flattened (note circular cross-sections). Note the near absence of sediment in the tufa. The tripartite sequence shown here is a typical "cycle" repeated many times over in the inland marsh sediment (Fig. 52). Scale bar is 1 mm.

distinct cap of denser mud, a cemented "paper-crust" commonly shattered by wispy, horizontal sheet-cracks and tubular filament molds such as described above for the "disrupted flat lamination."

The tan to brown-gray beds have a rather porous, massive, "earthy" look in slabbed sections. When rubbed between the fingers, fresh samples of these tan interbeds have a fragile, crunchy texture quite unlike the friable sandy or muddy feel of the white sediment layers. Apart from the roots and "floating" pebbles of white sediment (the fragmented paper-crusts mentioned above) these tan earthy layers appear homogeneous in hand specimen (Fig. 52). In thin section it is immediately obvious that these beds are porous *algal tufa* layers: like the fibrous, "palisade"-structured, laminated heads described above, they are seen to be crowded with beautifully preserved molds (30 to 75 μ across) of filaments and filament bundles (without doubt, *Scytonema*), which mostly are pressed nearly parallel to bedding and smothered here and there by lime mud, ovoid silt and fine-sand peloids, and tiny chips of dense "paper-crust" (Fig. 53B). Tiny irregular to amoeboidal fenestral pores between the molds make up as much as 50% of the volume of these tufa layers. The molds are fine, granular, crystalline sheaths

(10 to 20 μ wide) that mimic the tubular filament form, but leave no trace of trichome or other internal organic structure. Under the highest power of the microscope the granular sheaths are seen as a microspar of tiny (3 to 7 μ), clear crystals[4] that lightly weld the tubes together where they touch. This seems to be an *in situ* carbonate precipitate generated locally around the filaments, presumably aided by the metabolism of the *Scytonema* colony and/or the bacterial decay of the filaments (see Chapter 7 for detailed discussion). The filament sheath network of tufa (including fenestral pores) represents an entirely autochthonous carbonate deposit which accounts for 50 to 60% of the total volume of the entire inland marsh sediment! Although the bulk of the tufa layers consist of a porous mass of flattened and twisted filament envelopes that look remarkably like *Girvanella*, there are isolated, small, domal stromatolite heads of upright algal tufa.

There are, then, three basic elements to the algal tufa-peloidal mud interbeds: (1) fenestral algal tufa layers, (2) well-sorted peloid sediment or clotted mud layers, and (3) thin, brittle, fragmented paper-crust layers. These three kinds of layers are commonly, but not invariably, organized into a well-defined "cycle" of algal tufa layer → peloidal mud layer → thin paper-crust "cap." The tufa layers record long periods (several years) of *Scytonema* growth (*post-mortem* calcification on burial produces the tufa, see Chapter 7) which are abruptly interrupted by catastrophic inundation by sediment-charged seawater (hurricane floods, see later) that smothers the mat with a peloidal mud layer. After subsidence of the floodwaters, the sediment surface becomes covered with a *Schizothrix* mat that rapidly "grows" up through the sediment (calcification of this *Schizothrix*-bound surface produces the paper-crust). Slow but persistent recolonization of the *Schizothrix*-covered surface by *Scytonema* restarts the "cycle." Mud-cracking of the sediment layers and upgrowth of *Scytonema* along the cracks causes fragmentation of brittle paper-crust and "capture" of the flat chips by the *Scytonema* mat (Fig. 52B). These cycles are beautifully demonstrated in Figure 52B.

B. Disrupted fenestral bedding. Beneath the meadow of thick *Scytonema* "pincushions" of the low algal marshes fringing the ponds of the channeled belt (Fig. 18, and Chapter 7), the soft peloidal mud is layered into discontinuous, disrupted, lenticular laminae and thin beds (up to 3 cm in thickness). The layering stands out strongly in polished slabs as whitish and brown stripes (Fig. 54A). The whitish layers are peloid sand beds and mud laminae. They typically thicken and thin, pinch-out, end abruptly, bend, and even bifurcate; they commonly appear only as strings of lenticular fragments (Fig. 54A). The brown layers are mud-peloid mixtures that appear more like a massive matrix in which

---

4. X-ray analysis of this sheath material in samples free of detrital carbonate show it to be Mg-calcite with 12 to 16 mole % $MgCO_2$, but no detailed work was done to see if composition varies from one part to another in the marsh. This is to be expected in fact, because Monty (1972) reports tufa calcite compositions of 1 to 8% $MgCO_3$ in freshwater interior marshes and lakes that rarely if ever are flooded by seawater.

Fig. 54. Disrupted fenestral bedding: (A) Slab of sediment below *Scytonema* algal marsh at pond edge, showing discontinuous, pale peloidal sediment layers alternating with darker layers of massive fenestral mud. Note the upturned sediment-charged algal "pincushion" at the surface, bordered by desiccation cracks. Note also the circular hair-root-holes and dark roots. The underlying sediment is completely bioturbated pond sediment. Scale in mm. (B) Thin section photomicrograph of part of the slab shown above in (A) between 2 and 3 cm marks. The upper three-quarters of the photo shows the fenestral clotted mud typical of the massive dark layers seen in (A) above. At the top a horizontal sheet-crack joins a vertical but step-like mud-crack. The lower quarter of the photo shows a part of a discontinuous, peloidal sand layer, with a dense coherent (but unlithified) cap. This latter combination is one of the pale colored streaky layers in (A) above. Note the absence of tufa or any clear record of the *Scytonema* mats that dominate this depositional subenvironment! Scale bar is 1 mm.

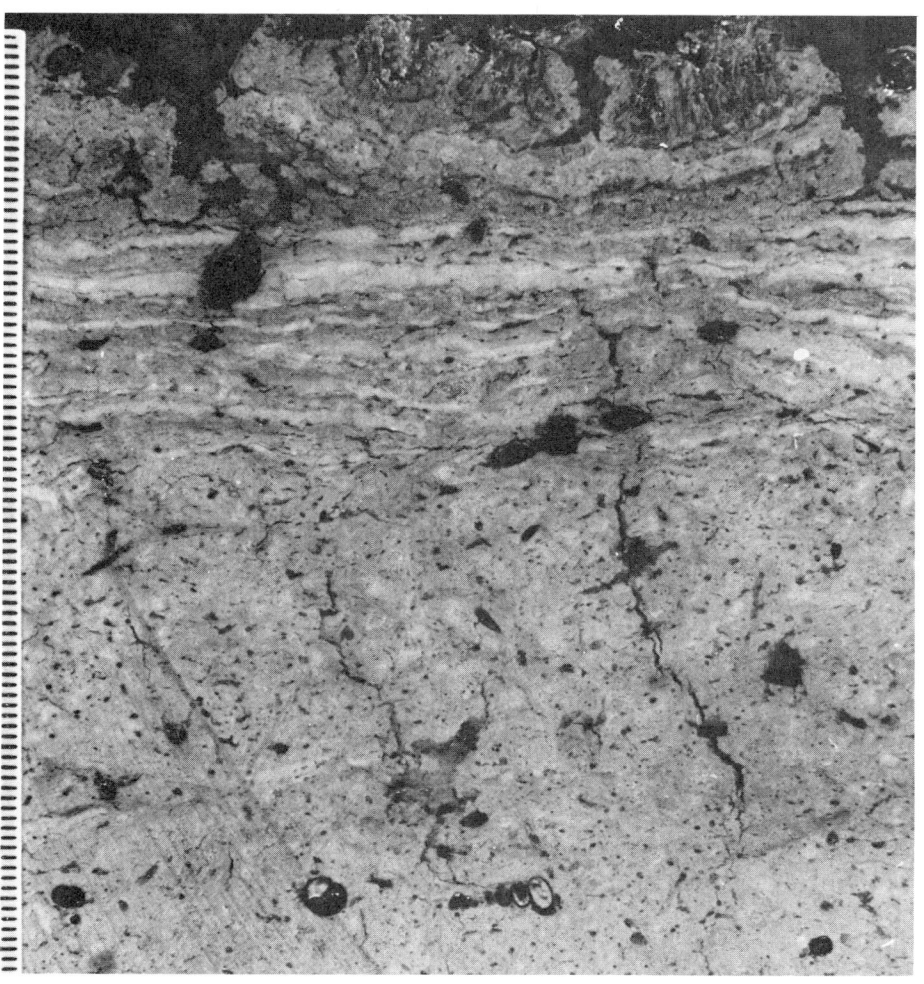

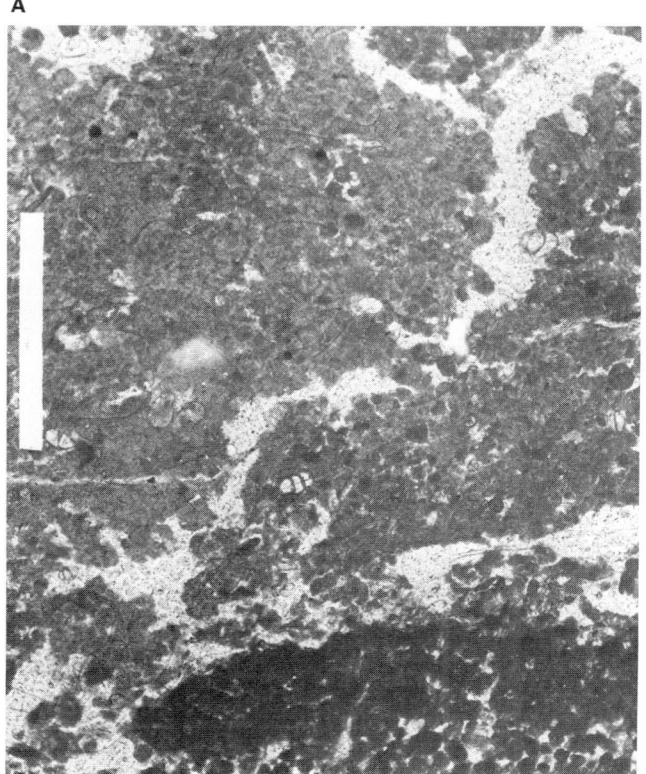

the sharp-bordered, fragmented, white layers are embedded than like individual beds. These brown layers typically are shot through with lenticular fenestral pores (up to 5 mm long) elongated in the plane of bedding, horizontal sheet-cracks (up to 1 cm long), and irregular to tubular fenestral pores and cracks (up to 5 mm in length) that are steeply inclined or vertical. The horizontal fenestral pores are commonly concentrated near the contacts between brown and pale-tan layers. The brown fenestral beds seem to be muddy sediment that has infiltrated and completely smothered a layer of lush, long, thick, upright-filamented *Scytonema* cushions, such as can be seen at the top of the slab shown in Figure 54A. Drying of such a mud-smothered *Scytonema* cushion layer produces polygonal cracking, upward curling of polygon edges, and sheet-cracking; this bends and disrupts the underlying white sediment layer, as can be seen to have happened just below the cushion layer in Figure 54A. A thin white sediment layer deposited on top of such a *Scytonema* layer will drape over and infill between these smothered, squashed, and polygonally cracked algal cushions, producing wavy, discontinuous lenticular bedding. Mud-cracking of this new, irregular, white sediment layer will cause additional disruption and fragmentation. The fenestral pores of the brown "algal" layers seem to be compacted remnants of primary air pockets between filament bundles (see top of slab in Fig. 54A) and openings left by bundles of filaments after complete decay of the organic matter. These complex processes probably account for the distinctive disrupted fenestral bedding of the low algal marsh sediment.

In thin section, the whitish layers are seen as sharp-bordered, discontinuous, thick laminae and thin beds of well-sorted peloid sand (dark, dense, ovoid peloids and light-brown, irregular to ovoid peloids ranging in size from 30 to 200 μ, and scattered tests of foraminifera) or thin laminae of fine, dense mud. Some of the thin beds are internally laminated, with alternations of peloid and mud laminae. These compound beds may simply represent a series of depositional events spaced closely enough in time that *Scytonema* could not colonize the surface between events; or they could represent a single sustained event during which pulses of sediment were introduced, such as a hurricane flood with tide cycles or wind cycles superimposed. A few individual peloid laminae show filament molds much like the crinkled fenestral lamination of the high algal marsh, but these are not typical.

The brownish layers under the microscope are simply a loosely packed, homogeneous mass of clotted mud (with subordinate dark, ovoid peloids up to 200 μ and isolated rare, large gastropod shells) pierced by large (up to 5 mm long), irregular amoeboidal and tubular fenestral pores and irregular cracks (Fig. 54B). Remarkably, there is no mass of filament molds preserved nor is there any trace of the algal tufa typical of the inland *Scytonema* algal marsh sediment. The significance of this latter point is discussed in Chapter 7.

Finally, the scattered, low red mangrove shrubs and the fiddler (*Uca*) and marsh (*Sesarme*) crabs that inhabit the pond marshes leave their record in the underlying sediment as roots, rootholes, and open burrows that cut across and into the layering.

C. Flat-pebble gravel. Scattered flat pebbles and sheets of flat-pebble gravels are found at the base of the beach cliff, on beach terraces, beach-ridge washovers, levee crests and backslopes, high and low algal marshes and the inland algal marsh, and in channels. The flat pebbles are intraclasts (Folk 1962), derived by penecontemporaneous erosion and redeposited at most only a few tens of meters from their source. These intraclast pebbles are flat, irregular to discoidal in outline, angular to round-edged fragments of two main kinds:

(1) *Breakable clasts*, up to 15 cm across and 3 cm thick, made of flat laminated and crinkled laminated sediment. They are generally flexible when moist, but hard enough to be broken between the fingers when dry. They are made co-

herent by drying, with attendant shrinkage and "case-hardening" by salt, carbonate, and organic matter "cement." Algal mats, part of the original sediment surface, bind the "topsides" of the fragments and help keep the more moist and flexible clasts together. Drying of algal-bound clasts commonly produces *curled* pebbles that show desiccation cracks on the exposed unbound sediment "undersides" (usually the convex side, see Fig. 55A). Most of these breakable clasts show some signs of abrasion, mainly rounding off of the sharpest edges. The clasts are derived (a) by wave erosion of the smooth flat laminated sediment exposed in the beach cliff and subsequent partial stripping of the beach-terrace sediment far back from the cliff (Fig. 55B); (b) by sheetwash of *Scytonema* mat polygons of crinkled fenestral laminated sediment of the high algal marsh (Fig. 17A) and inland algal marsh; (c) sheetwash of the cracked domes and bubbles of the *Schizothrix*-bound levee crest sediment (Fig. 13B) and beach-terrace sediment (Fig. 25); (d) undercutting of channel banks and slumping of flat laminated levee crest sediment onto the sloping lower walls of the channels (Fig. 30).

(2) *Cemented clasts*, up to 60 cm across and 5 cm thick, made of indurated crusts showing palisade-structured lamination. These intraclasts are cemented enough to ring when hit with a hammer and to break with a sharp fracture. They all shows signs of dissolution by freshwater, such as smooth re-entrant cavities; in this feature they are distinct from the laminated breakable clasts. These cemented clasts are fragments of the cemented crusts (see "Lamination with palisade structure" above and Chapter 7) that make hard, rocky pavements on the surface at the high marsh-levee backslope boundaries and on high crowns in the inland algal marsh. These pavements are broken into rather irregular, loose plates by upward growth of mangrove roots (Shinn et al. 1969, Fig. 23). Buried cemented crusts outcrop in the steep eroding banks of some channels and the landward-eroding beach cliff (Chapter 7); undercutting of the vertical walls causes large fragments of the buried crusts to slump into the channels and to the foot of the beach cliff.

Once formed, intraclast pebbles may be moved and redeposited to make distinctive pockets of flat-pebble gravel. We have seen the accumulation of flat-pebble gravels in the following subenvironments on Andros Island:

(i) **Beach, at the base of the beach cliff.** In several places along the foot of the beach cliff, large (up to 30 cm across), angular plates of cemented crust and smaller flat pebbles of laminated sediment have been eroded out of the cliff face and sorted by waves into fans of grain-supported, flat-pebble gravel with a skeletal sand matrix; in some instances the flat pebbles are standing on edge (Fig. 56A) and look remarkably like the edge-wise conglomerates so common in the early Paleozoic carbonates of the world.

(ii) **Beach-ridge washover and beach terrace.** Flat pebbles, both breakable and cemented clasts, torn up from the beach terrace and undercut from the beach cliff are thrown up and transported (by storm flood and wave currents) across the beach terrace and piled onto the beach-ridge washover crest along with skeletal sand (Fig. 26A). These accumulations make magnificent, imbricate, grain-supported gravels, a mixture of rubbery flexible and hard, algal-bound laminated clasts and rock-hard weathered crust plates, all sloping down toward the beach area from where they were derived (Fig. 25B). Some of the soft sediment clasts never make it across the beach terrace, but become welded to the beach-terrace surface by overgrowth of *Schizothrix* mats (Fig. 25A and Chapter 4). These welded pebbles do not make a framework gravel, but act as isolated nuclei for domal protostromatolite accretion, much like the Precambrian "encrusting" stromatolites described by Hoffman (1967, Fig. 3). Along some beach fronts the beach cliff erosion has pushed back onto the beach terrace in a series of low steps (up to 10 cms high) rather than making a single sheer wall. On these small, stepped terraces there are local accumulations of rubbery, laminated sediment pebbles at the base of the steps (Fig. 55B).

Fig. 55. Flat-pebble gravel: (A) Curled pebbles with mud-cracks on the "undersides" derived by erosion of cracked, updomed, algal-bound levee crest sediment (Fig. 13B); these pebbles are in transport across the levee backslope (note the small cm-sized mud-cracks). (B) Algal-bound laminated flat pebbles at the base of a small erosion ledge on the beach terrace; these pebbles are derived in place from undercutting of the ledge during severe onshore storms. Scale bar is 30 cm. Vegetated beach-ridge crest in background.

Fig. 56.
Flat-pebble gravel: (A) Cemented clasts on the upperbeach at Three Creeks. *Top photo* shows a sheet of buried cemented algal crust partly exposed by storm-wave erosion; to the right can be seen the scattered plates washed up across the outcropping crust sheet. *Bottom photo* shows a close-up of the storm-wave-sorted, cemented-clast beach gravel seen in the top photo. Note especially the edge-wise fabric of many of the plates and their ragged, pitted shapes. Pocket knife for scale. (B) Cemented clasts on a low narrow levee at the edge of a channel in the Pumpion Cay area. This gravel (note the edge-wise fabric of some clasts) sits on top of the outcropping pavement of algal crust from which the clasts were derived by storm-wave erosion. For scale, the mangrove pneumatophores in the middle ground are 15 to 20 cm high, and some of the very large clasts are 40 to 60 cm long.

A

B

(iii) **Edges of ponds.** Curled saucer-like pebbles of algal-bound laminated sediment from the levee crest and high algal marsh are carried by storm sheetwash down the levee slope onto the low algal marsh at the pond edge, where they accumulate as a thin, flat-pebble gravel sheet (Fig. 57); most of these curled pebbles are oriented with their convex undersides uppermost. On levees where cemented crusts outcrop, cemented clasts are also moved downslope toward the pond edge to join the algal-bound, laminated pebble accumulation. Pond-edge gravels made almost exclusively of plates of cemented crust are extremely common on the tidal flats around the palm hammocks (Shinn et al. 1965) near Williams Island. In this area the levees are quite low and narrow and their backslopes are commonly covered with fragmented crust pavements that are the sources of almost in-place, flat-pebble gravels (Fig. 56B). Similar cemented clast gravels are found at the pond edge, where the pond slopes up to meet the crust pavements that fringe the palm hammocks (Chapter 7).

(iv) **Channels.** Slumped blocks of dried laminated levee-crest sediment, along with the underlying bioturbated sediment, produce a mass of breakable flat and round pebbles and boulders on the sloping, lower muddy walls of the channels (Fig. 30). In places where cemented crust outcrops beneath the levee sediment, large angular plates (some up to 60 cms across) are added to this mix. Only in the channel thalweg are these pebbles and boulders likely to become sorted. We saw this on a small scale in very shallow channels where tidal currents are quite strong: the thalweg sediment consisted of small, flat, cemented intraclasts mixed in with quite well-sorted, coarse skeletal sand and granules. Also, some of our cores through channel bars showed sandy gravels near the channel axis.

(v) **Inland algal marsh.** The thin sediment beds of the inland algal marsh (see "algal tufa-peloidal mud interbeds" above) are commonly broken into desiccation polygons 5 to 15 cm across and these polygons are occasionally reworked during rare hurricanes to produce a thin lense of loose abraded fragments. In cores these fragments appear to "float" in algal tufa (Fig. 52), presumably due to overgrowth and intergrowth of *Scytonema* mats following deposition of the clasts. Granule-sized, very thin chips are also produced in the inland marsh by cracking and disruption of paper-thin surface crusts. Commonly, however, the paper-crusts are not transported but simply intergrown and overgrown by *Scytonema* mats to produce a fragmented lamina that looks like a flat-pebble layer (see, for example, the many fragmented paper-crusts in Fig. 52).

The question of the origin and accumulation of intraclasts in carbonate deposits is a very important one, because flat-pebble conglomerates and intraclast calcarenites are abundant in ancient limestones. It needs far more intense study than the rather sparse set of observations we have presented here. The special problem of the significance of grain-supported gravels versus matrix-supported "gravels" needs much attention, as does the problem of distinguishing a widespread, transgressive, basal-lag gravel from a local environment gravel.

D. Round-pebble gravel. Rounded equant to irregular-shaped pebbles and boulders of unlithified unlayered muddy sediment are found at the base of the beach cliff and in channels. In both subenvironments the fragments are derived by erosion of crab-burrowed, unlayered sediment, the lack of internal layering and the randomness of the crab burrows ensures that the eroded fragments will have round to irregular, rather than flat, shapes.

Smooth, wave-abraded, cohesive but breakable roundish clasts (up to 15 cm across) have collected, along with coarse skeletal hash (mainly gastropod shells), in small re-entrants in the beach cliff (Fig. 24B). In these areas the laminated beach terrace sediment has been stripped off and unlayered, bioturbated "old" pond sediment full of gastropod shells has been exposed to wave action during storms. *Uca* has riddled the cliff with burrows and produced a weak "Swiss

Fig. 57. Flat-pebble gravel. Curled, algal-bound clasts, derived from levee crest and levee backslope, collecting at pond edge. Note that many of the clasts are overturned, with their lower convex sides uppermost. Pencil (foreground) for scale.

Cheese" surface layer easily broken into equant to irregular lumps by undercutting and slumping.

Undercutting of channel banks at outer bends and slumping of the undermined sediment into the channels is a prime producer of large and small intraclasts, both flat and "round" (Fig. 30). Where the dry and coherent bank sediment above MTL is unlayered bioturbated peloidal mud, undercutting, aided to no small extent by fiddler-crab burrowing, will break off chunks of sediment as large as 1 m across. These irregular, unabraded fragments will roll down, break up and founder on the muddy, sloping lower walls of the channels (Fig. 30). Eventually, channel migration may expose these irregular lumps to reworking by the strong currents of the thalweg, producing a basal, round-pebble lag gravel. More commonly, laminated levee sediment is also added at the same time (Fig. 30) so a mixed flat- and round-pebble gravel will result.

As with the flat-pebble gravels described earlier, more intensive study needs to be made of the origin and significance of round-pebble intraclasts. We particularly need to know what the channel lag gravels actually look like when preserved and how thick and abundant they are in Andros-type tidal flat regressive deposits.

III. Thin to Thick Cross-Bedding

We did not make a study of the cross-bedded sands of the Three Creeks area but we have included our few observations in this descriptive section for completeness.

There are two types of cross-bedded sands in the Andros tidal flat complex:

A. Rippled skeletal sands. Coarse skeletal sands and grits cover the bottoms of many very shallow secondary distributary channels. These sands consist mainly of a well-sorted lag of whole and broken tests of gastropods, pelecypods, and foraminifers and rounded intraclasts eroded from the pond sediment into which the channels cut. These loose sands move back and forth daily with the tidal currents as ripples and small dunes (amplitudes up to 10 cm).

Very small pockets of rippled skeletal sands also occur at the base of the beach cliff, associated with the intraclast gravels described above.

Thin sheets of rippled skeletal sands are also deposited along with intraclast gravels on beach-ridge washover crests (Chapter 4).

It is interesting to note here that the intertidal beach zone is commonly a beautifully rippled surface (Fig. 24A), but the sediment consists of *soft* peloids and fecal pellets of fine sand size that completely lose their integrity on burial (by squashing and bioturbation) to produce a massive clotted mud, showing no trace of ripple cross-bedding. In the geologic record this wave-agitated beach sediment would probably be classified as a micrite deposited in a low energy environment!

B. Festoon cross-bedded skeletal sands. The sediment of the pine hummocks of the beach ridge, the highest parts of the tidal flat, is coarse sand made up of skeletal debris (mainly gastropods) and cemented clasts. Trenches dug into this sand clearly show that it is thin to thickly cross-bedded (amplitudes as high as 80 cm, see Shinn et al. 1969, Fig. 15) with steep landward-dipping foresets (30° avalanche slopes) and gentler seaward-inclined sets. In places this cross-bedding is festooned. Trenching and cores show that this shoreline ridge of coarsely cross-bedded sand is simply a cap (up to 1 m thick) on top of muddy pond sediment.

IV. Thick to Very Thick Bedding

The bulk of the sediment accumulating at present on the Bahama platform, both offshore and on the tidal flats, is subtidal bioturbated peloidal aragonite mud up to 2 m thick with no internal layering. Churning, burrowing, browsing, and pelleting by marine invertebrates (Chapter 6) are constant activities that ensure that no fine layering will survive below mean tide level.

At Three Creeks, homogeneous bioturbated mud more than a meter thick underlies the beach, offshore, pond, and subtidal channel bars. In each of these subenvironments the textural characteristics of the sediment is a little different, so that it is possible to distinguish, with careful examination, one from another.

(i) **Beach and offshore sediment.** A core taken 4 km offshore of Point Simon, Figure 58A, shows very well the churned and burrowed fabric of the Bank sediment, the lack of layering. In thin section, the sediment texture is a random patchwork of clotted mud and peloids (Fig. 58B). The clotted mud is a dark and light mottled fine mass in which are suspended sand grains of dark, dense, smoothly ovoid mud peloids, irregular skeletal fragments, and whole tests of foraminifera. In places the clotted texture of the mud is strongly suggestive of squashed pellets. Cutting through the mud are irregular to tubular patches of an openly packed peloid sand. The peloids range from 30 to 750 μ long and are both dark, dense, ovoid grains and lighter, more irregularly shaped, clotted mud aggregates. The features that distinguish the beach and offshore sediment from the pond and channel bar sediment are the abundance and coarseness of dark, dense peloids (up to 750 μ long), the abundance of *Callianassa* burrows and the absence of mangrove roots.

(ii) **Pond sediment.** Slabs of pond sediment show a dense white to gray homogeneous mud with only roots and root-hairs of mangroves, the curving branching worm burrows, and scattered whole shells of gastropods, pelecypods, and foraminifers to break the homogeneity (Fig. 59A). At the very pond edge, remnants of thin beds (Fig. 59B) testify to the originally layered structure of the pond sediment.

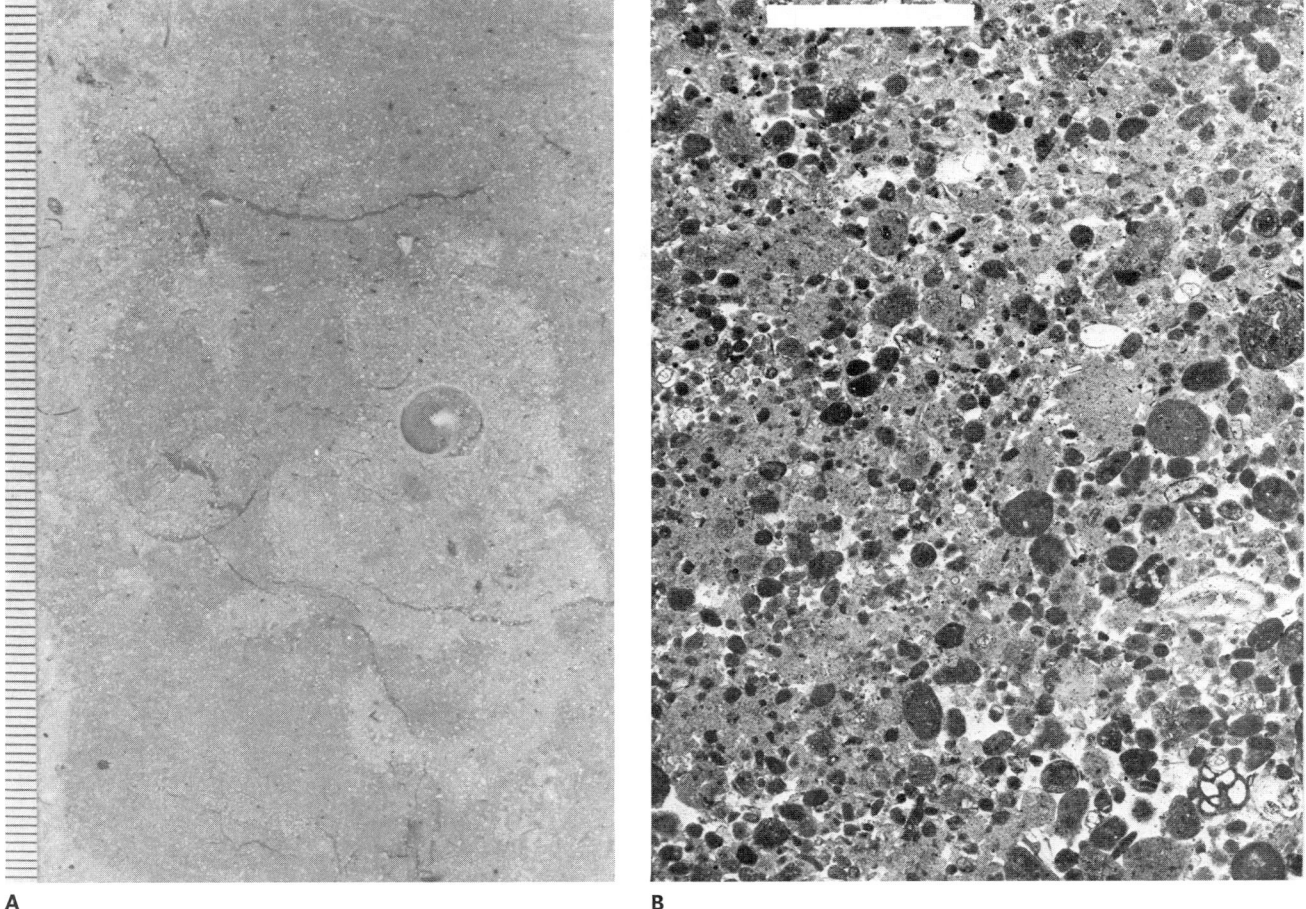

Fig. 58. Offshore sediment: (A) Vertical slab showing unlayered structure. White spots are firm, coherent peloids, a characteristic feature of the offshore sediment. Sample taken 4 km offshore of Three Creeks. Scale in mm. (B) Thin section photomicrograph of offshore sediment shown in (A). Note clotted mud, large dark peloids, and shell debris. Scale bar is 1 mm.

Sediment is added incrementally in thin layers, but bioturbation rapidly destroys each newly deposited bed. In thin section, this churned texture is readily recognized. The general pattern that one sees is an irregular patchwork of loosely packed, clotted fine mud (with suspended, dense, ovoid peloids, 50 to 150 μ, and skeletal grains) and pockets of an open framework of sorted dense and light peloids 50 to 150 μ in diameter (Fig. 75). Large foraminifera tests (50 to 1500 μ) are unevenly scattered through the sediment. Vague traces of fine layering remain in places. Zones are found crowded with whole gastropod shells, while others carry large, ovoid (up to 2 mm) mud aggregates, the fecal pellets of the "volcano-worm" *Marphysa*. For the most part, the deeper the sediment the more densely packed it is and the less recognizable is the pelleted fabric. At depths as little as a few tens of centimeters, compaction leaves a dense mud with a vague clotted texture, in which only the dark, smooth, indurated, ovoid peloids and skeletal grains have kept their original shape.

(iii) **Subtidal channel bars.** Cores taken through some of the larger permanently submerged bars typically show little or no traces of layering (Shinn *et al.* 1969, Fig. 31). For the most part, one sees in slabs only a massive, burrowed, pelleted mud with scattered gastropod and foraminifera tests, a sediment difficult to distinguish from burrowed pond sediment; this is true also of thin sections which show the same churned mass of clotted mud and dark, dense peloids and scattered

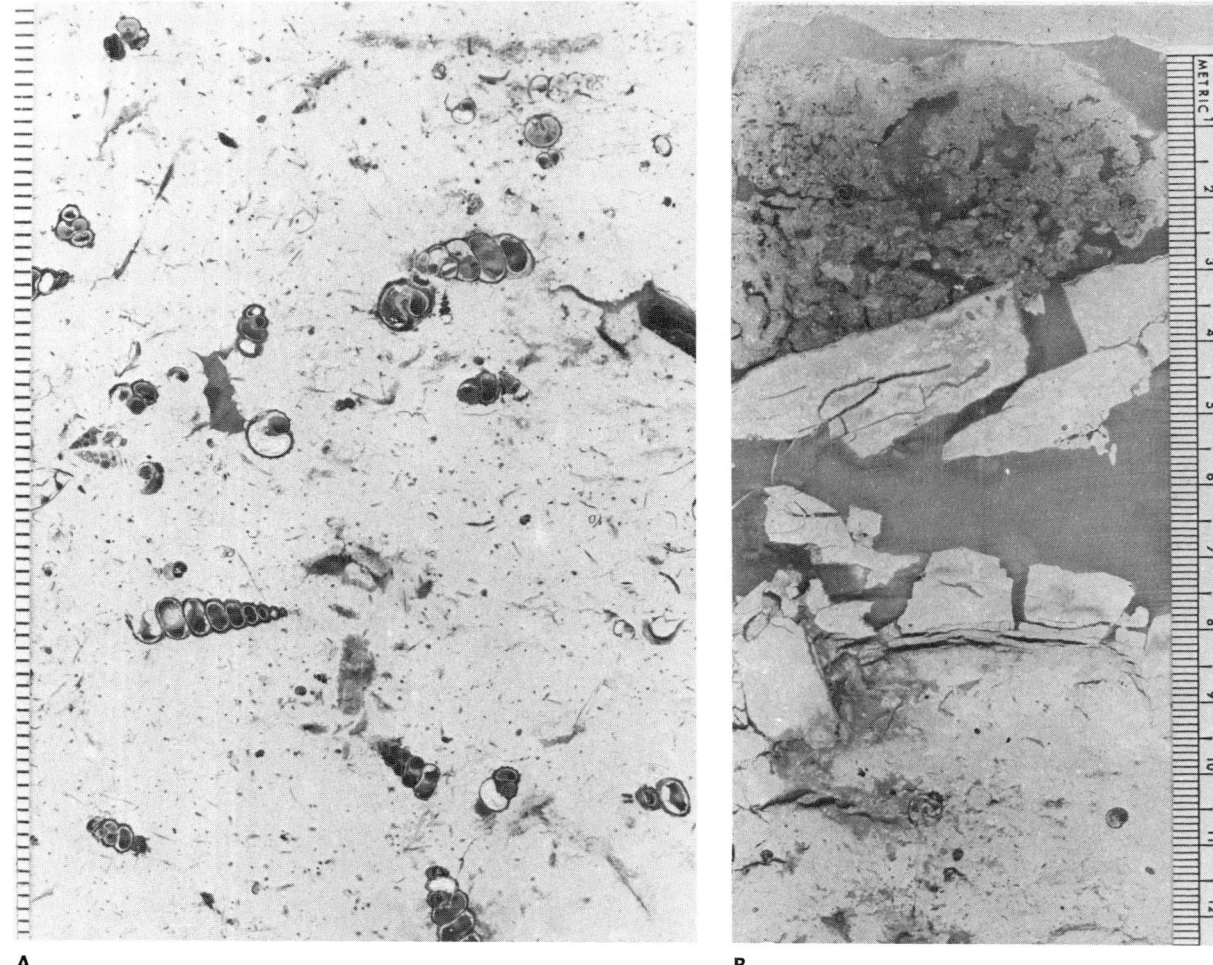

Fig. 59. Vertical slabs of pond sediment: (A) Note the unlayered structure and the cerithid gastropod tests. Small dark "threads" are hairroots of red mangroves. Scale in mm. (B) Dried coherent hurricane layer (between 3 and 8 cm marks) overlain and underlain by burrowed unlayered mud (disruption of the hurricane layer is a sample-preparation accident). Scale in mm.

gastropod and foraminifera tests, as does the pond sediment. However, at a few horizons there is a rough organization of the gastropod and/or foraminifera remains into vague but obviously inclined (ca. 30°) lines of strung out shells (Fig. 78A). Also, there is a lack of mangrove roots and roothairs, but instead one may find in some bars the roots of the seagrass *Thallasia*.

Much more data on channel-bar sedimentation and channel fills in general are clearly needed.

4. Time Value and Origin of the Millimeter Lamination: Rate of Deposition Experiments

Methods. To measure the rate of deposition of sediment on the Three Creeks tidal flat we used the simple technique of covering patches of the surface with marker dye and periodically returning to each site to observe the results (cf., Ginsburg et al. 1954, p. 22). The dye was finely ground hematite paint pigment dispersed in seawater and mixed with lime mud. It was poured onto the surface of the sediment as a thin film that covered a roughly circular area of about $\frac{1}{4}$–$\frac{1}{2}$ m², and the site was marked with a numbered stake. Each pigment film provided an easily recoverable and precise time horizon. In order to correlate deposition with weather and tides, we continuously monitored weather conditions (wind speed and direction, precipitation) and tidal regimen on the flats (Chapter 2).

In all we marked sediment surfaces at seventy-five sites which covered all subenvironments across the entire Three Creeks tidal flat complex from beach to inland marsh. Not all of these sites were examined at each visit to the island, instead we concentrated on several representative localities. Some of these are shown in Figure 2. In the three years of the study we made thirteen visits to check the pigment sites. Intervals between visits averaged three months, but the range was one to five months. Each site was examined for deposition, erosion, and algal growth. Samples were cut out and the holes filled, and new pigment was poured if necessary.

Results: The time value of a lamina. The results of the rate of deposition experiments made at critical locations are given in Table 5.

The first pigment patches were poured in March 1968 and were examined carefully for a week. No deposition occurred at any of the sites. The first return visit to these sites was made in late June of the same year. We found a thin lamina of sediment over the pigment on the channel banks and on some of the levees, but there was no deposition in the pond algal marshes or in the inland marsh. The pigment on the pond surfaces had been mixed into the underlying sediment by burrowing, so that we had no measure of any deposition in the pond. During this second stay (of four weeks duration), as during the first, there was no deposition of sediment at any of the sites. These early results—deposition of only a single thin layer on the levees in three months—clearly demonstrated that the depositional events producing the tidal flat lamination were infrequent, irregular, and certainly due to some unusual process. We felt, therefore, that we could safely space our visits several months apart.

As mentioned above, and as can be seen from Table 5, not every site was examined on each visit to the island. However, by piecing the data together we have been able to establish that in the period March 1968 to May 1971 there were *nine* major depositional events, *an average of three per year*, each one leaving a thin lamina on at least the higher parts of the tidal flats. There were at least two (and perhaps more) additional minor events that left a mere film of sediment on the channel banks and the beach terrace only.

Deposition was confined almost exclusively to the levees and levee backslopes, to the channel banks and channel bars, and to the beach-ridge terraces, washovers, and backslopes. There was little or no deposition in the pond algal marshes and none whatsoever in the inland marsh or on the beach-ridge pine hummocks. On the beach, mainly because of burrowing by polychaete worms, we had no sensible measure of sedimentation. We had the same result at most of the pond sites, but at the edge of a pond at station 8101 (Fig. 2) we found the pigment intact and uncovered after twelve months, indicating no deposition. At pond station 9101 we did find a sediment cover over small ceramic tiles put at the surface of a small channel delta built by inflow into the pond.

The laminae of the levees average about 0.5 mm in thickness, so we may compute a build-up rate of about 1.5 mm/year (with a range of about 0.3 to 3.0 mm/year). If sea level has remained the same over the last few centuries, then the highest levees would have been built from MTL to their present crest elevations (about 45 cms) in the last 300 years. This is only a rough estimate, because there is considerable local variation in sedimentation during a single depositional event and from event to event; the time span could have been as long as 1500 years, but almost certainly no more than this.

Origin of the lamination. There were three origins we had initially considered possible for the lamination. These are: (a) chemical precipitation, perhaps seasonal to account for the uniformity of lamination thickness; (b) normal diurnal tidal flooding (astronomical tides), particularly high spring tides for the levee and beach-ridge laminae; (c) flooding by storms (meteorological "tides").

Table 5. Rate of deposition observations at selected sites on the Three Creeks tidal flats

Site (see Fig. 2)	Subenvironment	3/26/68	6/23/68	11/16/68	1/1/69	2/1/69	6/18/69	10/12/69	1/1/70	3/2/70	4/2/70	9/12/70	12/8/70	2/3/71	5/11/71
SHORELINE AREA															
8269	Beach terrace							*	F,P	F,P	—	O	P	F,P	O
	Washover crest							*	P	F,P,C	—	O	P,C	P	O,C
	Washover backslope							*	P,C	P	O	O	O	P,C	O,C
	High agal marsh							*	O	—	O	O	O,C	F	O
8274	Beach terrace							*	F,P	F,P	—	O,C	F,C	F	—
	Washover crest							*	F,C	P,C,M	—	O,C	O,C	P,C	—
	Washover backslope							*	O,C	P,C	—	O,C	O	F,C	—
	High algal marsh							*	O	F	—	O	—	O	—
8268	Beach-ridge hummock							*	O	O	O	—	—	O	—
	Hummock backslope							*	O	O	O	—	—	O	—
NEARSHORE AREA OF CHANNELED BELT															
8013	Channel bank	*	L	L	—	F	F	—	—	—	TL	M	M	L	F
	Levee crest	*	F	L	—	F,C	P	—	—	—	L	M	M	P	O,C
	Levee backslope	*	F	L	—	F,C	O	—	—	—	L	M,C	F,C	M,C	O,M,C
	High algal marsh				*	O	O	—	—	—	L	C	F	O	O
8270	Levee crest							*	P	—	F,P	O	—	P,TL	—
	Levee backslope							*	O,C	—	L,M	M,C	—	P,M,C	—
	High algal marsh							*	O,C	—	O,C	O	—	O	—
70-35	Levee crest										*	O,C	O	P	O
	Levee backslope										*	O,M	C	P,M,C	O,M
	High algal marsh										*	M,C	M,C	P,C	O,C
9101	Pond					*	B	—	—	—	B	—	—	—	—
CENTRAL AREA OF CHANNELED BELT															
8101	Levee crest	*	F	L	—	O	O	—	L	—	—	P	O	P	—
	Levee backslope	*	P	L	—	O	O,M	—	F	—	—	—	M,C	F,C	—
	Upper pond				*	—	—	—	—	—	—	O	—	—	—
8105	Channel bank				*	F	F	F	—	F	O	F	F	—	
	Levee crest	*	—	L	—	F	F	L	F	—	L	O	O	O	—
	High algal marsh	*	—	P	—	B	B	O	O	—	O	O	—	O	—
INLAND ALGAL MARSH															
8055	High algal marsh	*	O			O	O	O	—	—	O	O	—	O	—
	Low algal marsh (edge of channeled belt)		—	—	*	O	O	F?	—	O	O	—	O	—	
8272	Interior of marsh					*	—	—	—	—	—	O	—		

Legend:
- * pigment first poured
- — site not visited
- O no deposition of sediment
- M surface lamina cracked, chipped, and eroded
- C isolated intraclast pebbles or chips over pigment
- F thin film of mud over pigment
- L thin lamina of sediment over pigment
- TL one thick lamina or several thin laminae over pigment
- P patchy deposition of fine sand, mainly in depressions
- B pigment mixed into sediment by burrowing

The first mechanism is quickly ruled out by the well-sorted, open framework peloid fabric of many of the laminae. The second mechanism, the astronomical tides, we found by direct observation not to be responsible for sedimentation, and this applies equally well to high spring tides. The reason is that the normal day-to-day tidewaters, including those of springs and neaps, carry little or no suspended sediment. This was confirmed again and again on each visit to the area. Indeed, the clarity of the water across the entire Great Bahama Banks is the most impressive aspect to anyone flying or sailing over it. In the Three Creeks area

Table 6. Suspended sediment in the waters of the tidal flats and offshore of northwest Andros Island, Bahamas

Sample location	Suspended solids (gm/liter)	
Main channel, Three Creeks, near the mouth. Sequence collected over tidal cycle at approx. 2 hr. intervals.	1)	0.00072
	2)	trace
	3)	trace
	4)	0.00397
	5)	0.00291
	6)	0.00554
11 km offshore of Three Creeks	1)	0.00042
	2)	trace
Surface water, middle of Great Bahama Bank (25°21'N 78°49'W)		0.00056
Channel north of Loggerhead Point (Shell Oil Co. sample)		0.00225
2.4 km offshore of Loggerhead Point (Shell Oil Co. sample)		0.00177

one cannot "see bottom" in the main channels of the tidal creek systems, even though their maximum depth is only 2 to 3 m. The lack of clarity, however, is due more to a yellow-brown organic stain than to suspended sediment particles. This becomes quite apparent in the shallower secondary channels and in the ponds where the water, although pale yellow in color, is notably transparent. To check this, water samples were collected in June–July 1968 from the open Bank and from the main channel at Three Creeks and were filtered for suspended solids greater than 0.45 μ. The data were compared to similar measurements made in the same general area by the Shell Oil Company in 1963. The results are given in Table 6. They clearly show a general paucity of suspended sediment in both Bank and tidal flat waters.[5]

From the beginning of the study we had been aware of the importance of storm deposition, because Black (1933, p. 176) had observed, for the lakes and bights of central and northern Andros, that "tidal action by itself was not responsible for any appreciable sedimentation but may become quite important during violent storms, when large quantities of sediment are stirred up into the water." Our first results, when we found that only a single thin lamina or film had been deposited on the levees in three months, certainly pointed to the laminae as storm layers. One of our team, Harold Wanless, was able to confirm this when he followed a "norther" from Miami to Andros in November 1968. We can reconstruct the event from Wanless's observations, from those of local inhabitants, from the weather records, and from our later experiences. The scenario goes something like this: a heavy squall line marking the turbulent cold front approached Andros from the northwest on November 10 bringing lightning, rain (1 cm in a few hours), and strong westerly, southwesterly, and southerly winds about 18 to 38 km/hr (10 to 20 knots), with gusts up to 44 km/hr (24 knots). The front moved slowly across the island on November 11 dumping more rain (another 1 cm) and maintaining the heavy gusty W and SW winds; gusts were now up to 70 km/hr (38 knots). By November 12 the front had passed through, temperatures had dropped 6°C into the low 20's, the winds had shifted to the westnorthwest, and the sky had begun to clear. The wind remained strong and somewhat steadier at 30 to 40 km/hr all through that day. During the following two days the winds diminished considerably and shifted first to the north and then came out of the northeast and east as the normal 10 to 20 km/hr trade winds.

---

5. If all the suspended solids in one meter depth of these tide waters were to be deposited, a layer only 2 to 3 μ thick would accumulate! About half of this would be degradable organic matter.

During and for almost two days after the passage of the cold front the strong winds had generated wind waves[6] that stirred the bottom sediment of the shallow Great Bahama Bank, making the water white and turbid with suspended sediment. The onshore storm winds piled this turbid Bank water up against the western shore of Andros and flooded the low-lying tidal flats with sediment-charged seawater. Waves attacked the beach cliff at Three Creeks and on the flood tide overtopped the beach-ridge washover crests with milky water a few centimeters deep which drained landward down the beach-ridge backslope toward the ponds. These floodwaters quickly dropped their sediment load on the beach terrace, the washovers, and the ridge backslopes, leaving these areas covered with a thin (< 2 mm) white layer of pelleted mud. The load must have been depleted by the time the floodwaters (now probably only a film of water) reached the algal marshes fringing the ponds, because these subenvironments were not covered at all by sediment.

The muddy water also poured through the openings of the channel mouths and spread upstream, overflowing the channel banks and smearing a thin layer of white sediment across the levees. The high seaward levees were perhaps under no more than a few centimeters of water at maximum flood.

After the floodwaters subsided, Wanless observed that the channel banks, the levee crests, the levee backslopes, and the channel bars were left covered with a sediment lamina up to 1 mm thick. Sediment was patchily deposited on the high algal marshes behind some of the levees, but in none of the low algal marshes around the ponds nor in the vast inland algal marsh was there any deposition. Sediment did reach the ponds through the small distributory channels, but collected at the channel inlets, building up the small inflow deltas.

This scene, we believe, describes the general *modus operandi* of the lamination-producing events. *Each lamina, then, is a storm layer produced by sporadic onshore storms which are generated for the most part by southward-pushing cold fronts.* In the three years of our study, the scene was repeated, sometimes with lesser and sometimes with greater violence, nine times. From our pigment-marker observations, from weather records, and from our tide gauges we have been able to determine with some precision when deposition of each sediment lamina by storm flooding took place. The results are given in Table 7. Figure 60 shows a short sediment core that records four of these depositional events. As Table 7 shows, the sediment-depositing storms are few and irregularly spaced through the year, but tend to be most common in the winter months and to a lesser extent in late summer-early fall, during the "hurricane season." This is in keeping with the long-term weather patterns of the Bahama area. In an average year perhaps forty southward-pushing cold fronts will reach the northern Bahamas from the U.S. mainland. However, very few of these have the strong northwesterly winds that make a violent winter "norther." Nassau, for example, averages less than four days a year with gale force winds from any quarter. In the Andros area winds over 50 km/hr (27 knots) are recorded on no more than 10 to 12 days a year. Our three-year records, then, are probably a reasonable sample of the long-term weather. Table 7 also shows that the storm waters generally do not flood the tide flats to any great depth or for any great length of time. Even during a particularly violent winter storm, such as that of February 3–4, 1970, the highest levees are covered at most by only a few centimeters of water and for only a few hours.

Consistent with the overall picture of deposition by onshore storms is the independent evidence that most of the sediment deposited on the tidal flats has

---

6. The open sea east of Andros was classified as "moderate to rough," with waves up to 3 m high (AUTEC records). The Bahama Bank, being shallower, must have had considerably smaller waves.

LAYERING

Fig. 60. Short sediment core from beach terrace at locality 8269 (see Fig. 2), where pigment patches were laid for rate-of-deposition field experiments. *Dark lines* are the pigment horizons (dates pigment poured on the left), *white layers* are laminae deposited by onshore storms (dates of deposition on the right). Scale in mm.

been derived from offshore on the Great Bahama Banks. There are two arguments for an offshore source of sediment. First, the levees, which are formed by channel-bank overflow, building the levees upward and pondward, progressively decrease in height and width eastward (upstream) away from the shoreline. Clearly, the sediment source for the levees must have been from the west, that is, from the Great Bahama Bank. Second, the peloids (particularly the hard, smooth, dense ovoid grains) in the offshore are remarkably similar to those in the tidal flat sediment.

With this storm-layer model in mind we have tried to make as quantitative an estimate as possible of the minimal storm conditions required before deposition of laminae will occur on the Three Creeks tidal flat: the calculations are given below.

Estimate of minimal conditions required for storm deposition on the Andros tidal flats. It is possible to make some reasonable speculations about the conditions necessary before sedimentation on the tidal flats can occur. The two main requirements are (1) to get the bottom sediment of the offshore Bank into suspension, and (2) to then move this sediment-laden water onto the tidal flats. Both of these are achieved by wind stress. The second phenomenon, that of shoreward mass transport and piling-up of water onshore, is well known to occur along coastlines during a severe storm. It is the so-called "storm surge" (Harris 1959, 1963; Welander 1961). The water level reached at the coast by these storm surges depends on wind velocity, duration, fetch, air pressure, angle of approach of the storm, the shape of the coastline, the bottom topography, and the prevailing astronomical tide. If we use a very simple model with a uniform wind stress operating on a shallow body of water of uniform depth, then it is possible to make an approximate calculation of the storm-surge amplitude. Welander (1961, p. 321) gives the following relationship:

Table 7. Major onshore storms and storm flooding at Three Creeks tidal flats in the period March 1968 to May 1971

Date of storm	Weather conditions	Flooding of channel levee crest			Remarks
		Date of flooding	Duration crest submerged (hours)	Max. depth of water over crest (cm)	
1968:					
4–6 Jun	SSW to WSW winds, strong and persistent, averaging 7.7–8.7 m/sec (15–17 k), gusts to 21 m/sec (40 k) for almost 3 days. 13 cm rain on 3 Jun. Sea state 3.				Hurricane Abby passed 450 km to west on 2–3 Jun. Widespread deposition of a thin lamina of sediment.
10–12 Nov	WSW to NW winds, 7.7–10.3 m/sec (15–20 k), gusts to 19.6 m/sec (38 k) for at least 30 hr. 2.1 cm rain. Sea state 4. Strong cold front.				Widespread deposition of a thin lamina of sediment.
15 Dec	WNW winds 7.2–10.3 m/sec (14–20 k), gusts to 13.4 m/sec (26 k) for 21 hr. Sustained 9–10 m/sec winds for 12 hr. No rain. Sea state 4. Cold front with 7°C temp. drop.				Thin film of sediment deposited.
1969:					
16–17 Feb	WSW to WNW winds 7.2–12.4 m/sec (14–24 k) for 30 hr. 2.6 cm rain on 16 Jun. Sea state 4. Strong NW winds continued through 19 Jun. Cold front.	15 Feb. 16 Feb. 17 Feb	1.0 2.0 3.0 2.5	1.3 2.5 5.1 3.8	Either this storm or the one that followed on 4 Mar, or both, left a film of sediment and patches of fine sand on some of the levees.
4 Mar	WSW to NW winds 7.2–11.8 m/sec (14–23 k), gusts to 14.4 m/sec (28 k), for 30 hr. No rain. Seas rough, state 5. Cold front. Strong W to NW winds continued on and off for next week, particularly persistent on 9–10 Mar.	4 Mar	5.3	7.6	As above.
25–26 Aug	WSW to WNW winds 3.1–8.2 m/sec (6–16 k), gusts to 14.9 m/sec (29 k) for two days. Gusty, but no periods of sustained high winds. 2.8 cm rain in 24 hr. Gusty N and NE winds continued through 30 Aug with further 2.7 cm rain. Low pressure cell.	24 Aug 25 Aug 26 Aug* 27 Aug 28 Aug 29 Aug. 30 Aug	1.3 3.7 3.8 4.8 3.9 2.7 1.3 1.6 1.1 1.8	1.0 7.6 12.7 11.4 11.4 5.8 1.8 2.0 1.8 2.5	One to three thin lamina of sediment deposited. Not a strong storm, but persistent onshore winds and heavy rain caused widespread flooding. On 26 Aug the entire channeled belt was under water for several hours.
26–27 Dec	WNW winds 7.2–11.3 m/sec (14–22 k), gusts to 17 m/sec (33 k), for 24 hr. Trace of rain. Seas rough, state 5. Cold front.	27 Dec	3.2	5.6	Film of sediment and patches of fine sand.
1970:					
7–10 Jan	NW to N winds 6.2–12.4 m/sec (12–24 k), gusts to 14.4 m/sec (28 k), for 70 hr. 0.05 cm rain. Seas rough, state 5. Cold front.	7 Jan	2.0	1.3	Blush of sediment on beach terrace and some channel banks. No deposition on levees or beach ridges.
3–4 Feb	SW to NW winds 8.2–17.5 m/sec (16–34 k), gusts to 21 m/sec (40 k), for 30 hr. 0.9 cm rain. Seas rough, state 5. Very sharp cold front, 12°C temp. drop.	3–4 Feb*	16.5	22.9	Deepest and most prolonged flooding we recorded. Widespread deposition of thin sediment lamina.

Table 7. Major onshore storms and storm flooding at Three Creeks tidal flats in the period March 1968 to May 1971 (continued)

Date of storm	Weather conditions	Flooding of channel levee crest			Remarks
		Date of flooding	Duration crest submerged (hours)	Max. depth of water over crest (cm)	
9 Mar	SW to WNW winds 5.1–9.3 m/sec (10–18 k), gusts to 14.9 m/sec (29 k), for 24 hr. Persistent 7.2–9.3 m/sec (14–18 k) winds for 12 hr. Trace rain. Weak cold front.	9 Mar*	4.5 6.0	10.2 17.8	Film of sediment? No positive data for this storm.
4–5 Nov	NW winds 5.1–11.3 m/sec (10–22 k), gusts to 13.4 m/sec (26 k), for 24 hr. Persistent 9.3–10.3 m/sec (18–20 k) winds for 12 hours. 0.1 cm rain. Sea state 4. Sharp cold front.	5 Nov	0.3	0.5	Either this storm or the following one on the 15–17 Nov, or both, left a patchy deposit of sediment on the beach terrace and washovers.
15–17 Nov	WNW to NNW winds 5.1–10.3 m/sec (10–20 k), gusts to 14.4 m/sec (28 k), for 36 hr. Persistent 7.2–10.3 m/sec (14–20 k) winds for 18 hr. 6.4 cm rain on 15 Nov. Sea state 4. Cold front.	15 Nov	1.0	0.8	Film of sediment on the channel banks, but no deposition on the levees.
1971					
19–20 Jan.	SW to NW winds 6.2–10.3 m/sec (12–20 k) winds, gusts to 14.4 m/sec (28 k), for 48 hr. 1.3 cm rain. Sea state 4. Sharp cold front.	19 Jan*	5.0	17.8	Film to thin lamina of sediment and patches of fine sand deposited. Widespread.
4 Mar	SW to NW winds 6.2–10.8 m/sec (12–21 k) for 18 hr. Cold front.	4 Mar*	4.5	10.2	Between 3 Feb and 11 May only a film of sediment was deposited on lower part of beach terrace and on some channel banks. No deposition on levees or beach ridges. No data on which of the 6 storms that occurred during this period (listed here) produced the sediment film.
27 Mar	SW to NW winds 6.2–9.3 m/sec (12–18 k) for 18 hr. Cold front.	27 Mar*	5.0	12.7	
30 Mar	SW to NW winds 6.2–9.3 m/sec (12–18 k) for 15 hr. Weak cold front.	30 Mar*	4.0	12.7	
6–7 Apr	SSW to WNW winds 3.1–10.3 m/sec (6–20 k), gusts to 13.4 m/sec (26 k) for 48 hr. Sustained winds 7.7–10.3 m/sec (15–20 k) on 7 Apr. 0.4 cm rain. Weak cold front.	7 Apr	0.3	0.3	
24–25 Apr	SSW to W winds 4.1–9.3 m/sec (8–18 k), gusts to 11.8 m/sec (23 k), for 30 hr. No sustained strong winds. No rain. Weak cold front.	24 Apr 25 Apr 26 Apr	1.5 1.0 0.3	2.5 2.5 0.3	
1–3 May	SSW to W winds 2.1–9.3 m/sec (4–18 k), gusts to 16.5 m/sec (33 k). No sustained strong winds. 0.05 cm rain. Light N to NE winds on 2 May.	1 May 3 May	3.0 2.0	3.8 1.3	

Notes:
(1) Flood measurements were made with respect to levee crest at location 70–35, Fig. 2.
(2) Weather data compiled from records at Nassau, Fresh Creek (east Andros), and our own records.
(3) Sea state from Fresh Creek AUTEC records; refers to the sea east of Andros (Tongue of the Ocean). State 3 = 0.63–1.26 m waves; state 4 = 1.26–2.32 m waves; state 5 = 2.32–4.10 m waves.
(4) Duration of submergence was measured as length of time the highest levee crest (location 70–35) was completely under water.
(5) * indicates that flooding extended across the entire levee from channel to pond. Floods not so indicated are really channel-bank overflows with the pond water level much lower than that in the channels.
(6) No water level records before 4 Jan 1969.

$$s \cong K\, T_w\, F / g\, d\, h \tag{1}$$

where
- s = surge amplitude
- K = constant of the order of 1, depends on bottom stress
- T_w = wind stress
- F = fetch
- g = acceleration of gravity
- d = water density
- h = water depth

Also following Welander (1961, p. 322) we can estimate wind stress as follows:

$$T_w = k\, d_a\, W^2 \tag{2}$$

where
- k = a constant of value about 2.5×10^{-3}
- d_a = density of the air
- W = wind speed.

Combining (1) and (2), and using $F = 100$ km, $h = 5$ m, $K = 1$, $d_a = 1.3$ kg/m³, $d = 10^3$ kg/m³, $g = 10$ m/sec², $k = 2.5 \times 10^{-3}$ and W in m/sec, we have, for the Bahama Bank model:

$$s = 6.5 \times 10^{-3}\, W^2 \text{ meters} \tag{3}$$

Equation (3) is plotted in Figure 61. From this figure we can see that wind speeds over about 8 m/sec (16 knots) should produce a surge of sufficient amplitude to flood the highest levees (43 cm above MTL). Speeds over 10 m/sec (20 knots) should cause a surge high enough to overtop the highest beach-ridge washover crests (66 cm above MTL). These values are in surprisingly good agreement with the order of speed of storm winds that produced the major floodings of the tidal flats during our period of record (Table 7). For example, on February 3–4, 1970 direct onshore winds (W to NW) averaging between 10 to 11 m/sec (20 to 22 knots) blew continuously for fifteen hours, and water level in the channeled belt reached 66 cm above MTL in the channels and 60 cm above MTL in the ponds. Figure 61 predicts a surge of about 65 to 80 cm for such winds! For the most part, the storms have been of lesser intensity than this one, usually with persistent winds averaging between 7 and 10 m/sec (14 to 20 knots), and flood levels have been commensurately lower, about 45 to 55 cm above MTL (Table 7).

Complete flooding of the channeled belt then would seem to require as a minimum, persistent (at least several hours) onshore storm winds that maintain an average speed of at least 8 m/sec (15 to 16 knots). Flooding of the beach-ridge washovers would certainly take place at wind speeds lower than that predicted by Figure 61, because we must add to the surge the effects of waves breaking on the shore. Wave run-up will certainly spill water over the beach ridge at elevations above the mean water level of the storm surge (Harris 1963, p. 6). Our data confirm this: sediment deposition has occurred on washovers 45 cm above MHW, even though highest flood level in the channeled belt behind the protecting beach-ridge barrier has been only about 30 cm above MHW (Table 7). Clearly, the main body of water during the storm surge reached the tidal flats through the channel openings rather than by overtopping the beach ridge. To put the entire tidal flats completely under water would need onshore winds of average speed above 14 to 15 m/sec (about 30 knots). Such persistently strong winds are only associated with tropical cyclonic storms. The great hurricane of September 25–27, 1929, with wind speeds near the eye of over 45 m/sec (90 knots), passed directly over North Andros and flooded the whole island to depths of 1 to 4 m. Such fierce cyclones pass over Andros about once every four years on the average (Chapter 4).

Getting the bottom sediment into suspension requires that (a) bottom currents generated by the wind-waves be strong enough to erode the bottom sediment, and (b) flow be turbulent enough to thoroughly mix the water from bottom to top and so put the eroded sediment into suspension. Turbulent flow is assured by

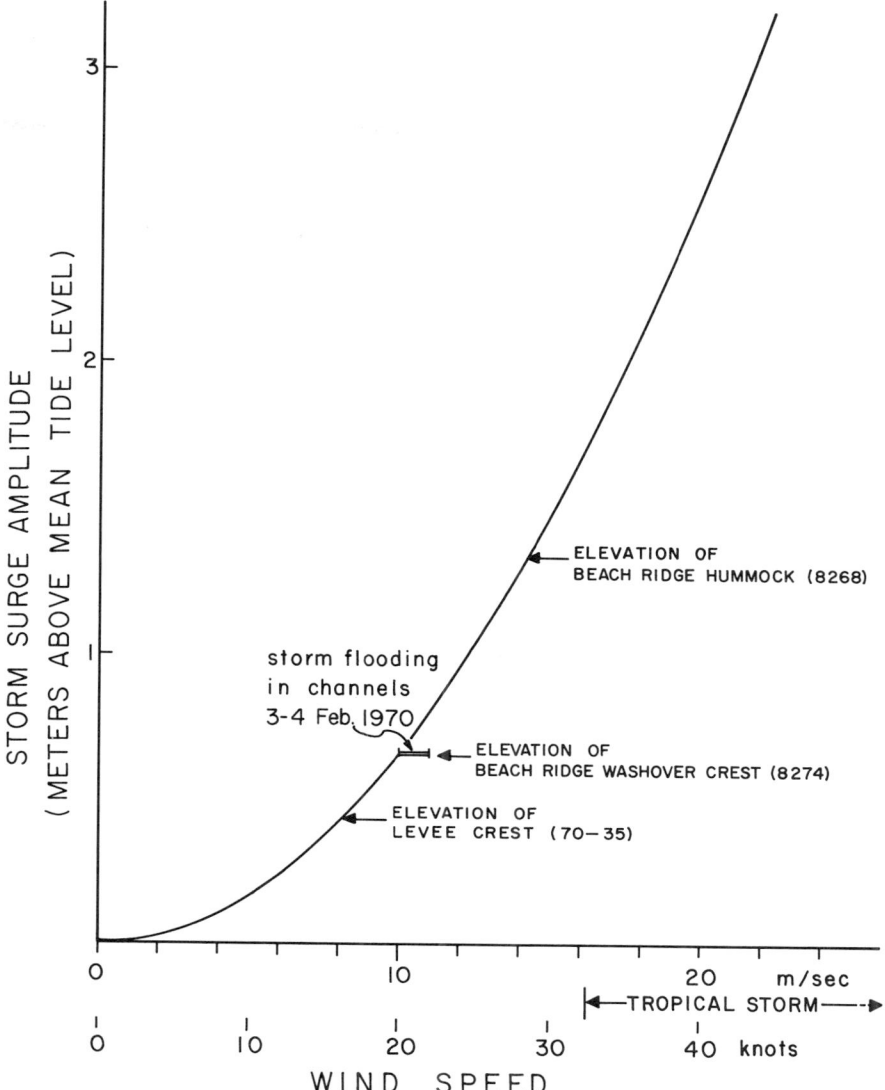

Fig. 61. Graph of surface wind speed against storm surge generated by the wind calculated from the equation Surge = $6.5 \times 10^{-3} \times$ wind speed (see text). Typical elevations of a levee crest, washover crest, and beach-ridge hummock (with their locality numbers in parentheses, see Fig. 2) at Three Creeks are indicated.

the very shallowness of the Bank waters. The problem then reduces to the bottom currents and their effect on the bottom sediment. Logan and Cebulski (1970, pp. 8–10) have calculated the horizontal component of orbital velocity of wind waves under varying conditions of wind speed, fetch, and water depth for short-period waves such as occur in shallow restricted bodies of water. From their Figure 4, for a water depth of 5 m and a fetch of 100 km, and using Hjulström's (1935) critical erosion velocity values, we have *roughly* the following:

Wind speed (m/sec)	Minimum duration (hrs)	Bottom velocity (cm/sec)	Maximum size sediment eroded (mm)
5 (10 knots)	2.4	10	0
7.5 (15 knots)	5.9	50	3.0
10 (20 knots)	10	75	6.5
15 (30 knots)	23	100	8.0

From Hjulström's (1935) data we know that the threshold erosion velocity is about 20 cm/sec; this agrees extremely well with the underwater flume measurements of Scoffin (1970) and Neuman et al. (1970), who measured velocities of 18 to 25 cm/sec for the initiation of fine-sand movement, and with those of Guy et al. (1966), whose data indicate a minimum erosion velocity of about 18 cm/sec for fine sand (Harms 1969, Fig. 9A).[7] If we use a threshold erosion velocity of about 20 cm/sec, then, from the data in the table, we would calculate that the minimum surface wind velocity to induce bottom erosion on the shallow Bahama Bank would be approximately 6 m/sec (12 knots). At wind speeds of about 10 m/sec (20 knots) the "erosive capacity" has increased dramatically, so that, in 5 m of water, granule-size material can be lifted. To move gravel such as the 10 to 15 cm flat pebbles we find on the beach ridges would require bottom currents of about 500 cm/sec! Wind speeds to generate this kind of velocity, even in very shallow water, could only be produced by storms of hurricane intensity. Now, in the off-shore, on the beach-ridge washovers, and on the levees the main range of size of detrital particles is 30 to 750 μ. To move such particles would require bottom currents of at least 35 cm/sec: in 5 m of water this translates to surface winds of no less than 6.5 to 7 m/sec (13 to 14 knots). However, as Neumann et al. (1970) have shown, the surface sediment may be bound by an organic mat that requires bottom currents as much as 120 cm/sec before any erosion can take place. Off Andros Island we found the bottom sediment to be covered by what is best described as a coating of organic scum rather than a cohesive mat. We found the sediment to be easily disturbed, so that binding is probably considerably lighter than in the subtidal mats of Abaco Sound where Neumann et al. worked. This suggests somewhat lower erosion tolerance for the Andros offshore than for Abaco Sound, but certainly greater than would be indicated by the Hjulström (1935) relation used in the table above. We think it likely, therefore, that winds of 7.5 to 8 m/sec (15 to 16 knots) generating bottom currents of about 50 to 55 cm/sec would be capable of eroding the Andros offshore and putting all grains up to 750 μ into suspension. Lower velocity winds around 6.5 to 7 m/sec (13 to 14 knots) could be equally effective where the water is very shallow (no more than 1 m deep) and would still be capable of moving 100 to 200 μ grains in 5 m of water. We have some confirmation of these rather rough calculations. In early January 1970, we observed that three days of WNW–NNW winds averaging 8.5 m/sec (17 knots), but ranging from 5 to 12 m/sec, made the Bank water turbid with suspended sediment. We also observed in April of the same year that even lower winds (southerly, 5 to 8 m/sec; 10 to 16 knots) stirred up the sediment in the very shallow offshore (1 m depth) of Point Simon, without, however, causing flooding of the levees and beach ridge.

All in all, then, we would estimate that for sedimentation to take place on the levees and beach-ridge washovers of the Three Creeks tidal flats, onshore winds (SW–NW) must blow for at least half a day at speeds at least 7 to 8 m/sec (14 to 16 knots). Producing a large enough storm surge is really the limiting factor. Therefore, the winds must be directly onshore (or nearly so), otherwise the surge (which will be propagated in shallow water about parallel to wind direction; Harris 1963, p. 4) will be partly dissipated by generation of longshore components as the coast is reached. This we found for the storm of January 1970 mentioned above. Although the winds were strong for almost three days (8.5 m/sec average), the direction (NNW) was oblique to the Three Creeks shoreline and the surge just spilled over the levees from the channels, but was still some 12 cm below the levee crest in the ponds.

5. Origin of the Thin Beds of the Marshes: Some Speculations

Thin beds and thick laminae are characteristic of the sediment of the vast inland algal marsh and the algal marshes fringing the ponds of the channeled belt (Fig. 52). The wide and random range in layer-to-layer thickness (the outstanding feature of these marsh sediments) makes the marsh bedding distinctly different from the uniformly thin lamination of the levees, beach ridges, and channel bars.

---

7. Sundborg's (1956, Fig. 13) calculated curves for erosion velocity are widely used, but his values appear to be somewhat high when compared to the flume data given above.

Clearly, the storms we have recorded are not responsible for the marsh layering, because in the three years of the study no sediment (except perhaps a blush of mud on some of the pond marshes) has been deposited in these marsh subenvironments, but, instead, slow growth of *Scytonema*—about ½ to 1 mm a year—has been observed. *Scytonema* does not have the capacity, as the Oscillatoriacea, like *Schizothrix*, do, to slip out of its sheath and migrate upward through a sediment cover and so avoid fatal burial. *Scytonema*, therefore, can only thrive in areas where sedimentation is absent or very infrequent (Chapter 7). Therefore, we interpret the alternations of thick (up to 5 cm) algal tufa with thin beds of sediment in the algal marshes (Fig. 52) as a clear record of long periods (measured in years) of algal growth free from sedimentation, punctuated by sudden influxes of sediment. Burial of the *Scytonema* mat and infilling between the "pincushions" suggests that the sediment came in single widespread pulses and covered the mat under a mud and mud-peloid layer that in some cases was as thick as 7 cm and in others as thin as a few millimeters (Fig. 52; see also Shinn et al. 1969, Figs. 21 & 22). *These thin beds and thick laminae of peloids and mud are, we believe, "hurricane layers" deposited during rare catastrophic flooding caused by storms of hurricane or near hurricane strength.* If this interpretation is correct then these algal marsh sediments preserve an exclusive record of the major tropical cyclones that have passed over, or very close to, Andros Island in the past several hundred years. The record is highly selective, an exclusive "filter," because small seasonal storms like the "northers" do not deposit sediment on the marshes (Table 5).[8]

No hurricanes have passed close to Andros in the three years of our study[9] so we have no direct evidence that the thin beds represent "hurricane layers." However, thin beds (from 2 to 10 cm thick) of lime mud and sand are known to have been deposited by hurricanes Donna (1960) and Betsy (1965) on Andros and in Florida Bay (Pray, 1968; Ball et al. 1967; Perkins & Enos 1968; see also, Ginsburg et al. 1954, pp. 27–28). A short core, Figure 62A, taken on Crane Key in Florida Bay from the sediment deposited above MSL shows the same thick algal layer-sediment layer structure seen in the marsh sediments of Three Creeks. In the Williams Island area of Andros, Perkins and Enos (1968, p. 716) report deposition of a layer of lime mud (about 2 to 5 cm thick) at the edges of palm hammocks (Shinn et al. 1965) by Hurricane Betsy (Fig. 62B); they suggest that deposition was due to the baffling effect of the vegetation fringing the hammocks. For the Three Creeks area these same authors (p. 714) report flooding of the tidal flats to depths of about 3 m during Betsy, but no deposition of sediment.

From our observations it seems most likely that deposition of thick sediment layers during major storm flooding will occur on the tidal flats only where there is either an effective vegetation baffle, or substantial ponding of the floodwaters so that particle settling is more rapid than floodwater subsidence. For example, in a few places at the pond edges we find "crag-and-tail" piles of sediment up to several centimeters thick and a few meters long in the lee of very low mangrove clumps and small bushes.[10] These deposits, now fragmented by wide mud-cracks (see, for example, Fig. 62B), invariably tail *away* from the pond upslope toward levee crest or beach ridge. Clearly, deposition took place on the ebb, as the deep, sediment-charged floodwaters rushed directly seaward off the flats. The sediment

---

8. It may be possible to read the marsh cores (Fig. 52) like tree rings and decipher the cyclone periodicity, since the thickness of the algal layers is certainly proportional in rough terms to growth time. Thickness of the sediment layers at first thought would seem to be at least roughly proportional to storm intensity, but this is by no means reliable as the observations of Perkins and Enos (1968) suggest.

9. Hurricane Abby, the eye of which passed about 450 km to the west of Andros in early June 1968, caused only minor flooding of the Three Creeks tidal flat and left but a thin lamina on the levees only (Tables 5 & 7).

10. These are the most recent thick layers on the flats and were probably deposited by Hurricane Betsy (cf. Perkins & Enos 1968).

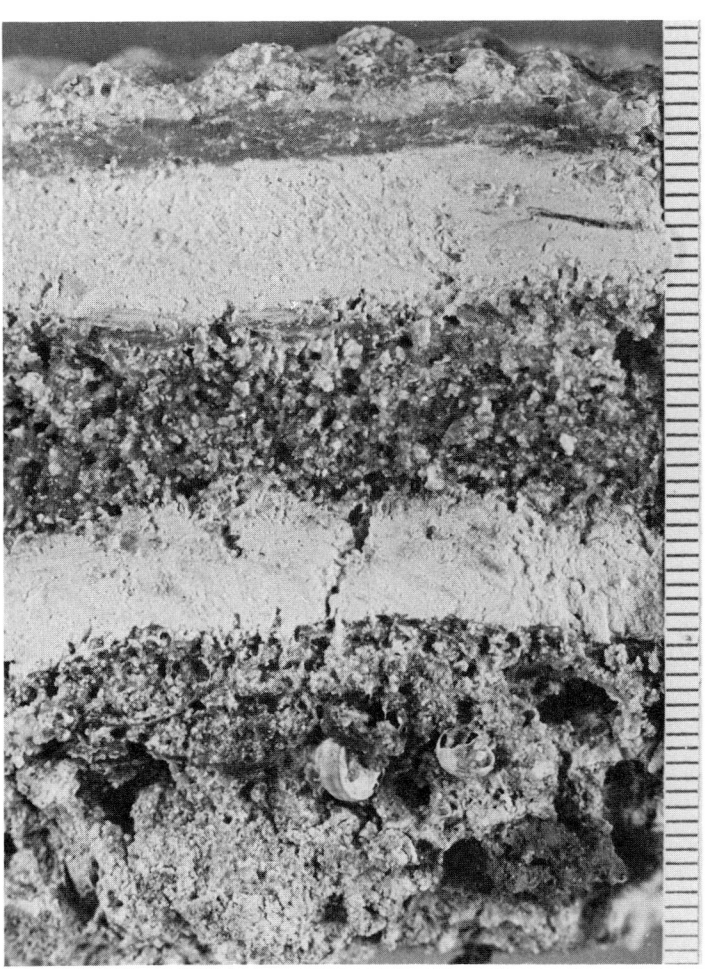

Fig. 62A.
Short core taken from supratidal pond on Crane Key, Florida Bay, showing hurricane layers (white, lower deposited by Donna 1960, upper deposited by Betsy 1965), alternating with thick, gelatinous algal layers (rough dark layers). Scale in mm.

Fig. 62B.
View of widely cracked and eroded hurricane layer on Snake Hammock, north of Williams Island, Andros (Fig. 88). This aragonite pelleted mud layer was deposited by hurricane Betsy of 1965 on the supratidal fringes of the palm hammocks. Note the intraclastic grit, derived from erosion of the polygons, filling in the mud-cracks. Trowel for scale.

Fig. 62C.
Lee-shadow accumulation of sediment mound behind a small halophyte tuft on a levee at Three Creeks (near locality 8256, see Fig. 2). Penknife for scale.

in these mounds was probably partly derived from the ponds themselves. More pervasive baffling is provided in the long, thick *Scytonema* filament bundles and "pincushions" in the algal marshes (Fig. 32).

Ponding of floodwaters must certainly occur in the ponds of the channeled belt when the ebb-water level drops below the levee crest. The water can now only escape from the ponds through the narrow distributory channels and subsidence is temporarily checked, enough at least to allow sand- and silt-sized mud peloids and shell fragments to settle to the pond and fringing algal marsh surfaces. With this mechanism one would expect the thickest layer to be in the ponds (where the water was deepest), with a feathering-out of the settled layer across the algal marsh zone and up the levee backslope. Remnants of such thick layers that have survived bioturbation are found in a few places at the pond edges (Fig. 59B). These layers thin toward the levees, supporting the picture outlined above.

In the inland marsh, which slowly rises to meet the pinelands of east Andros, both ponding and baffling are probably responsible for deposition. This huge tract, although relatively flat, does undulate (Fig. 31), with shallow depressions and low hummocks. Sediment layer thickness does vary both on a small scale, due to infilling between algal "pincushions," and on a large scale, due undoubtedly to local ponding in the depressions. Ponding of water across the *entire* inland marsh, although this area is slightly higher in elevation than most of the channeled belt, can occur. We have observed during heavy summer rains that the rainwater runoff can be trapped for days in the inland marsh by general high level of tidewaters that reach to the edge of the marsh. In the aftermath of a hurricane, abnormally high tides with a high "low-water" level will continue for several days, and so draining of floodwaters from the inland marsh will be considerably retarded, allowing settling of at least the coarser sediment to take place.

The frequency of hurricanes in the Andros area is rather high. As noted earlier, in the years 1900 to 1937, thirty-two tropical cyclones passed within 100 km of Andros Island; eleven of these passed directly over the island. However, not all of these cyclones were of full hurricane intensity when they tracked through the Andros area; some, too, were of small diameter. It is probable, therefore, that major flooding of the northwest Andros tidal flats has occurred at most once every four years on the average. Indeed, in the past decade only Betsy (1965), which passed directly over the north tip of Andros heading west across the Great Bahama Bank, caused widespread inundation of Andros's western shore.

The origin of the algal tufa—the magnesian calcite sheaths around *Scytonema* filaments in the algal interbeds—is discussed in detail in Chapter 7.

6. The Problem of the Uniform Thickness of the Millimeter Laminae

Uniformity of laminae thickness. Thinly laminated rocks ("mm-laminites") are typical of ancient tidal flat deposits, particularly carbonates (see, for example, Fischer 1964; Matter 1967; Laporte 1967; Roehl 1967; Schenk 1967; Tourek 1970; etc.). These laminites are remarkable for the very small range in thickness from lamina to lamina. This relative uniformity in thickness is rather puzzling because it appears to require that the lamina-producing events be all of about the same magnitude. Some regularity in time seems also to be implied, and, in fact, some geologists have interpreted laminites as varves (e.g., Fischer 1964, p. 143). Our observations on the mm-lamination forming today on Andros Island throw some light on the problem.

We have established that the levee and beach-ridge lamination at Three Creeks are produced by onshore storms. The storms we have recorded, although all rather small, have been of variable intensity and duration (Table 7), and, as might be predicted, this is reflected in a small but obvious variability in lamina thickness (Fig. 36). If we can extrapolate this relationship, then we would expect to find at least a few thick "hurricane layers" interbedded with the thin laminae such as occurs in the blocked (standing water) tidal channels of the Gladstone embayment tidal flats in Shark Bay (Davies 1970, Figs. 10 & 11). However, in all levee, beach-ridge, and channel-bar sediments where layering persists with depth, the range in thickness of the laminae is always small, no more than an order of magnitude with a maximum at about 2 mm (Fig. 36). Thick lamination and thin beds, such as characterize the sediments of the ponds and marshes, are notably absent, a characteristic these Three Creeks "laminites" share with many ancient laminites. The problem, very simply then, is why doesn't a huge storm leave any thicker layer on the levees, beach ridges, and channel bars than does a small winter "norther"? The answer, we think, must lie in the different depositional dynamics that occur on the *mound-like* levees, beach ridges and channel bars as against the *basin-like* ponds and plant-baffled inland marsh.

Some possible explanations. In looking for some control on layer thickness other than "intensity" of the storm, we turned to the role played by organisms. Workers such as Black (1933), Ginsburg (1960), Logan (1961), Monty (1967), Gebelein and Hoffman (1968), and Gebelein (1969) have presented evidence for the sediment trapping role of blue-green algae. Since the entire surface of the Three Creeks tidal flat is covered with a blue-green algal mat, our first thought was that the mm-laminae of the levees, beach ridges, and channel bars are simply planar algal proto-stromatolites, each lamina being produced by the sediment-trapping action of the surface mats. We envisioned the wet, slimy algal mat (and its tenant micro-organism film) as a "fly-paper," capturing and holding fast sediment suspended in the floodwaters streaming over the levees, beach ridges, and channel bars during a storm. Once this "fly-paper" was choked with sediment, it would no longer be able to act as a trap. Regardless, then, of the amount of sediment passed over the "fly-paper" only one thin lamina could be formed during a storm. This is certainly a very attractive hypothesis and has strong support from a field experiment carried out on the Lake Ingraham, Florida, tidal flats by P. Hoffman and C. Gebelein (personal communication; see also Bathurst 1971, p. 229). These workers killed with formalin a small patch of the algal mat that covered the surface of a channel levee. They found after overbank flooding with sediment-charged water that no deposition occurred where the mat had been destroyed. The implications of this experiment are that (1) tiny filamented algal mats do trap sediment like a "fly-paper," (2) the flow over these levees was of such a velocity that purely mechanical bed-load deposition could not take place. At Three Creeks we have some evidence that such algal trapping to produce laminae does occur (we will present this evidence in the next section) but, on the other hand, we also have observations that demonstrate that at least some of the laminae are *purely mechanically deposited*. Traction-load deposition of

LAYERING

Fig. 63. Isolated patches of peloid sand and silt filling shallow depressions in levee crest surface after an onshore storm. The dark underlying surface is a coat of red pigment we poured on the levee to act as a time line. Pocketknife for scale.

laminae was very obvious after several storms when we found patches of *loose*, well-sorted, fine sand strewn across the levees and beach-ridge washovers as very thin depression fills (Fig. 63) and as low, flat, starved ripples. In addition, the common presence of "current lineations" (Fig. 12B, Chapter 4) clearly demonstrates bed-load deposition (see Wolman & Brush 1961, p. 196, Fig. 123; Vincent 1967, Fig. 6). In cores it is not very difficult to recognize these mechanically deposited layers—they are the thin, discontinuous, open-framework peloid sand laminae so typical of the levee, beach terrace, and beach-ridge washover sediments (see, for example, Fig. 36). The lack of any really thick sand layers then becomes puzzling because in purely mechanical deposition, layer thickness is certainly a function of the nature of the fluid suspension and the characteristics of the flow. The most obvious consideration is the concentration of suspended sediment: very low concentrations can only produce a thin accumulation of sediment, whatever might be the flow regime. On the other hand, no deposition will occur if the flow velocity exceeds the settling velocity of the suspended sediment, whatever is the concentration of suspended sediment. We do not have any measurements of floodwater velocities during onshore storms at Three Creeks, but we do know that low sediment concentration seems to be a characteristic of all the winter-storm deposition we have so far recorded. This is inferred from the fact that (1) we find "starved" ripples, and (2) little of the overbank sediment load, except a little mud, reached the marshes at the pond shores (Table 5). It appears then that a relative "deficiency" of sediment is at least one of the major reasons for the thinness of the layers on the levees, beach terraces, beach-ridge washovers, and channel bars.

Now, low suspended sediment concentration in floodwaters of small winter storms (and hence deposition of only thin laminae) is understandable because of both the moderate wind velocities and the short storm duration. But what of a major tropical storm such as Betsy of 1965 that churned the Great Bahama Bank waters for several days and flooded the Andros Island tidal flats with up to 3 m of water (Perkins & Enos 1968)? Betsy produced floodwaters that clearly were highly charged with sediment, because we find a layer *several centimeters thick* left at the grassy edges of the palm hammocks near Williams Island (Fig. 62B). Yet, from Williams Island to Three Creeks, the sediment record shows that on the levees, beach ridges, and channel bars, Betsy did not leave any layer thicker than a few millimeters (if any sediment was deposited there at all!—see Perkins & Enos 1968). The best explanation, we think, lies in sediment "by-passing" during a major catastrophic storm. We can imagine two kinds of by-passing processes: (1) bottom current velocity exceeds bedform washaway velocity, and (2) the storm is of insufficient duration to overcome the induction period for establishment of stable bedforms. We can roughly calculate the requirements for the first kind of "by-passing." The data of Kondratev (1962) (see Graf 1971, pp. 282–83) suggest that for grains of 750 μ diameter, washaway of bedforms occurs when the flow has a Froude number above about 1.1. In water depths of 3 m (such as Betsy's maximum flood depth on Andros) this Froude number translates to a mean flow velocity of 596 cm/sec.[11] For shallower depths the washaway velocity would be commensurately lower, for example, for 1 m water depth the velocity would be 344 cm/sec. We have not found any data on floodwater velocities at shorelines during hurricanes, but 3 to 6 m/sec does seem very high, although we know that mean flow velocities of 2 to 3 m/sec are average for rivers in flood and extreme values up to nearly 10 m/sec have been measured (Leopold et al. 1964, p. 167); normal rip current velocities on beaches may reach 1 m/sec. However, earlier we had computed that a velocity around 5 m/sec would be needed to move the flat pebbles we find deposited on the beach-ridge washovers (Fig. 26A).

Perhaps the answer lies in the fact that levee and beach-ridge crests would be sites of local wave action, where breakers would move as localized currents of very high velocity. This would suggest that during catastrophic storms such as Betsy, local wave-generated flow over the levee and beach ridges is of so high a velocity that, in the *absence of current baffles*, mechanical deposition of a sand layer cannot occur, but, instead, the sediment is swept into the pond and marsh basins. This idea is supported by the observation that small isolated "crag-and-tail" mounds of sediment are found on the levees and beach-ridge washovers in the lee of scattered low bushes and grassy tufts (Fig. 62C), which have obviously acted as current baffles. This demonstrates that not only was sediment available but that it passed over the levees and beach-ridge washovers.

Another possible explanation for the lack of thin-bed deposition on the levees and beach ridges during hurricanes was suggested to us by John Southard. Southard (personal communication) has found experimentally that for new bedforms to stabilize on a smooth, *firm* (nonerodable) surface there is quite a long induction period (several hours), during which time sediment continues to be transported without building ripples or dunes. Such a firm surface, free of loose sediment, is provided by the smooth, *Schizothrix*-bound surface of the levee crests, beach-ridge washover crests, beach terraces, and intertidal channel-bar crests. By-passing by very high velocity currents would then not be needed to explain the absence of thick hurricane sand layers on the surfaces of these subenvironments. However, this does not explain how the coarse, flat-pebble gravels on the beach-ridge washovers were transported.

---

11. This was calculated from the relation N (Froude number) $= \bar{u}/g^{1/2} D^{1/2}$, where $\bar{u}$ is mean velocity in cm/sec, g is acceleration due to gravity (980 cm/sec²), D is flow depth in centimeters.

If only uniformly thin laminae are deposited on the levees, beach ridges, and channel bars, what then of the ponds and algal marshes, where wide and random variation in layer thickness is the characteristic feature? In contrast to the "positive" topography of the levees, beach ridges, and channel bars, the ponds are basin-like features. As described in the preceding section, when storm floodwaters subside below the level of the bounding levee crests, the only outlets from the ponds are through narrow channel ways. Ebbwater drainage from the ponds then will be considerably retarded and the settling rates of at least the coarser sediment may exceed drainage rate, and thick sediment layers could accumulate as *settle-outs*. The layer thickness will depend on sediment concentration and ponding duration, and hence will vary from storm to storm. The algal marsh zones that fringe the pond shores will also accumulate sediment by settle-out of ponded floodwaters, but, in addition, the *Scytonema* carpet will act as an effective and *pervasive* baffle, collecting a sediment layer by *entrainment* in the complex network of large filaments and filament bundles. These same processes of ponding of sediment-charged floodwaters and baffling by the long-filamented *Scytonema* carpet account for the accumulation of thick sediment layers in the inland algal marsh.

We conclude then that baffle-free, rapidly drained depositional surfaces like the levees, beach ridges, and channel bars will allow only thin laminae to accumulate, whatever the magnitude of the depositing storm. In the absence of ponding and current baffles, big storms do not leave any thicker layers (if any at all) than do small storms. Such "laminites" are more revealing of the physiographic setting than they are of the energy of the depositional events: they are, in fact, a highly "filtered," and hence very deceptive, record. They are an exclusive record of *small* onshore storms, just as, in contrast, the thin beds and thick lamination of the marshes are an exclusive record of major tropical cyclones (see above).

The role of blue-green algae. We outlined above a possible mechanism in which blue-green algal mats act as a "fly-paper" to trap sediment and so produce uniformly thin lamination. Our primary evidence for agglutination of sediment particles onto sticky *Schizothrix*-type mats at Three Creeks is the steeply draped (even "overhung") lamination of the knobby domal structures of the channel banks (Fig. 34). These "gravity-defying" laminae carry detrital peloids, yet they cannot be mechanically deposited layers—what is clearly required is some kind of surface glue to which the passing sediment particles will adhere. Such a sticky substrate is provided by the surface *Schizothrix*-type mats which, when wet, become coated with a slimy mucus (this algal surface mat on the levees, beach-ridge washovers, and beach terraces at Three Creeks is so slippery when wet that one can slide bare-footed many meters as though on ice!). For the *flat* lamination of the levees, beach ridges, and channel bars we think the same trapping process occurs. Our evidence here is in the fabric of the lamination. In thin section, the laminae are of two kinds: discontinuous lenticular layers of well-sorted peloid sand and thin, more continuous muddy layers (Figs. 37A and B). The well-sorted peloid layers and some of the thicker, coarser, clotted mud layers are the mechanically deposited traction-load depression fills and starved ripples we have discussed above. Many of the muddy layers, on the other hand, are fine-grained (max. peloid size 70 μ), uniformly very thin (a fraction of a millimeter), laterally continuous, and drape any irregularities, *even steeply inclined slopes*. These features strongly suggest that such thin, muddy layers are algal "stick-ons." Our pigment-marker observations support this interpretation: after storms which left only patches of sand over the levees and beach-ridge washovers (Fig. 63), the pigment-marker surfaces were found to be lightly coated with a blush of mud, giving the red dye a milky hue. Clearly, a mud film will collect where sand will not: in an event where sheet flow predominates, selective algal trapping of the finer sand, silt, and mud seems to be the most obvious explanation. In this regard, Black (1933) and Gebelein (1969) emphasized that trapping by the tiny filamented *Schizothrix*-type mats was size-selective, that the smaller particles were preferentially collected from the passing sediment load by the mat. Our interpretation of the thin, draped, finer-grained muddy laminae as algal "stick-ons" is certainly consistent with the observations of Black and Gebelein. However, it is well to note that the Andros channel-bank proto-stromatolites have trapped grains up to 150 μ, so that, for Andros at least, "smaller particles" means grains up to fine-sand size and not just mud.

A vital role played by the surface algal mats at Three Creeks is that of protecting the underlying sediment from erosion. On the levees, beach terraces, beach-ridge washovers, and "exposed" channel bars the firmness of the smooth, *Schizothrix*-type, algal-covered surface (even the heaviest footstep leaves no imprint), our direct observation that even mud particles are not dislodged by the heaviest rain-splash, and our inability to peel the mat from the topmost sediment, are more than ample demonstrations of the effectiveness of the very thin (less than a millimeter) but complex meshwork of tiny filaments in tightly binding together the particles of surface sediment. By quick upward growth through newly deposited sediment Oscillatoriacea like *Schizothrix* can colonize a fresh sediment surface in a matter of hours (Chapter 7). Therefore, the sediment at Three Creeks is virtually never without a protective coating. This surface algal binding, then, by protecting even the most delicate sediment layer from erosion by storm floods or rainwash, preserves in splendid detail each and every lamina deposited on channel banks, levee and washover crests, beach terraces, and "intertidal" channel bars. We judge that the algally bound sediment can resist even hurricane flooding (with flow velocities perhaps in the order of several m/sec at peak flood ?). We certainly find no record of deep scouring of the laminated sediment, nor have we been able to induce any artificially by violently pouring large barrel loads of water onto levee surfaces. On the beach-ridge washover crests, where large, flat pebbles have been deposited (Fig. 26A), presumably by very high energy currents, the underlying thin laminae are not scoured nor disrupted. Only where blistering (Fig. 13B) and cracking (Fig. 17A) occurs, or along the exposed beach-cliff edge (Fig. 24B), or the undercut channel banks (Fig. 30) does erosion of this algally bound surface sediment take place. It is this pervasive algal binding then that ensures the continued and uninterrupted vertical accretion, the upward growth, of the levees, beach-ridge washovers, and channel bars.

Of all the subenvironments of the Three Creeks tidal flat the ponds appear most susceptible to erosion, because browsing by gastropods and rapid sediment turnover by polychaetes in these ponds (Chapter 6) does not allow the development of cohesive algal mats like those of the levees and beach-ridge washovers. Although daily tidal currents, and even winter storms (to judge by the pigment-markers), do not appear to erode the pond sediment, we have observed that heavy rain-splash will dislodge sediment from the pond surface and turn the run-off waters milky. It seems likely, therefore, that very high energy floodwaters, such as those produced by hurricanes, would erode the ponds, while the algally bound levee, beach-ridge, and "intertidal" channel-bar sediments remain intact. Perhaps such hurricane floodwaters remove from the ponds during peak overflow stages about as much sediment as they deposit there at low ebb stages and thereby maintain the pond sediment build-up rate below that of the enclosing beach ridges and levees. Some such process is certainly needed, because if a single hurricane can deposit over a centimeter of sediment in the ponds, and hurricane deposition recurs once every four years or so on the average, then the ponds would soon fill up, because the accretion rate of the levees is probably no more than 3 mm per year. The lack of a resistant algal cover on the ponds would seem, therefore, to be a prime factor in the long-term maintenance of the existing pond-levee physiography.

7. Criteria for Recognizing Andros-Type Tidal Flat Layering: A Review

Lamination. More often than not, laminites of all compositions (carbonate, siliciclastic, evaporitic) in the geologic record have been interpreted as varves, usually with the implication that they are "settle-outs" in a relatively deep body of water in a region of strong seasonal climatic changes. Yet there exist several other viable (and perhaps more common) mechanisms for producing laminites, for example, *nonseasonal* (random storm or flood-induced) "settle-outs" in either deep or *very shallow* "standing" water; selective sediment trapping by algae from a passing suspension load (planar stromatolites); traction-load, plane-bed flow in stream channels, on flood-plains, or on beaches; sheetwash on tidal flats and playas, particularly overbank sheetwash on levees, bars, beach ridges, such as we have described in this paper; chemical precipitation, such as produces laminated caliche or travertine; bacterial or algal precipitation. It is very clear that what we need are diagnostic criteria for distinguishing among these different laminite-types. We offer here such criteria for recognizing the kind of laminite being deposited today on the Three Creeks tidal flats.

On the Three Creeks tidal flats the millimeter-lamination is produced by storm-induced overbank flooding of subaerially exposed channel levees, channel bars, and beach ridges. The particular combination of purely mechanical traction-load deposition and algal trapping leaves a readily recognizable fabric in these laminated sediments. In brief review, the Three Creeks lamination is characterized by the intimate interlayering of two kinds of layers: (1) *discontinuous* flat-lenticular layers of quite well-sorted peloid sand,[12] deposited mechanically as thin depression fills and as starved ripples (see Fig. 63), thickness varies from a few tenths to several millimeters, lateral persistence generally on the order of a few to tens of centimeters; (2) laterally *continuous*, uniformly thin muddy layers, deposited as muddy films (most likely as algal "stick-ons"), thickness seldom greater than a few tenths of a millimeter, lateral persistence at least on the order of many meters. These features can, with close examination, be seen in hand-specimen (Fig. 36); in thin section, they are unmistakable. The photomicrographs, Figures 37, 38, and 42, provide the best single illustrations of the typical "Andros-type" laminite fabric. This fabric, we think, is the basic criterion for recognizing an "overbank flooding" or sheet-wash mechanism in a low energy tidal flat environment.

We can go a step further. As we pointed out above and summarized in Table 3, on the basis of the morphology of the individual laminae we can identify distinctive depositional subenvironments in the Three Creeks tidal flat complex. The major distinctive lamination types, modified from Table 3, are worth re-stating here:

1) smooth domal lamination (SH protostromatolites) = channel bank (unlikely to be preserved in the geologic record as a widespread lithology);
2) smooth flat lamination = levee crest, beach terrace, beach-ridge washover crest, intertidal channel-bar crests (Table 4 gives further criteria for distinguishing among these four subenvironments);
3) disrupted (mud-cracked) lamination = levee and beach-ridge backslopes;
4) crinkled fenestral lamination and lamination with "palisade" structure (LLH protostromatolites and SH algal tufas) = high algal marsh and inland freshwater algal marsh.

Each of these variants reflects differences in local conditions, such as microtopography, frequency of sediment influx, type of algal mat, etc. The reader is referred to the particular descriptions in the text for detailed criteria. The point to be made here is that it is not only possible to recognize an Andros-type

---

12. Many of the peloids are *soft* and will easily deform on drying and slight compaction, so that they are likely in the geologic record to end up as a *muddy matrix*. One must then rely on the good sorting of the skeletal grains and firm peloids and on the thickness and lenticular geometry of the layer to identify these laminae as mechanically deposited sands.

laminite in the geologic record, but very refined interpretations of the environment in which lamination forms are within our grasp.

Four additional features of the Three Creeks lamination should be noted. First, all lamination types are pervaded by algal filament molds (10 to 30 μ wide). These molds are most obvious as thread-like voids in the muddy laminae, but they are also recognized, with some difficulty, by the vertical orientation of sand grains and pores in peloid layers. We do not regard the presence of filament molds as a primary criterion for Bahama-type lamination, except for the crinkled fenestral lamination and lamination with palisade structure, because: (1) algal mats are ubiquitous on all kinds of tidal flats and are common even on nonmarine playas where similar laminated sediment accumulates; (2) the thin "algal stick-on" mud laminae are not produced by the algae (*Scytonema* sp.) that leave the large molds in the Three Creeks lamination!

Second, although some laminae without doubt appear in thin section as graded from sand up to mud, and perhaps would be so described by many workers, the significant feature of the Bahama-type lamination is that if these apparently graded lamina are traced laterally the sandy "basal" layer pinches-out, while the "muddy" upper layer continues without change in thickness (Fig. 38A). The mud and sand are deposited during a *single event*, but the order of deposition is, we think, quite the opposite of a graded bed: during a storm, mud is deposited *first* as the wetted algal mat selectively traps the finer particles in the early stages of overbank flooding, and then in the later stages, when flooding is at its peak, the algally bound mud film is partially covered by patches of sand. A mud layer and a sand layer, then, make a "couplet," with the thin, persistent mud film at the base and the thicker, lenticular sand layer on top. However, not every storm lifts sand-sized particles over the levees and beach ridges, so that the mud-sand "couplet" is only randomly deposited in time. The overall vertical succession of mud and sand layers reflects this random deposition of sand—commonly several successive mud layers separate the "couplets" or packages of "couplets" (Fig. 42B). Some of the sand laminae are themselves internally graded but the grading is from dense, indurated sand peloids up to light, soft sand peloids, or, in a few cases, the sand laminae show *reverse* grading (Fig. 42B). We think that the presence of both kinds of graded sand layers points to waning and waxing traction currents as the cause of the sand grading, rather than settle-out from slack water. All in all, sharp separation of sand from mud laminae is the norm in Andros-type lamination. We would urge sedimentologists looking at laminites in the geologic record to make the distinction between laterally persistent, sand-mud graded laminae and the Andros tidal flat lamination, because their respective origins are quite different; certainly algal trapping cannot produce an internally graded lamina.[13]

The third important additional feature, as pointed out earlier, is that mud-cracking is absent from the laminites collecting in the *highest and driest* environments, the levee crests, beach terraces, and beach-ridge washovers. In ancient laminites, then, the absence of mud-cracks should not be considered as an inviolate criterion against a "subaerial" origin.

The fourth additional feature of significance is that a single storm can deposit a lamina over a very wide area. On northwest Andros Island a "norther" can smear sediment on the tidal flats over a longshore distance of at least 60 km and an onshore distance of up to 4 or 5 km. At Three Creeks, a lamina, either a

---

13. Davies (1970a, p. 186), for example, notes that "graded bedding is a common feature in the algally-laminated sediments of the Gladstone area." In fact, the only internally graded bedding, strictly defined (see *AGI Glossary*), that Davies illustrates involves thin beds (> 1 cm) which surely are not algally trapped layers. All the fine lamination pictured by Davies shows sharp separation of sand and mud, and, indeed, Davies says "grading . . . may be represented simply by an alternation of coarser and finer laminae" (!).

"couplet" or a muddy layer, although widespread will not be a completely continuous sheet because of the patchwork distribution of levees over the tidal flat area. However, on the beach ridges the along-strike "continuity" is broken only by narrow tidal channels. South of Williams Island (Fig. 1), where a coastal plain is being built, a single lamina could be deposited as a completely continuous widespread sheet, as for example occurs on the silici-clastic tidal flats of the northwest Gulf of California. Here, at the southwest corner of the Colorado River delta, the "high flats" above normal high water (Thompson 1968) is a vast, unbroken, featureless coastal plain covering about 500 km^2. A single onshore storm will flood the entire high flat, depositing a silty lamina over the whole area. Therefore, the existence of areal "continuity" of laminae—the ability to correlate packages of laminae over tens of kilometers (Davies & Ludlam 1971)—should not be indiscriminately used as a criterion to rule out a tidal flat (particularly a "supratidal") origin for laminites, whether carbonate, silici-clastic, or evaporitic.

A quick look at thin sections of a small sampling of ancient laminated limestones shows that the "Andros-type" laminite is quite common. We have seen the "couplet" fabric in the Cambro-Ordovician Conococheague Limestone and the Ordovician St. Paul's Group of Western Maryland, and also in the Cambrian Mauv Limestone of the Grand Canyon (Wanless 1971, 1973). In addition, the laminites of the Macumber Formation (Mississippian) of the Maritime provinces of Canada described by Schenk (1967, Figs. 8, 9, 10, 12) have a "couplet" fabric with a thin (0.1 mm), basal bituminous quartz rich micrite (algal) layer overlain by a thicker (1 mm) calcite mosaic layer showing "inverse grading" (Schenk 1967, p. 368). Perhaps the best ancient analog we have seen is that of the upper Miocene carbonates of the Appennines, Italy. The photomicrographs presented by Rabbi and Ricci-Lucchi (1968, Table LVIII, Fig. 1 and Table LIX, Figs. 1–4) show a precise parallelism of fabric, even to the presence of peloids in the discontinuous sand layers. We suggest that on careful examination many other ancient carbonate laminites might prove to have a similar fabric.

Thin Bedding. We want here to briefly review only the algal marsh thin bedding which is so distinctive of the Andros-type tidal flats.

In summary, the main diagnostic features of the freshwater marsh bedding are the algal tufa layers, the wide and random range in thickness of succesive beds from millimeter laminae to thin beds up to 10 cm thick, the fragmented discontinuous pinch-and-swell nature of individual beds, and the small pockets of flat-pebble and granule matrix (tufa)-supported gravels. Figure 52 is a good summary illustration of what marsh bedding looks like in cores.

The highly variable thickness of successive layers records only that deposition occurred in an environment where water could be *ponded* during infrequent storm-flooding. Similar variable thickness bedding occurs in shallow depressions on the wide sabkha of the silici-clastic tidal flats of the Colorado River delta (Hardie, unpublished data), and also in partly blocked tidal channels of the Gladstone Embayment in Shark Bay (Davies 1970a, Fig. 10 & 11). The desiccation features (fragmentation and flat-pebble gravels) record the extensive periods of exposure of the marsh surface; they indicate the essentially "supratidal" setting of the marsh.

The one truly unique feature of the marsh bedding, the single unequivocal criterion of the marsh subenvironment, is the *Scytonema algal tufa*. This tufa, and its associated laminated, "palisade"-structured crusts and lithified algal "heads," betrays the dominant influence of freshwater in the inland marsh. It is only in the inland marsh that the magnesian calcite sheaths are precipitated around *Scytonema* filaments; it does not occur in the seawater-bathed channeled belt (see Chapter 7 for a discussion of this freshwater lithification process). This algal tufa is an unmistakable record of the nonevaporitic, *rainy* tropical climate of the Andros Island tidal flats. Analogous algal tufas in the geologic record (like *Girvanella* structures) should carry the same climatic message.

8. Criteria for Recognizing Andros-Type Algal Stromatolites and Crinkled Algal Lamination

Descriptions and discussions of layering influenced by algal mats have been presented in some detail above, but we want here to pull together in summary the major types of "algal bedding structures" analogous to ancient stromatolites that we found at Three Creeks, and to briefly discuss their special importance and criteria for their recognition. We will consider here only those structures that are clearly "gravity defying" and hence would be immediately classified as stromatolites by most geologists.

Some clarification of our terminology is needed. First, we would distinguish between algal structures that are discrete centimeter-scale domes or heads, and essentially horizontal algal lamination that is crinkled on a millimeter-scale. While the crinkled algal lamination would fit into the classification of Logan et al. (1964) as LLH-C stromatolites, it is the main internal lamination type of the most widespread of the domal algal structures on Andros Island. Second, as pointed out in footnote 5 of Chapter 4, the term "stromatolite" is used rigorously for *lithified* structures, and so to avoid confusion we will use the term "proto-stromatolite" for unlithified algal structures on Andros Island. Third, by "algal tufa" we mean carbonate material precipitated directly around algal filaments by chemical or biochemical processes.

Domal proto-stromatolites and stromatolites are not spectacularly developed on the Three Creeks tidal flats, but their internal architecture and modes of construction have much to tell us about genesis of stromatolites and their environmental significance. We recognized the following types of domal algal structures at Three Creeks:

I. *Proto-stromatolites formed mainly by sediment trapping*
   A. small (< 10 cm) SH-C structures with smooth domal lamination—channel bank subenvironment (Figs. 11, 34, 35).
   B. small (up 15 cm across) LLH-S structures with smooth domal and flat lamination—beach terrace (see Fig. 25A).
   C. "raised disc" SH-C structures (5 to 15 cm across) with crinkled fenestral lamination (Black's type C algal heads)—high algal marsh of channeled belt and inland algal marsh (Fig. 17A; Black 1933, Figs. 4, 8).
   D. very small (few centimeters), thrombolite-type, unlayered fenestral clots in a matrix of smooth flat or crinkled fenestral lamination—beach terrace and high algal marsh-levee backslope boundary (Fig. 45).

II. *Stromatolites formed mainly by sediment trapping*
   A. small (< 10 cm), lithified SH-C knobs and mounds with "palisade" lamination—upper high algal marsh of channeled belt (Figs. 46A, 47A, 50).

III. *Stromatolites formed mainly by in situ chemical or biochemical precipitation (algal tufa)*
   A. small (< 10 cm) lithified mushroom-shaped heads and rounded knobs and flat-topped mounds of *Scytonema* algal tufa with "palisade" lamination (SH-C type mainly)—inland algal marsh (Figs. 46B, 48, 49).

None of these domal algal structures are formed below mean tide level, because burrowing and browsing by marine invertebrates destroys all layering in the lower intertidal and subtidal zones (Garrett 1970, and Chapter 6). In this respect the contrast with the spectacular development of club-shaped and ellipsoidal columnar stromatolites throughout the tidal range in the hypersaline embayments in Shark Bay is outstanding (Logan 1961; Logan et al. 1974). This is because the high salinity in Hamelin Pool (> 53‰) in Shark Bay excludes the bioturbators and so allows healthy development of algal mats in the subtidal and intertidal zones. At the same time, desiccation and evaporite growth discourages algal growth in the supratidal zones, so that supratidal stromatolites are weakly developed or absent. By contrast the rainy climate of Andros provides an exceptional environment for growth of freshwater-loving and freshwater-tolerant algal mats in the supratidal zones and, hence, development of supratidal stromatolites and proto-stromatolites. The Persian Gulf tidal flats (Kendall & Skipwith 1968) fit in between Andros Island and Shark Bay because the subtidal lagoons of the Persian Gulf are not salty enough to discourage bioturbators, while the supratidal zone is a salt-encrusted sabkha. So, in the Persian Gulf, algal mats are restricted to a fairly narrow zone in the upper intertidal.

On Andros Island the distinction between stromatolites formed by sediment-trapping processes and stromatolites made of precipitated algal tufa is of major importance because the algal tufa is found *only* in the inland, freshwater-dominated *Scytonema* algal marsh. These domal algal tufas of the inland marsh record the truly significant role of freshwater in the supratidal of the rainy Andros tidal flat complex (see Chapter 7 for more detailed discussion), and are characterized by their mushroom shape, fibrous palisade internal structure, and *Girvanella*-like radiating mass of micrite-encrusted filament molds (see detailed descriptions above, under "lamination with 'palisade' structure"). The morphology of these algal tufa domes reflects the growth form of *Scytonema*, which is a large-filamented terrestrial alga with a limited tolerance to seawater that grows in isolated, tufted knobs where sedimentation is infrequent and wetting is moderate to low (Exposure Index range for this particular mat form is 85 to 99). A full discussion of the distribution of *Scytonema*-dominated mats and the restriction of filament calcification to freshwater settings is given in Chapter 7.

In strong contrast to the tufa domes are the SH-C and LLH-S proto-stromatolites of types IA and IB. These type I algal structures, by their small size and stacked smooth laminae of mud and peloids, record sediment-trapping by tiny, filamented Oscillatoriaceae mats in an actively accreting environment of low-wave energy and tidal currents. These smooth laminated proto-stromatolites have precisely the same laminae architecture as the smooth flat lamination (compare Figs. 35 and 37) and the laminae have the same basic origin. Like the smooth flat lamination, the smooth domal lamination consists of alternating thin mud laminae and peloid sand-silt lenses. However, in the domal structures the mud laminae, representing algal "stick-ons," are predominant and the peloid laminae are relatively thin. These discontinuous peloid laminae are best developed over the flattish crests of the domes and seem to represent *mechanically deposited* layers almost like micro-ripples (see discussion of the surface ridges on the domes, in the section on "smooth domal lamination" above). It is in fact the preferential accumulation of these peloid lenses on the crests that gives the pronounced dome shape to the proto-stromatolites. So the criteria for recognizing Andros types IA and IB stromatolites in the geologic column are exactly those used to recognize the smooth flat lamination as outlined in the preceding section. A look at thin sections of a number of domal stromatolites from the lower Paleozoic carbonates of the central Appalachians showed that the mud-sand couplet fabric is a common feature. What then can the Andros domal proto-stromatolites with smooth internal lamination tell us about their ancient analogues? First, there can be little doubt that stromatolites with this internal fabric have been formed by the sediment-trapping action (agglutination) of algal mats of the *Schizothrix*-type, so that a caliche, travertine or tufa origin can be ruled out. The presence of filament molds is not necessary to this interpretation, because *Schizothrix*-type algae do not leave any filament record as they move upward to recolonize the surface of a newly deposited lamina; the filament molds that are present in the Andros type IA and IB proto-stromatolites belong to isolated tufts of *Scytonema* and *Rivularia* which are not responsible for making the smooth domal structures. Second, unlike the algal tufas, the domal shape of the smooth laminated proto-stromatolites is not a mimic of algal growth forms, but instead depends on substrate morphology. In fact, the only real difference between the smooth flat laminated sediment and the domal structures at Three Creeks is the shape of the surface that the *Schizothrix*-type mats coat and bind: where the surface is sharply irregular the high spots are built up to form domes, but where the surface is flat only flat lamination (a planar proto-stromatolite!) is produced. For example, the type IA domes are formed by draping over sediment mounds between *Uca* burrow entrance holes, while the type IB are formed over flat pebbles strewn on the beach terrace. Third, the lateral asymmetry of these domal structures reflects the depositional current direction. In vertical cross-sections of the SH-C channel-bank structures, the steepest slope is on the downcurrent side and the gentlest slope faces directly into the depositional current direction (Fig. 34B). As explained in the section on "smooth domal lamination," this asymmetry simply reflects the preferential deposition of sandy laminae on the upcurrent slope and crest of the domes. These sandy layers are thickest on the downcurrent edge of the crest, where they end abruptly with a steep terminal slope (exactly like a flat "micro-delta" ripple). For the LLH-S pebble-draped structures the same kind of picture holds, but trapping of sediment against the leading edge of the pebble has smoothed out the sharp relief, making a ramp-like structure facing into the current (Fig. 25A). This relation of shape to current direction in these Andros proto-stromatolites is the opposite of that reported by Hoffman (1967, Fig. 3) for draped pebble LLH-S stromatolites in the Precambrian Pethei Group of Great Slave Lake, N.W.T., Canada.

The value of type IA proto-stromatolites on Andros Island as environmental indicators is moot. These domal structures of the crab-burrowed channel banks will not be preserved as a widespread layer, because the channel banks are ephemeral features destroyed by cannabalistic migration of the channels. So precise analogs of such channel-bank algal drapes are likely to be rare in the geologic record. The type IB LLH-S pebble-drape structures are much more useful environment-indicator models. On the beach terrace of Three Creeks these proto-stromatolites are the product of wave erosion of the beach cliff, due to *local* transgression (induced by deficiency of sediment rather than sea-level changes). These structures will be preserved in a matrix of smooth flat lamination, if the long-term trend is actually a net regression. So ancient analogs will indicate areas of *local* erosion. But even more than this we think that the small size and laterally linked structure reflect that the prevailing conditions are low turbulence ("low kinetic energy") ones. In a stable substrate environment, normally free from scouring currents (except during infrequent severe storms), such as exemplified by the beach-terrace zone, the algal mat is free to cover the entire surface, and the initial high relief provided by isolated flat pebbles will soon be deemphasized by the smoothing trend of the draped mats and the continued build-up of the substrate between the pebbles (see Fig. 25A, where it is obvious that a smoothing-out of the sharp relief of the pebbles is being achieved by the draped-over mat). It is difficult to visualize large, discrete columnar heads developing in such an environment. By contrast, in a prevailing "high kinetic energy" setting, where daily scouring by waves or tidal currents produces a mobile lag-sand substrate, it is impossible for blue-green algal colonies to establish a pervasive coherent mat. Only where a stable "island" exists in the shifting sand can small patches of algae thrive and trap and bind sediment laminae. Such a situation exists in the hypersaline lagoon of Hamelin Pool in Shark Bay. On wave-scoured headlands, algal mats are established only on stable fragments, such as cemented clasts, shells, etc. embedded in the mobile rippled sand substrate. These mats have built discrete SH-V columnar stromatolites as high as 2 m; their height decreases onshore as the intertidal surface rises so that their size is proportional to the tidal "depth" (Logan et al. 1974).

The "raised disc" proto-stromatolites of type IC were first recognized and described by Black (1933) in his pioneer work on algal sediments of Andros Island. Black observed this type of algal head (his Type C) to cover wide areas bordering interior "lakes" (partly tidal but mainly freshwater influenced). The heads are formed by draping of *Schizothrix-Scytonema* mats over mud-crack polygons with upturned edges (see Black 1933, Figs. 4 & 8). At Three Creeks this type of proto-stromatolite is found at the upper boundary of the high marsh zone of the channeled belt and over wide areas of the inland algal marsh. In the inland marsh the heads consist mainly of crinkled fenestral lamination typical of the flat *Scytonema* mats (see descriptions above), whereas, in the channeled belt it is alternations of crinkled fenestral laminae (*Scytonema*-bound layers) with smooth laminae (*Schizothrix*-bound layers) that make the heads. In the inland marsh, incipient calcification of *Scytonema* filaments is found in many layers so that the heads may carry some features of the algal tufas. This "raised-disc" type of proto-stromatolite is by far the most widespread domal algal structure at Three Creeks. It is a sensitive subenvironment indicator, being restricted to marine supratidal and to freshwater settings, because the essential alga is *Scytonema*. Like the algal tufas, these "raised-disc" structures are a record of the strong influence of freshwater on the tidal flat complex.

Scytonema is also responsible for the formation of the remaining two types of domal structures. Type IIA is characteristic of the cemented surface crusts that partly pave the upper levels of the high algal marsh zones of the channeled belt (Chapter 7). Here internally laminated domes with a striking fibrous palisade structure encrust flat, horizontally laminated crust surfaces, mimicking the patchy *Scytonema*-mat growths of the high marshes. These lithified domal structures are different from the algal tufas in that they are made of laminated and bioturbated detrital sediment pierced by large tubular *Scytonema* filament molds (see description above, under "lamination with 'palisade' structure"). The molds, which impart the fibrous palisade structure, are not tufa envelopes, but simply sediment-supported voids. Nonetheless, like the algal tufa domes, these indurated stromatolites owe their domal form to the patchy growth habit of *Scytonema* mats in dry supratidal environments, where sedimentation is relatively infrequent. The cementation seems to be an accident of their particular environmental setting and not the result of any algal process (Chapter 7). The type ID domal structures are found as small isolated clots, no more than a centimeter or two across, in a matrix of smooth flat laminated sediment (on the beach terraces) or in crinkled fenestral laminated sediment (on the high algal marshes). These clots are produced by entrainment of sediment (mud and peloids) in high standing tufts of coarse filaments of *Scytonema* (and *Rivularia* on the beach terrace). They typically are unlayered masses of poorly sorted sediment shot through with tubular fenestrae marking where bundles of filaments stood upright. Apart from the fact that they record that large filamented alga were only minor constituents of the mat colony, these clots are strong evidence of the predominance of sheet-flood, traction-load currents and lack of settle-out deposition.

Finally, to complete this review, we must give special attention to the crinkled algal lamination, because not only is it the major internal structure of the widespread "raised-disc" stromatolite (type IC) first reported by Black (1933), but it is the one flat lamination type that clearly signals its algal origin. This is the "crinkled fenestral lamination" produced by sediment entrainment in the flat *Scytonema* mats of the high algal marshes. Drying and wrinkling of these fleshy, sheet-like *Scytonema* mats forms a mass of arch-shaped hollow fenestrae and horizontal sheet-cracks as the algal mat separates from the underlying sediment substrate (Figs. 43A & 44A). A thin sediment lamina newly deposited on a flat *Scytonema* mat will be contorted and crinkled along with the mat and so produce mud layers sharply crinkled beyond the normal angle of repose of detrital sediment (Fig. 43A). This crinkled morphology of the sediment laminae and the attendant hollow fenestrae survive long after the organic matter of the mats has been destroyed and so record the algal influence as convincingly as any

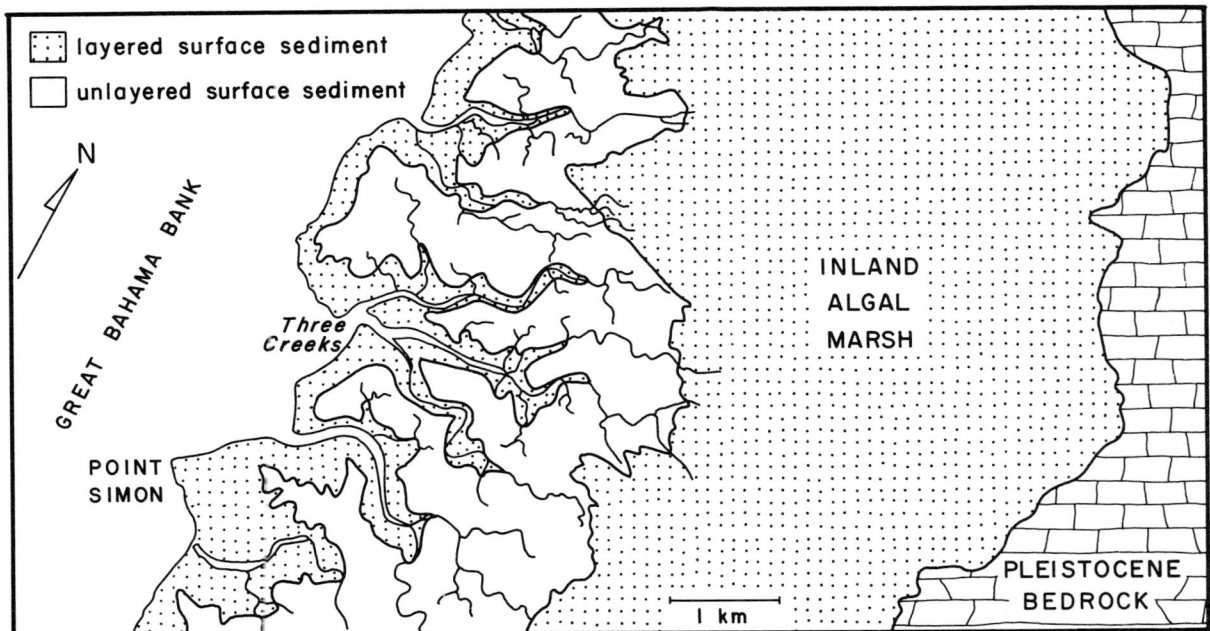

Fig. 64. Map of Three Creeks area showing distribution of layering in surface sediment.

domal stromatolite. It is the fleshy, large-filamented nature of the mat that produces this special crinkled fenestral fabric; the mat does not bind the surface sediment like a *Schizothrix* mat does, but simply covers the surface as an essentially separate layer. Therefore, on drying and shrinking, the mat will *separate* from the sediment substrate and so produce the characteristic crinkled surface with protected hollows (open fenestrae) beneath the ridges.

9. Vertical and Lateral Distribution of Layering: Some Implications

The areal distribution of layering in the *surface* sediment of the Three Creeks tidal flat complex is summarized in Figure 64. As the figure shows, more than 50% of the surface of the onshore area is underlain by layered sediment, most of this being found in the vast inland marsh. The vertical distribution of layering is shown in Figure 65, as a series of typical cores taken in each major subenvironment, and in Figure 66, which is a schematic vertical section across the tidal flats.

The significant features brought out in these diagrams are (1) the pond and offshore sediments are completely unlayered; (2) on the levees and beach ridges the layered sediment is only a thin "laminite cap" over unlayered sediment; (3) in the inland marsh, layering persists all the way down to the Pleistocene bedrock; (4) thin-bedded marsh sediment separates the bedrock from the channeled belt and offshore sediment. These features emphasize two major aspects of the distribution of layering that need to be discussed: (1) preservation and destruction of layering, and (2) stratigraphic relations among layer types and their implications for reconstructing depositional environment and history.

Preservation and destruction of layering. The existing distribution of layered sediment across the tidal flats is entirely a "survival" distribution. Discrete laminae and/or thin beds are deposited spasmodically in *every part of the complex* from the offshore to the landward edge of the inland marsh, but, in all save the few highest "supratidal" subenvironments (levees, beach ridges, inland marsh), the layering is relentlessly destroyed by browsing and burrowing marine organisms (Garrett 1970, 1971; & Chapter 6). In the vast inland marsh, layering survives because the low salinity of this freshwater "pond" keeps out burrowing

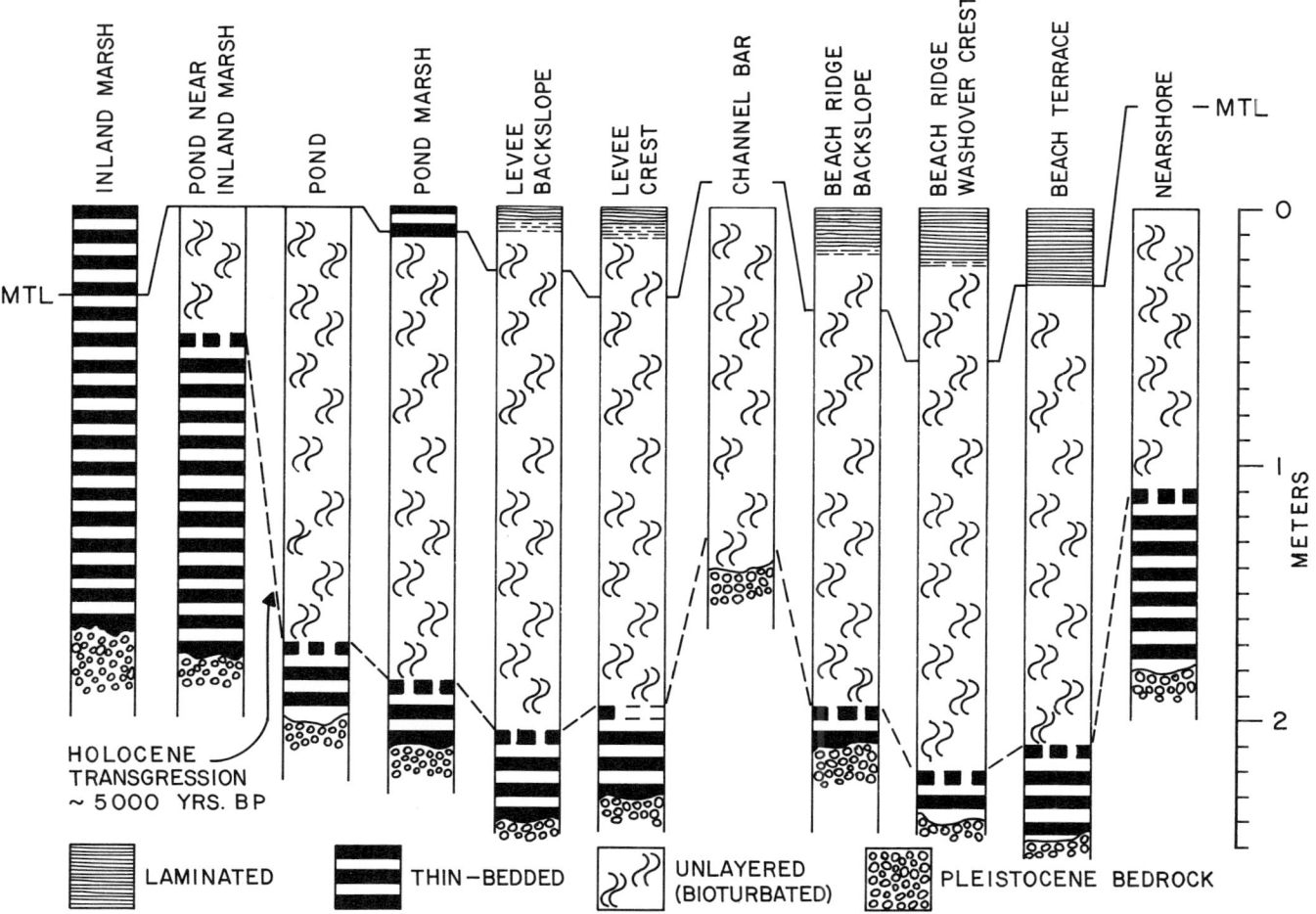

Fig. 65. Schematic diagram of typical cores showing distribution of layered sediment in the different subenvironments at Three Creeks.

and browsing marine organisms. In the levee and beach-ridge sediments it is the dryness that excludes all the marine infaunal burrowing organisms except the fiddler and marsh crabs. However, in all these "supratidal" subenvironments the layered sediment that survives "homogenization" does not necessarily escape post-depositional modification, but may be disrupted by mud-cracking, by rooting, and by burrowing of terrestrial organisms.

All of the processes that destroy or disrupt layering have been mentioned above, and organic processes are dealt with by Garrett in Chapter 6, but they are worth reiterating and are summarized in Table A.

Without question, homogenization of originally layered sediment by marine organisms is the single major sedimentary process that operates almost continuously day-in and day-out on those parts of the tidal flats that are regularly wetted by seawater. The efficiency with which marine burrowers and browsers turn over sediment in the ponds, channels, and offshore has been emphasized by Garrett (1970, 1971; & Chapter 6). Garrett has shown that *Apseudes* (a tiny tanaid shrimp) alone can completely churn the upper 2 cm of pond sediment in less than three months, not to mention the work of polychaete worms. In the offshore sediment the burrowing shrimp *Callianassa* (Shinn 1968) is incredibly abundant; its burrows may extend to a depth of over 1 m, so that the volume of

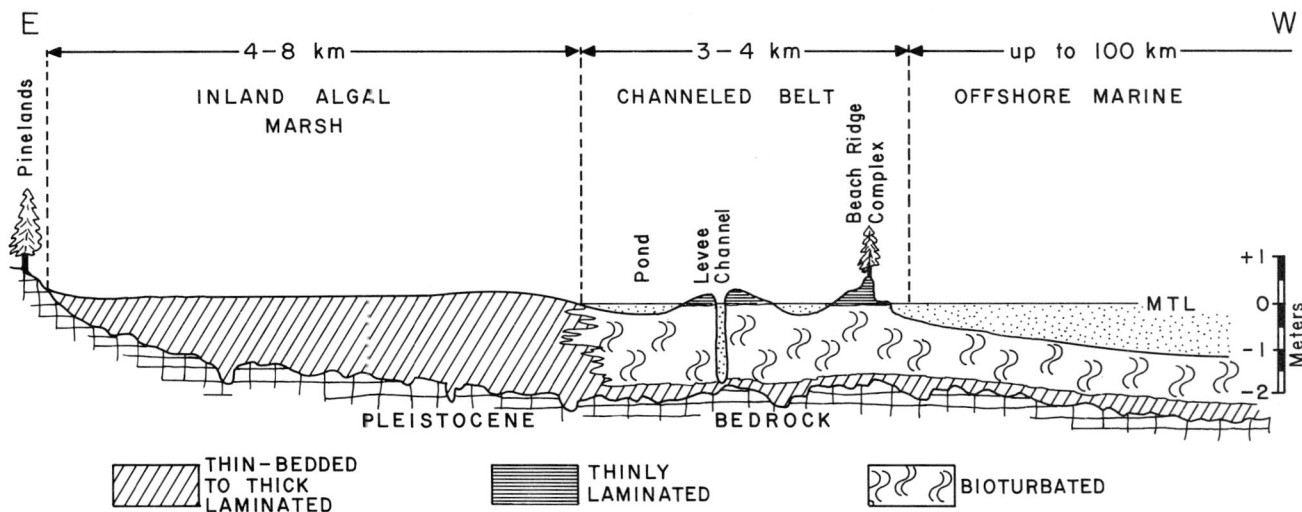

Fig. 66. Highly schematic cross-section through the Three Creeks tidal flats to show the thin "laminite" cap over bioturbated sediment in the channeled belt and the thin-bedded landward edge of the flats (the inland marsh sediment).

sediment worked by this shrimp in burrow construction and repair must be enormous. Across the Three Creeks tidal flat complex, then, it is not surprising that all the sediment below the mean tide level (and in places up to MHW) is unlayered—the rate of sediment reworking by organisms (at least 10 cm/year) is more than an order of magnitude greater than the rate of sediment deposition (less than 1 cm/year).

Table A. Types of layer-destroying processes

I Mechanical Processes (mainly disruptive)
—— shallow mud-cracking by desiccation, producing erodable intraclasts (levee backslope, beach-ridge backslope, algal marsh)
—— deep polygonal (prism) cracking by drainage of porewater at low tide (ponds, channel bars)
—— air forced to surface by rising watertable, producing sediment domes and bubbles at surface (levee crest, beach-ridge washovers)

II. Organic Processes (mainly destructive)
—— burrowing by worms
  —— polychaetes (offshore, beach, ponds, channel bars)
  —— oligochaetes (levees, beach ridges)
—— burrowing by crustaceans
  —— shrimp (offshore, channel bars, ponds)
  —— crabs (channel banks, high algal marsh, beach cliff, beach ridge)
—— burrowing by insects (levees, beach ridges)
—— browsing by gastropods (ponds, beach)
—— rooting by mangroves (levees, marshes, ponds, beach ridges)
—— rooting by halophytes (levees, marshes, beach ridges)
—— rooting by sea-grasses (offshore, channel bars)

If we include the entire offshore shelf lagoon with the Andros tidal flats, then the overwhelming sediment record being compiled today on the west side of Andros Island is bioturbated, burrowed, unlayered pelleted mud and silt. Layered sediment makes up only a small part of the existing record.

The homogenization by burrowers of sediments below the mean tide level is not confined to the Bahama shelf setting, but is typical of most modern shallow marginal marine environments, both carbonate and silici-clastic. Such an abundance of thick beds of unlayered sediment contrasts strongly with the record of Precambrian and early Paleozoic tidal flat-shelf lagoon complexes where layering is predominant. Perhaps this has to do with a rather late radiation of the burrowing shrimp and crabs, a group (the malacostraca) known to go back to the Ordovician (Beerbower 1960, p. 316), but with a very "dim" past, because they are not easily fossilized.

Stratigraphic relations: the "laminite cap". Shinn et al. (1969) made a study of the Holocene stratigraphy at Three Creeks, based on cores put down to the Pleistocene bedrock, and have worked out the basic stratigraphic relations. Our work was not aimed in this direction, but several ideas emerged from our layering study that may add to the picture presented by Shinn et al. (1969).

First of all, we want to reiterate a point we have made several times: the type of layering and its associated structures are a very sensitive record of the depositional subenvironment in which they are formed. It is easy to distinguish the layered sediments of levees, beach ridges, marshes, and their subdivisions, from each other and, of course, from the unlayered sediment of the offshore, pond and channels (see our detailed description above). On the basis of these distinctions, and despite the ravages of burrowers, we were able to see in our cores very small-scale stratigraphic changes, such as levee crest over levee backslope, levee backslope over algal marsh, beach terrace over pond, that reflect local "progradation" (lateral migration really) of levees and beach ridges. We have not made a detailed analysis of such stratigraphic relations—a special study of this is needed—but we simply want to make the point that we found that where layering is preserved it is an unmistakable indicator of the subenvironment.

Second, the Holocene record in the study area begins with a freshwater marsh accumulation unconformably overlying Pleistocene limestone bedrock (Figs. 65, 66; & Shinn et al. 1969). This marsh covered a wide area of the Great Bahama platform; Traverse and Ginsburg (1966, and personal communication) found freshwater "swamp peat" beneath marine sediment over 24 km bankward from the Andros shoreline. A major transgression of the sea over this wide marsh about 5,000 years ago (Traverse & Ginsburg 1966) initiated the present phase of sediment accumulation. This transgression left no recognizable deposit but eroded the early Holocene marsh sediments down to the thin (few tens of centimeters) remnants we find at the base of the Three Creeks sediment pile (Figs. 65 & 66). The advance of the sea stopped before it reached the Andros Island mainland, so that the area of the existing inland algal marsh has remained a freshwater marsh environment without interruption to the present day.

Third, the overall stratigraphy of the Three Creeks tidal flat clearly shows that accumulation since the latest Holocene rise in sealevel is not due to continuous seaward progradation of the shoreline. There is no simple regional progradation pattern of supratidal over intertidal over subtidal sediments, such as is found in the tidal flats of the Persian Gulf (Butler 1970) and the northwest Gulf of California (Thompson 1968). Instead, the accumulation mechanism seems to be mainly *vertical* accretion in a complex of environments behind a protecting barrier beach ridge, much as occurs on the tidal flats along the Atlantic coast of the United States from New Jersey to Georgia. We envision tidal flat accumulation beginning when an accreting offshore bar emerges above mean high water to become a barrier (by whatever process, see Schwartz 1971). Infilling behind the barrier is very complex with channels, channel bars, levees, ponds, and marshes making an intricate spatial pattern of subtidal, intertidal, and supratidal depositional environments. This is the present stage of development of the Three Creeks tidal flats. If continued sedimentation causes plugging of the main channels, then the ponds will soon become choked with sediment and the entire channeled belt behind the beach ridge will become a vast supratidal coastal

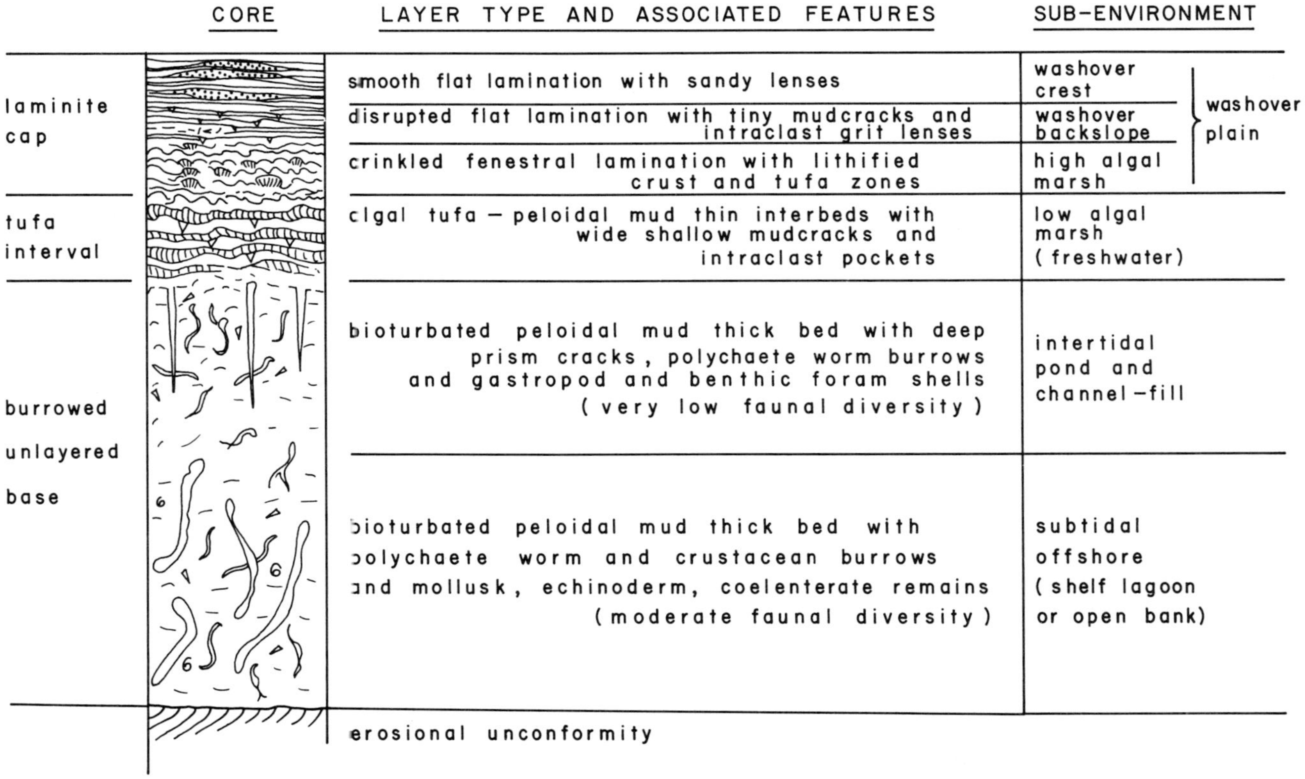

CORE	LAYER TYPE AND ASSOCIATED FEATURES	SUB-ENVIRONMENT	
laminite cap	smooth flat lamination with sandy lenses	washover crest	washover plain
	disrupted flat lamination with tiny mudcracks and intraclast grit lenses	washover backslope	
	crinkled fenestral lamination with lithified crust and tufa zones	high algal marsh	
tufa interval	algal tufa — peloidal mud thin interbeds with wide shallow mudcracks and intraclast pockets	low algal marsh (freshwater)	
burrowed unlayered base	bioturbated peloidal mud thick bed with deep prism cracks, polychaete worm burrows and gastropod and benthic foram shells (very low faunal diversity)	intertidal pond and channel-fill	
	bioturbated peloidal mud thick bed with polychaete worm and crustacean burrows and mollusk, echinoderm, coelenterate remains (moderate faunal diversity)	subtidal offshore (shelf lagoon or open bank)	
	erosional unconformity		

Fig. 67. Schematic drawing, showing what we think a vertical core taken through the Andros tidal flats at the completion of a regressive cycle might look like. (The similarity with Fischer's Lofer cycle (Fischer, 1964, and Fischer, 1975, pp. 235–42, in Ginsburg, 1975) is, we think, rather striking although Fischer interpreted the boundaries of his cycle in a different way, putting the prism-cracked unit as the exposed *top* of the package. Our Andros-type cycle may offer a simpler interpretation.)

plain. The gross stratigraphic record compiled will be a complex basal unit (1 to 2 m thick) of bioturbated unlayered subtidal and intertidal sediment overlain by a thinner (about 1 m) well-layered unit, a *"laminite cap"* deposited by the most severe onshore storms. The supratidal plain would start as a complex of blocked ponds and channels that would be reached by sediment-charged seawater only when hurricane floodwaters overtopped the barrier beach ridge. These now isolated depressions would become freshwater algal marshes which would slowly fill, first with algal tufa and peloidal mud thin beds and then crinkled fenestral laminated sediment and lithified crusts. When filling of these marshy "lakes" with sediment was complete (this would occur first at the seaward edge) the coastal plain would become a *Schizothrix*-covered subaerial flat, a vast "beach-ridge washover crest" many kilometers wide. We imagine a typical vertical core through the sediment beneath this high and dry coastal plain might look something like that shown schematically in Figure 67. Continued seaward accumulation of sediment presumably would be by the growth of a new offshore bar and subsequent infilling behind this new barrier, and so on.[14]

Repeated regressions of this type, coupled with continuous subsidence, will produce a package of vertical cycles in which the "repeat" would be a supratidal laminite cap over unlayered bioturbated subtidal and intertidal pelleted mud, such as is shown in Figure 67. This basic pattern of a laminite cap seems to be typical of most modern tidal flat sequences, as, for example, in Florida Bay, the Gulf of California, the North Sea, the Persian Gulf, and Shark Bay. In the

---

14. At the present time the beach ridge is undergoing erosion. Our probing offshore shows that the shoreline has been slowly advancing landward: for at least 100 m offshore of Three Creeks

geologic record, then, the "laminite cap" should be a primary mark of tidal flat accumulation.

---

the subsurface sediment is old pond sediment, complete with mangrove roots and root-hairs. This transgression is probably due to a recent local decrease in sediment supply (related to Bank water circulation) rather than to a sea-level change, because concurrent regression appears to be occurring south of Williams Island (Shinn et al. 1969, pp. 1225–26).

6 BIOLOGICAL COMMUNITIES AND THEIR SEDIMENTARY RECORD

Peter Garrett

1. Introduction

This biological part of our study was designed in part to complement the sedimentological aspects treated in other chapters. It was also aimed at extending the coverage of Newell et al. (1959), who in their study of the communities of the Great Bahama Bank, stopped short of the tidal flats. But mainly this study was conceived as an opportunity to follow the biota and biological processes of a modern carbonate environment through to final burial, in order to get some feeling for the means and biases of preservation, and to provide one possible model for comparing with the communities of ancient carbonate tidal flats, such as those described and compared so beautifully by Walker and Laporte (1970).

The communities of organisms are first described, and as might be expected, their boundaries correspond quite closely with the boundaries of the physiographic divisions outlined by Hardie and Garrett (Chapter 4). Then the range, abundance, and habits of the organisms are tabulated. Next, the contribution of the several skeletal organisms is discussed, in an attempt to show the difficulties of interpreting a skeletal record. Finally, the various types of sediment-modifying activities of organisms are described, in order to show how a trace record is created and how biased its preservation can be.

This study thus attempts to cover at least some aspects of the biology and ecology of the tidal flat, to follow through with observations and deductions on the taphonomy and early diagenesis of the skeletal material, and on the preservation of a record of sediment-modifying activities. Similar types of studies have been made, especially by German workers, most notably Schäfer (1962), the field having recently been reviewed by Reineck (1972).

2. Communities of Organisms

Following Johnson's (1964) definition of a community as a group of organisms related spatially, though not necessarily related by any interaction between species, the following communities are defined:

1. Nearshore Community

This is a semirestricted shallow water, sand-mud bottom community, and is described by Newell et al. (1959, p. 223) as the "Didemnum Community" after a characteristic tunicate. It includes blue-green algae, diatoms, green algae, marine grasses, foraminifera, sponges, a coral, an ectoproct, polychaetes, nematodes, molluscs, crustaceans, echinoderms, and a tunicate (Table 8).

No one species is particularly abundant, and because of the paucity of the macro-flora (usually less than 10 plants per m^2) one gets the impression of a low biomass, which is supported by the difficulty of finding members of the infauna. This community has in fact a lower diversity and abundance than communities in more open parts of the Great Bahama Bank, e.g., the *Strombus samba* community of Newell et al. (1959).

Table 8. Characteristic species of the nearshore community

Blue-green Algae 　"Schizothrix" sp. sensu lato Diatoms Green Algae 　Halimeda spp. 　Penicillus pyriformis 　Acetabularia crenulata 　Dictyosphaeria cavernosa 　*Rhipocephalus phoenix 　*Caulerpa paspaloides 　Batophora oerstedi Grasses 　*Thalassia testudinum 　Cymodocea manatorum Foraminifera 　see Streeter (1963) 　and Todd & Low (1971) Sponges 　*Verongia fistularis 　Ircinia fasciculata Coral 　Manicina areolata Ectoproct 　Schizoporella pungens	Polychaetes 　*Myriochele sp. 　*Armandia maculata Nematodes Gastropods 　Cantharus multangulus 　Cerithium eburneum 　Tonna maculosa 　Fasciolaria tulipa 　Strombus costatus 　? Atys sp. 　Vermicularia knorri Bivalves 　Pitar simpsoni 　*Chione cancellata 　Laevicardium laevigatum 　Aequipecten gibbus Crustaceans 　Ostracods 　Callinectes sapidus 　Callianassa sp. Echinoderms 　Echinaster sentus 　Clypeaster rosaceus Tunicate 　Didemnum candidum

Source: Data from this study combined with data from Newell et al. (1959) and Cloud (1962).
* Most abundant forms.

2. *Pond Community*

This is an intertidal muddy bottom community occurring in the ponds and intertidal portions of the beach and of channel banks and bars from 0 to 75% Exposure Index. It is a restricted marine community in the sense that though the groups present here are also present offshore, they are reduced in number. Groups which are notably absent from the pond community include the sponges, corals, ectoprocts, and echinoderms. Species diversity is also generally lower than in the nearshore community.

The organisms present include red mangroves, and blue-green algae, but only *Batophora* among the green algae, foraminifera, deposit-feeding polychaetes, nematodes, molluscs (mostly grazing gastropods), and several crustaceans (Table 9). Common pond organisms are illustrated in Figure 68 and their vertical distribution is sketched in Figure 69.

A few of these organisms are restricted to the intertidal, and have special adaptations to cope with the special conditions. For instance, the red mangrove *Rhizophora* has aerial prop roots with a layer of air-filled spongy tissue extending down into the submerged roots, enabling them to breathe. It also has various physiological adaptations, enabling it to cope with its unusual salt balance.

The two cerithid gastropods, *Cerithidea* and *Batillaria*, are also restricted to the intertidal, though the reason is not clear. They only feed, by grazing on surface algal mats, when submerged or at least when wet, and yet when submerged on an algal mat in a bowl of water for periods in excess of their normal tidal submergence, they invariably crawl out above the water surface to rest, with operculum closed. The same intolerance to conditions different from their normal habitat applies to their ability to withstand desiccation. Several specimens of *Batillaria minima* collected from a beach at the lower end of their distribution (Exposure Index 25%) were compared with speci-

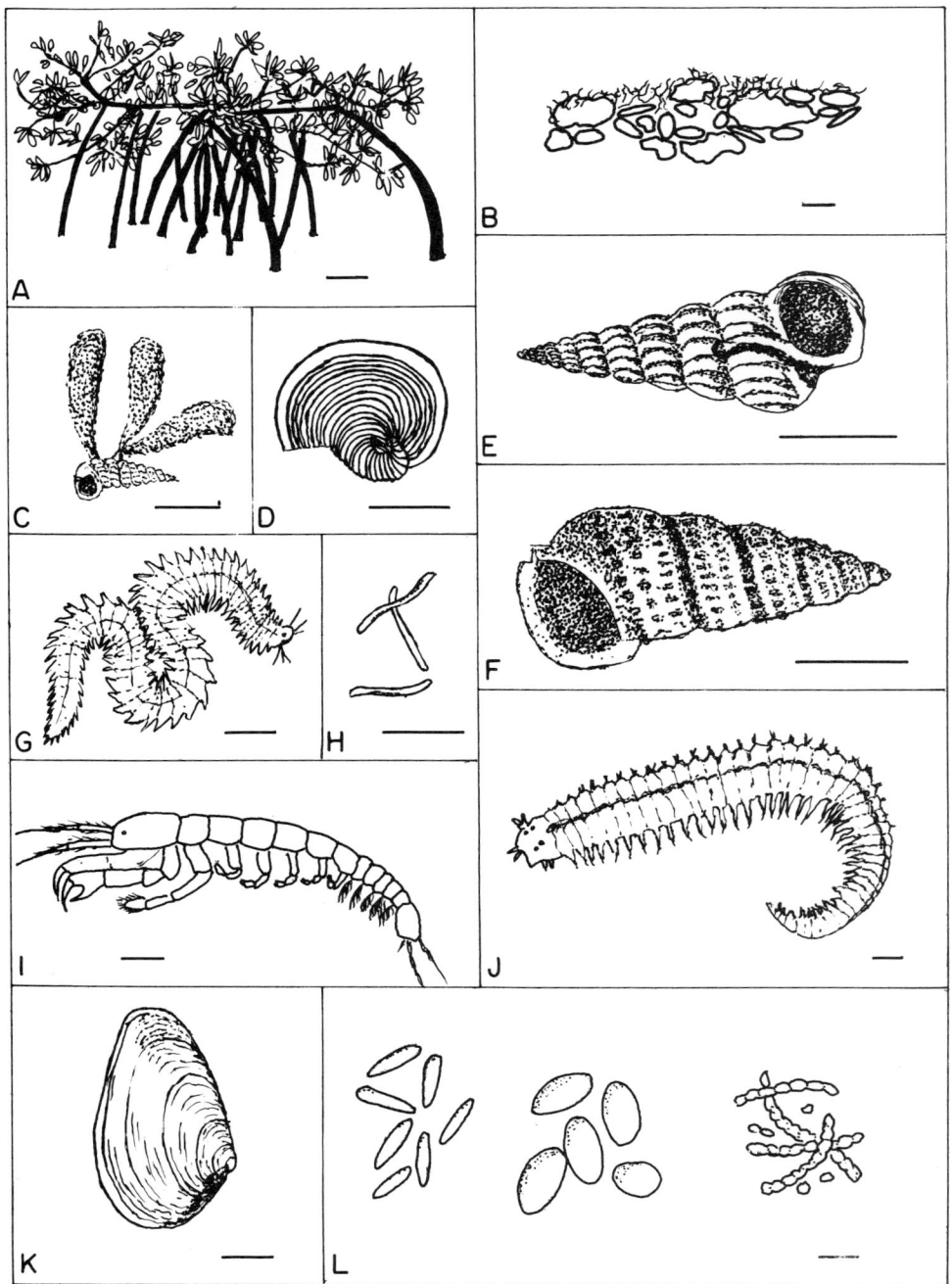

Fig. 68. Common pond organisms. (A) Red mangrove (*Rhizophora mangle*) in its typical stunted growth habit. These plants abound at the margins of all ponds and extend across the shallower ones. Scale bar 15 cm. (B) Diagrammatic "*Schizothrix*" algal mat. Fine filaments intertwine with sediment particles, but do not form a cohesive mat because of the activities of the grazing and burrowing fauna. Scale bar 50 μ. (C) The green alga *Batophora oerstedi* with holdfasts attached to a gastropod shell. These plants can survive desiccation, but vary greatly in abundance from pond to pond. Scale bar 1 cm. (D) The foraminifer *Peneroplis proteus*, a common surface organism. Scale bar 1 mm. (E & F) The cerithid gastropods *Cerithidea costata* (E) and *Batillaria minima* (F), both abundant and important surface grazers in intertidal zones. Scale bar 5 mm. (G) Nereid polychaetes burrow mainly in near surface sediments and graze on the surface at night. Scale bar 2 mm. (H) Nematode worms—the commonest members of the meiofauna. Scale bar 50 μ. (I) The tanaid shrimp *Apseudes* sp. is sometimes a common near-surface burrower. Scale bar 0.5 mm. (J) The eunicid polychaete *Marphysa sanguinea*, a common and important deep burrower. Scale bar 1 mm. (K) The bivalve *Geloina* sp. is an infaunal suspension feeder, but does not usually attain its adult size in pond sediments. Scale bar 1 mm. (L) Common pond surface pellets. The rods of cerithid gastropods, the ellipsoids of *Marphysa*, and the cylindrical strings of nereid polychaetes. Scale bar 1 mm.

Table 9. Characteristic species of the pond community

Blue-green Algae *"Schizothrix" sp. (s. l.)	*Nematodes
Green Algae *Batophora oerstedi	Gastropods *Cerithidea costata *Batillaria minima Prunum apicina
Grasses Diplantheria wrightii	Detracia bullaoides Sayella cf. bahamensis
Other Angiosperms *Rhizophora mangle Avicennia germinans	Syncera succinea Tornatina candei
Foraminifera *Peneroplis proteus others	Bivalves *Geloina sp. Brachidontes citrinus
Polychaetes *Marphysa sanguinea Dasybranchus lumbricoides? *Perinereis cf. anderssoni *?Ceratonereis tridentata	Crustaceans *Apseudes sp. Callinectes sapidus Alpheus heterochaelis Ostracods

* Most abundant forms.

Table 10. Characteristic species of the levee community

Blue-green algae *"Schizothrix" sp. (s.l.) †*Scytonema spp.	*Avicennia nitida Salicornia bigelovii Borrichia arborescens Conocarpus erecta Rachicallis americana
Grasses †*Distichlis spicata †*Monanthochloe littoralis *Fimbristylis sp. *Spartina cf. patens Juncus sp. Cladium jamaicense	Oligochaetes Pontodrilus bermudensis
Other angiosperms †*Rhizophora mangle	Crustaceans †Uca sp. †Sesarma curacaoense ants other insects

* Most abundant forms.
† Species characteristic of algal marsh subcommunity.

mens of the same species collected from a pond margin (Exposure Index 65%). Both groups were left dry on a levee surface for an extended period. After fifteen days the beach specimens were all dead and being eaten by ants, but after twenty-five days the pond margin group quickly emerged when put on an algal mat in water, but spent only a few hours feeding before retreating behind their opercula above the water surface.

Not all pond community organisms are restricted to the intertidal however. Some, such as *Batophora* and *Diplantheria*, extend their range from their normal subtidal habitat, though the plants are smaller and are killed by prolonged exposure. Others, such as the polychaete assemblage, are probably controlled more by substrate requirements than by other factors.

3. *Levee Community*

This is effectively the supratidal community, with a range upward from 75% Exposure Index and a spatial distribution covering the levees, beach ridges, and algal marsh. It is a low diversity marginal-terrestrial community with several halophytic angiosperms, including mangroves, grasses, sedges and others, and blue-green algal mats, but few animals—only two brachyuran crabs, an earthworm, ants, and a few other insects (Table 10 & Figs. 69 & 70).

The pond and levee communities show changes with elevation and physiographic location, and in certain areas gradational subcommunities can be defined. For instance:

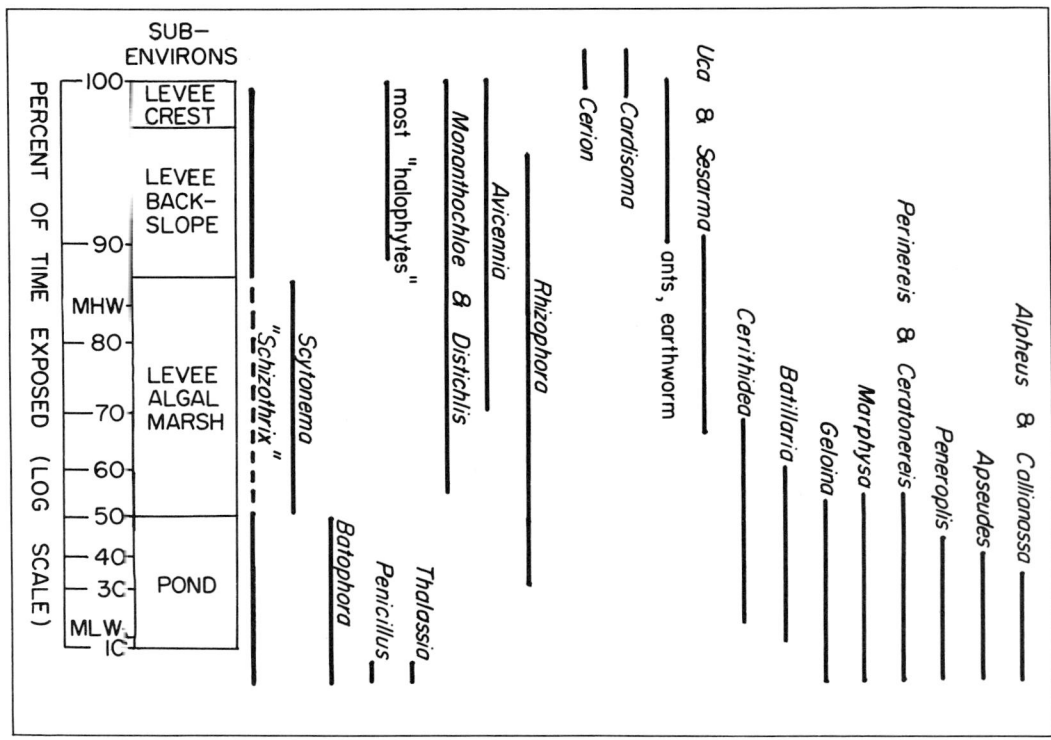

Fig. 69. Vertical distribution of the common organisms of ponds and levee, plotted against percent of time exposed (Exposure Index), and subenvironments.

i) The subtidal floors of the channels and subtidal portions of the ponds have some elements in common with the nearshore community, such as the grass and green-algal assemblage, and some in common with the muddy bottom pond community, such as the polychaete assemblage.

ii) The algal marsh has a very low diversity subcommunity of the levee community, with only one principal blue-green algal genus, *Scytonema*, and a much reduced halophyte assemblage. The inland algal marsh is practically devoid of animal life, though on the levee marsh, fiddler and marsh crabs (*Uca* and *Sesarma*) are usually present, and some of the pond animals extend into the lower levels of the marsh.

iii) The highest points on the beach ridge are colonized by most of the plants and animals characteristic of the levee community, and by many other plants, most conspicuously the casuarina tree. There is thus a far greater diversity than on the levees, and even several distinctly terrestrial animals, such as mice, birds, and a plumonate gastropod, are present. This then, is an assemblage transitional between levee and truly terrestrial communities.

There are definite physical causes for the division of the biota into communities. Tidal flats lie at the important boundary between the marine and terrestrial habitats, and thus to some degree share the physical and biological characteristics of each. The lower levels of the flats are basically an extension of the marine habitat, because of the regular day-to-day flooding by marine tidal waters. The upper levels can be considered as basically terrestrial, because exposure to air and dowsing with rainwater are more common occurrences than flooding by seawater, and the beach ridge even has fresh groundwater. However, tidal flat organisms must to some extent be able to escape from or withstand the climates

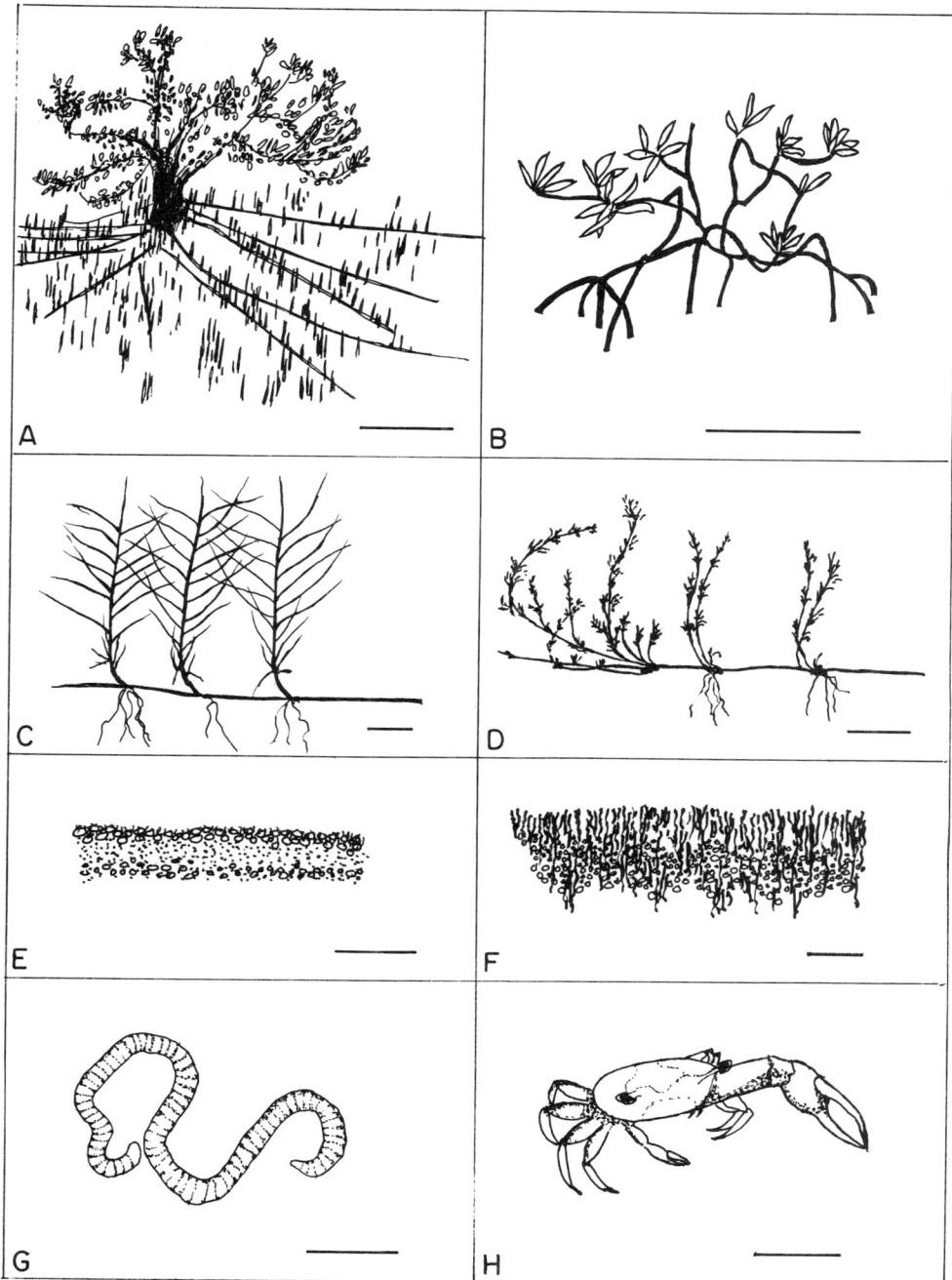

Fig. 70. Common levee and marsh organisms. (A) Black mangrove (*Avicennia nitida*) with root system eroded (on the beach) to show horizontal runners and short vertical pneumatophores. Scale bar 1 m. (B) Red mangrove (*Rhizophora mangle*) in straggly growth habit typical of marsh and levee subenvironments. Scale bar 50 cm. (C) *Distichlis spicata*, and (D) *Monanthochloe littoralis*, common small grasses with shallow horizontal runners. Scale bars 2 cm. (E) Diagrammatic "*Schizothrix*" algal mat. These mats develop the flat laminations characteristic of levee sediments. Scale bar 1 cm. (F) Diagrammatic *Scytonema* mat with coarse filaments enclosing very little sediment. Scale bar 1 cm. (G) Earthworm *Pontodrilus bermudensis*, an uncommon but important burrower and bioturbator of levee sediments. Scale bar 1 cm. (H) *Uca sp.*, the fiddler crab, which excavates subvertical shaft burrows in levee marsh zones. Scale bar 1 cm.

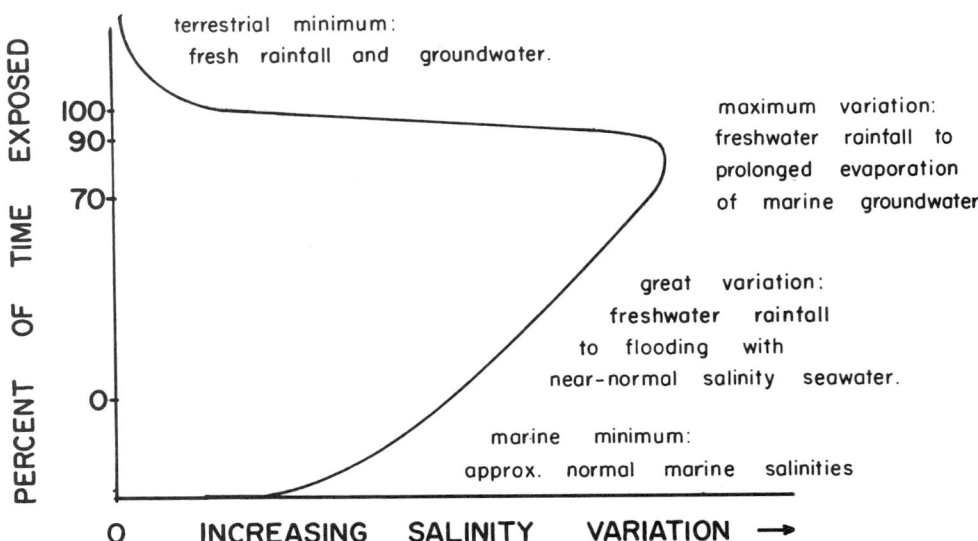

Fig. 71. Diagrammatic illustration of the relationship between degree of exposure and salinity variation. The terrestrial minimum salinity variation is only rarely disturbed by hurricane flooding of low-lying land. Hurricane Betsy (1965), for instance, flooded the west coast of Andros, killing large tracts of pine forest by temporarily salting the groundwater. Between 90 and 100% Exposure Index the salinity variation climbs steeply. The lower levee and levee algal marsh zones have maximum variation, the highest value recorded being 61.4% from a hard, levee-crust surface, the corresponding low salinity value being fresh rainwater. In the zone of everyday tidal flooding (channels and ponds), salinity may range between 45% (recorded in a channel in January 1970) and 5% (recorded after three days of torrential rain in June 1969). Open-bank waters have a relatively small variation slightly elevated above normal marine at between 38 and 45% (Smith 1940; Cloud 1962).

of both habitats, and, indeed, greater extremes than are usually experienced in either habitat. Figures 71, 72, and 73 illustrate diagramatically that salinity and surface temperature fluctuations reach a maximum within the tidal zone, and that desiccation is always greater than in marine habitats. So it should be no surprise that at the most intolerable level, approximately between the 70 to 90% Exposure Index levels, the diversity of organisms should be at its lowest (Fig. 69). This generalization appears to be true for the fauna, though the flora is least diverse nearer to the 10 to 30% Exposure Index level, a fact which suggests that the flora is controlled more by desiccation and flooding frequencies than by temperature and salinity fluctuations.

The greatest change in flora and fauna occurs across the boundary between the pond and levee communities. The other two major community boundaries, pond/nearshore and levee/terrestrial, are less pronounced, because the physical conditions change less abruptly.

3. The Skeletal Record

With many organisms secreting calcareous skeletons, the skeletal record preserved from carbonate environments should be unusually rich and easily decipherable. Such is not always the case, however, because of a number of factors, some of which can be illustrated in examples from this tidal flat. In particular; (1) some skeletal particles are frequently difficult or impossible to identify because of their similarity to nonskeletal particles, even before the processes of diagenetic change set in; and (2) skeletons or fragments of skeletons act as sediment particles and may be transported to and deposited in different environments from those in which they grow.

Codiacean algae. Most of the sediment on these tidal flats and nearshore consists of needles of aragonite a few microns in length, most of which are aggregated into soft and indurated pellets. The origin of these needles has been debated at

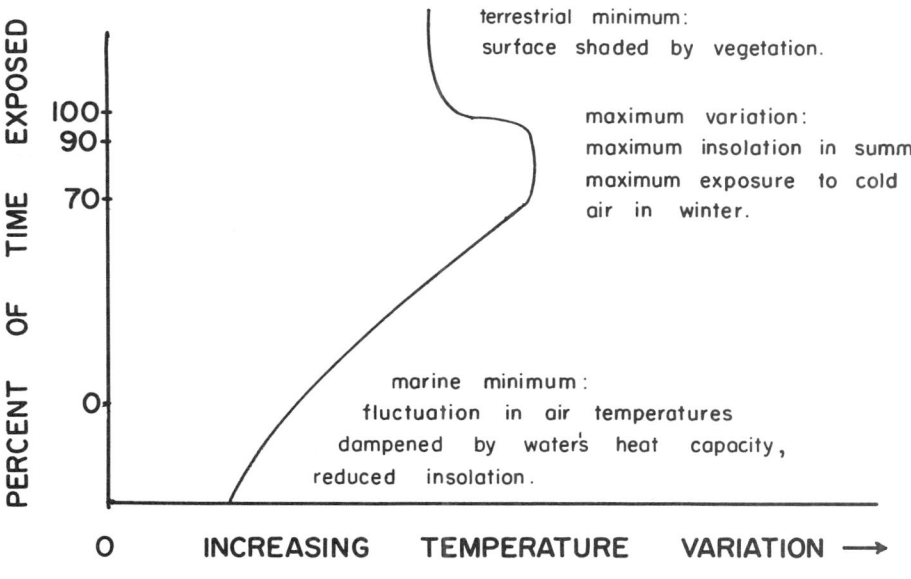

Fig. 72. Diagrammatic illustration of the relationship between degree of exposure and temperature variation. Maximum temperature variations recorded were 40°C in summer, from marsh waters overlying dark algal mats, to 8°C air temperature, and 10°C water temperature in the channels, during the passage of a winter cold front in January 1970.

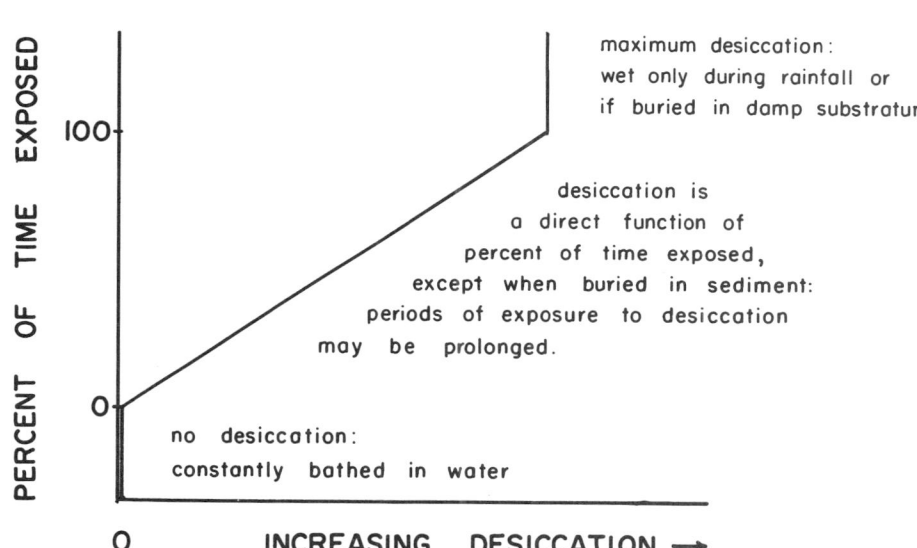

Fig. 73. Diagrammatic illustration of the relationship between degree of exposure and desiccation.

length elsewhere, the debate having been summarized recently by Bathurst (1971, pp. 276ff.). It appears that the needles may have a multiple origin, as inorganic precipitates and as skeletal spicules of codiacean algae, particularly *Penicillus* and *Rhipocephalus*.

The evidence for inorganic precipitation is based on changes in the chemistry of the seawater as it is evaporated during its period of residence on the Great Bahama Bank, though the "whitings," patches of turbid water, which many workers have assumed to be the sites of new precipitation, are known to local fishermen as areas where schools of such fish as mullet and bonefish are stirring up the bottom sediment in search of food. Evidence for the production of the needles by the algae rests on questionable isotopic data, though the populations of these algae are in any case apparently insufficient to account for the total volume of mud offshore, let alone the volume of tidal flat sediment, which is mostly derived from offshore (Hardie & Ginsburg, Chapter 5, this volume).

The question of the origin of the needles may never be resolvable, however, since inorganic and algal needles once formed cannot be distinguished with any degree of certainty, even using isotopic and electron microscopic methods. And being fine-grained aragonite, such particles are of course most susceptible to diagenetic change. Thus their identification in older sediments and rocks is fraught with further problems.

Foraminifera. Mg-calcite is the next most abundant component to aragonite, comprising 5 to 15% of the sediment. This Mg-calcite is believed to be derived entirely from foraminiferal skeletons, which have the same chemical composition (11 to 14 mole % Mg), and which occur in about the same quantities (as judged from point counts on sediment thin sections).

The smaller foraminifera, especially the triloculinids, are numerous in sediments from all environments, including the beach ridge, levee, and marsh. This distribution strongly suggests that the majority if not all the forams are transported as dead tests from offshore, together with the pelleted aragonitic mud. Only in the ponds and channels could these forams live, but in those environments, a few trials with the protoplasmic stain Rose Bengal, failed to identify any alive. Streeter (1963) and Todd and Low (1971) report on the foram faunas of the Bahama Banks west of these tidal flats, but no detailed studies have been made in the tidal flat area.

Ostracod carapaces have a similar distribution to that of the smaller foraminifera, though they are far less common, both offshore and in the tidal flat sediments.

The one large foraminifer, *Peneroplis proteus*, has a flattened disc-shaped test, 1 to 3 mm in diameter. Its tests are abundant in pond and channel sediments and are also present offshore. But because they are rare or absent in levee and marsh sediments, it seems unlikely that they were all washed in from offshore with the rest of the sediment, though fragments of their tests undoubtedly were.

In pond sediments, whole and broken *Peneroplis* tests occur randomly scattered throughout, though they are frequently more abundant at some horizons than others. In channel sediments, *Peneroplis* tests occur either randomly or lying parallel in distinctive cross-beds. These channel tests may be indigenous to the channels or they may have been eroded and redistributed from pond sediments like the cerithid gastropods (see below).

Peneroplis is also responsible for the production of small "dense peloids," which range in size from 50 to 250 μ, but are mostly in the size range 125 to 150 μ. All dense peloids are approximately ovoid, though there is great variability and most show slight irregularities in outline. All have smoothly rounded corners and a porcellaneous rather than a polished luster (Fig. 74A). They are all hard, and split or jump, rather than crush, under pressure of a needle.

In thin section, these dense peloids (Fig. 74B) are colored either a dark grey or dark brown in plane polarized light. With crossed polarizers they go very dark, though a fine microgranular texture is usually apparent.

Their exact mode of origin is in doubt. According to Tappan and Loeblich (in Moore 1964) they may be a type of foraminiferal fecal pellet. I quote page C87: "waste products (e.g., empty diatom frustules) may remain in the cytoplasm until just before reproduction occurs, or the waste may be condensed into small pellets (stercomata) as in *Peneroplis*. The tiny brown xanthosomes also appear in the protoplasm after feeding and are excreted from time to time." The abundance of dense peloids in all sediments, but especially pond and channel sediments (Fig. 75), supports this view. However petrographic and mineralogical evidence suggests that they may be recrystallized and broken internal walls of the chambered tests. Thus (1) as shown in Figure 74B, the dense peloids inside the *Peneroplis* test appear to form an integral, though disintegrating, part of the chamber walls, and (2) the fine-sand fraction of pond sediments, which consists mostly of dense peloids, has a higher than average Mg-calcite to aragonite ratio, suggesting that the dense peloids are in fact a part of the *Peneroplis* skeleton.

Grains like these are not described in the texts of Illing (1954), Purdy (1963), or Cloud (1962), though all discuss at length pellets of larger sizes and more

Fig. 74A.
Fine, sand-sized fraction of pond sediment, showing mostly "dense peloids" with slightly irregular, tough, smoothly rounded outlines and a porcellaneous luster. A few hyaline foraminifera and broken shell fragments are scattered throughout.

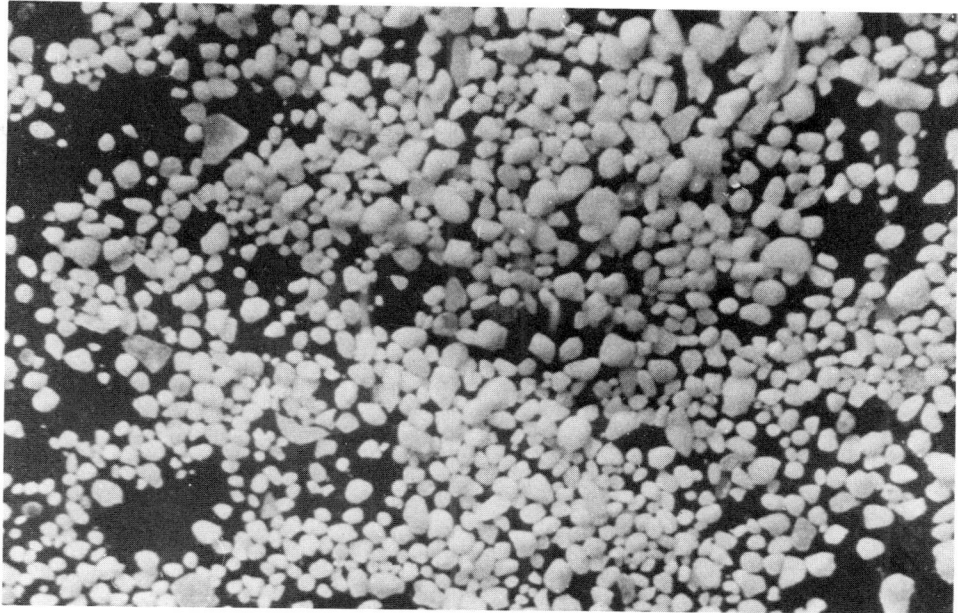

Fig. 74B.
Photomicrograph of test of the porcellaneous foraminifer *Peneroplis proteus*, showing fecal pellets or micritic skeletal struts within the test, which closely resemble dense peloids outside the test.

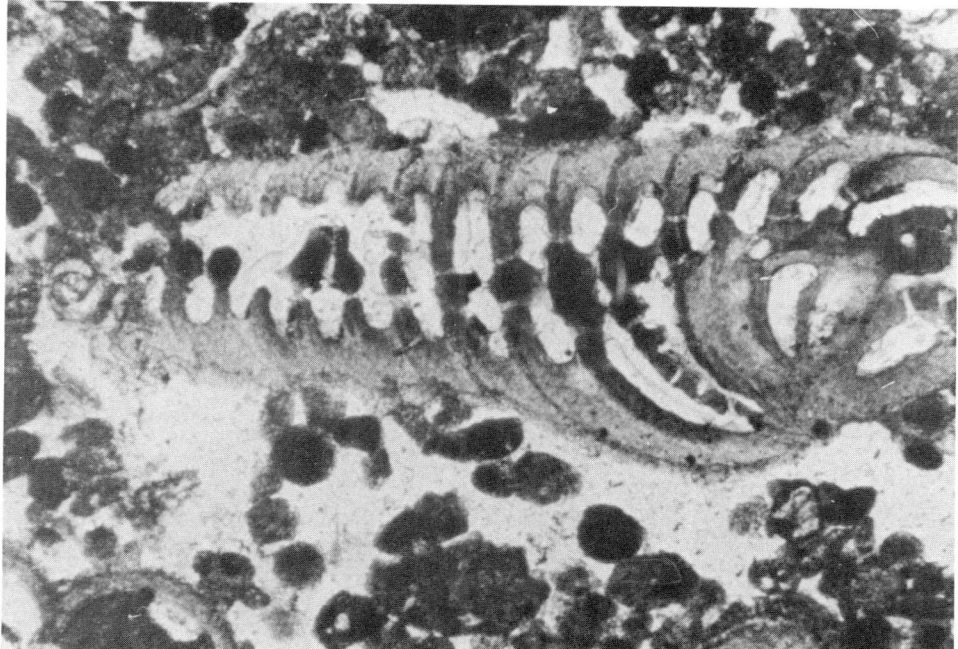

regular ellipsoidal outlines. Fecal pellets are usually distinguishable by having smaller particles, such as tunicate spicules and small forams (Purdy 1963), included in them. Cloud however, illustrates in his Plate 9G, particles which are very similar to dense grains, though they have a polished luster—he calls them pellets.

Matthews (1965) describes similar particles in the sediments of the southern British Honduras Shelf, though apparently he did not consider anything but a skeletal origin for them.

If these grains are indeed of skeletal origin, they illustrate how difficult it can be to distinguish a skeletal from a nonskeletal peloid. With further recrystallization the problems are compounded (Purdy 1968).

Fig. 75A. Compacted pond mud, showing faint outlines of probable pellets near pores and oblivion of pelleted texture away from pores. Darker ovoid grains are dense peloids. Scale: × 100.

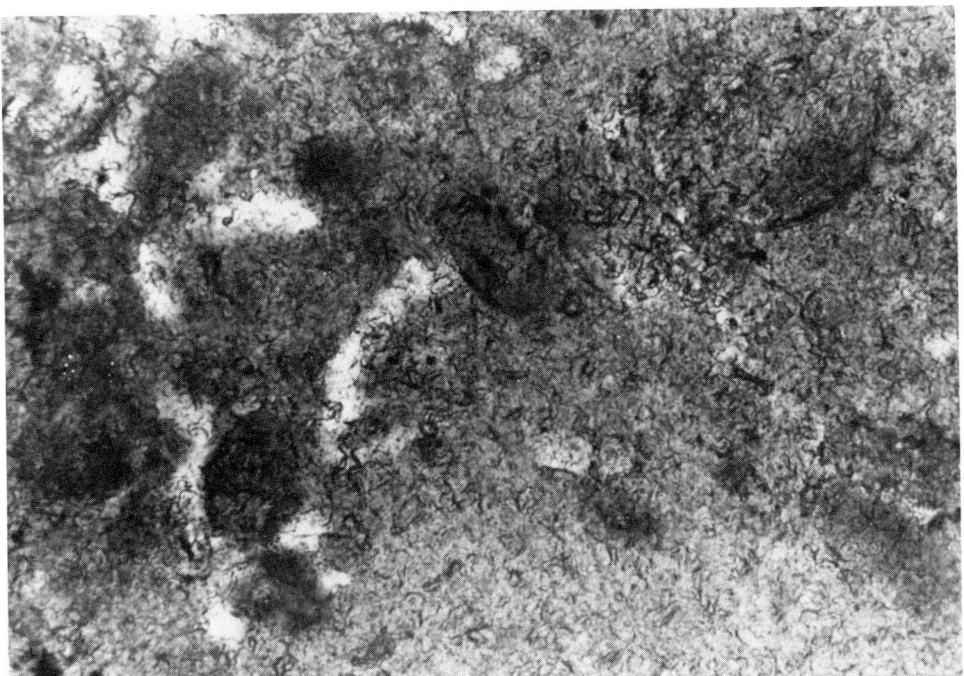

Fig. 75B. Thin section photomicrograph of a core from an abandoned channel. Clearly visible are large ellipsoidal pellets, probably of *Marphysa*, which are composed of silt matrix as well as dense grains and small foraminifera. The rest of the sediment is a porous mixture of the same components together with skeletal grains, mostly foraminifera. Scale: × 15.

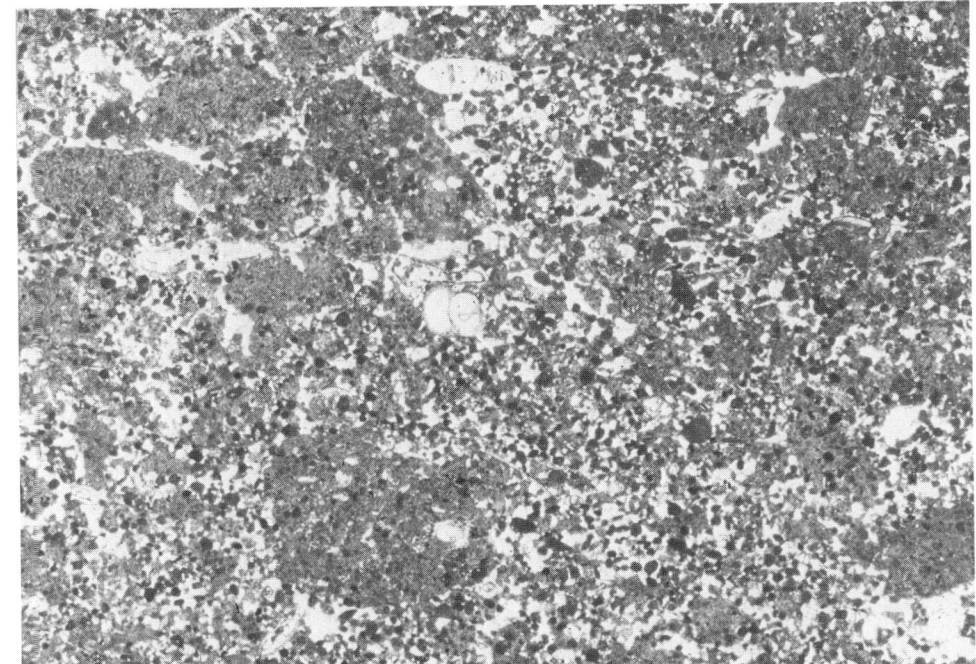

Molluscs. Several molluscs are present on the tidal flats, but only three species (*Cerithidea*, *Batillaria*, and *Geloina*) are common enough to provide a significant skeletal record. Most of the remaining species (*Cerion, Prunum, Detracia, Sayella, Brachidontes*, and a few scaphopods) are sporadic in occurrence and few in number, though there are two species (*Syncera* and *Tornatina*) which are common, but so small that their skeletal record is more like that of a small foraminifera.

Because the gastropods *Cerithidea* and *Batillaria* are abundant and easy to observe, they were chosen for special study of their live and dead distribution.

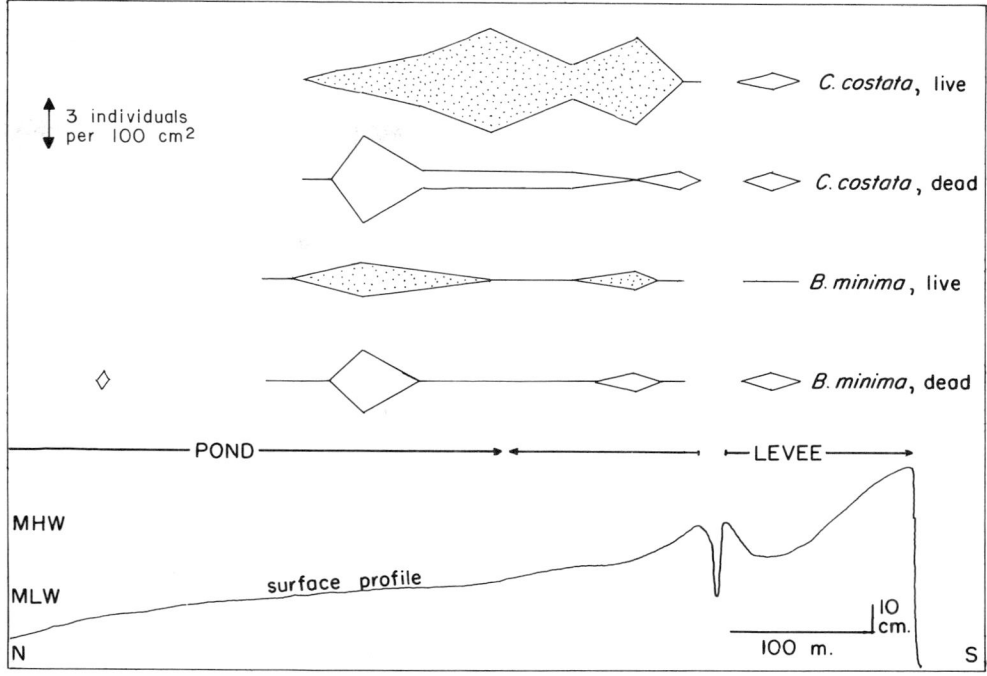

Fig. 76. Levelled profile and abundance of live and dead cerithids, transect 8255. See Fig. 2 for location.

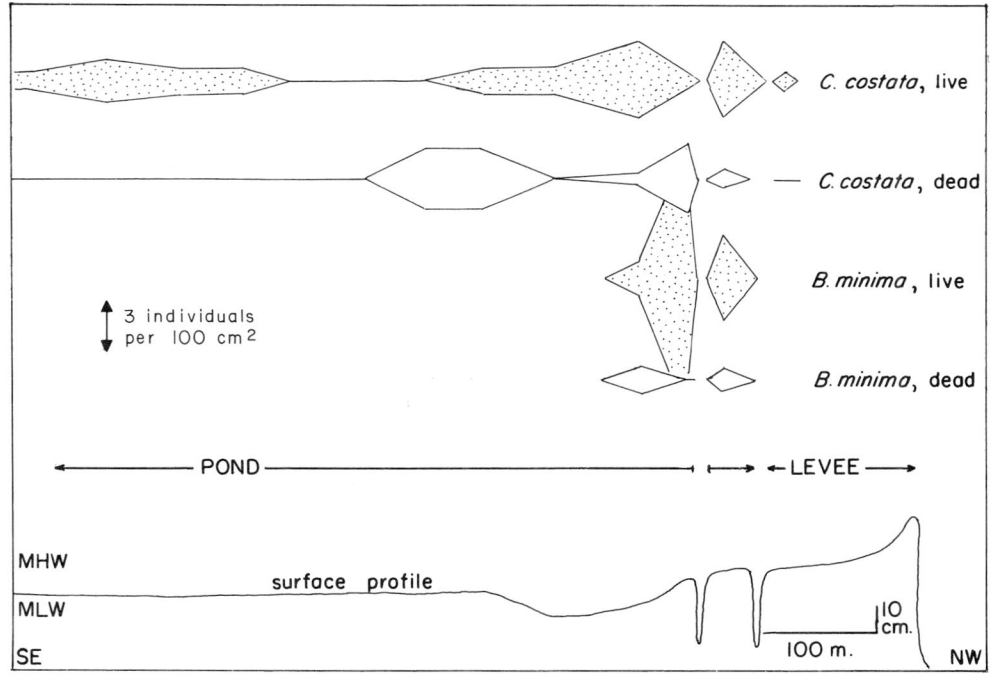

Fig. 77. Levelled profile and abundance of live and dead cerithids, transect 8256. See Fig. 2 for location.

This was done by counting their shells along two levelled transects (Figs. 76, 77), using as a unit of measure, a 100 cm² wire grid thrown on the ground in two places at the point of observation. The number of gastropods enclosed within the grid, both live and dead, were picked up and counted: later these counts were averaged and plotted with the profile of the transect. Many other observations of cerithid abundance were made in other parts of the tidal flat.

Observed facts concerning live cerithid distribution are as follows:

1) The limits of the gastropods are: Upper—just above the lower limit of the *Scytonema* mat, approximately at the 75% exposure level. Lower—at or

below the lowest limit of red mangrove occurrence, approximately at the 20% exposure level.

2) Within these limits, there is a tendency for *C. costata* to be commoner higher on the pond shores and to extend its range into dryer zones than *B. minima*, and vice versa. *B. minima* is the only cerithid to occur in the beach zone, where it frequently lives gregariously (Garrett 1970, Plate 1A).

3) Live cerithids are absent from ponds which are at present subtidal and from those which are tideless (because they lack a connection to the sea). They are also absent from the algal marsh, which lies within their requisite flooding-frequency range, but which can be flooded by freshwater, or dried out completely for several months at a time.

4) Live cerithids are not always present where they would be expected, and sometimes one species is present alone or in a single-year size-class. Such a spotty distribution is no doubt due to the vagaries of their larval settling.

The distribution of dead cerithid shells on the pond surfaces (Figs. 76, 77), approximately follows the distribution, though not the abundance, of the living gastropods. There are two possible causes for this discrepancy: irregular larval settling and redistribution of shells during storms. Clear evidence of the latter is not common, though cerithid lag "beaches" are developed on the margins of some ponds, and the floors of some subtidal ponds are strewn with dead shells. Such ponds show evidence for having been scoured out to subtidal levels from a previous intertidal level, because neither cerithids nor mangroves (whose roots are also exhumed) live in subtidal ponds. Lag deposits are recognizable in pond cores as layers of unusually great concentrations of shells. Most pond cores, however, show a more or less random distribution of cerithid shells.

In channel and beach subenvironments, live cerithids are absent (except occasionally on intertidal beach mounds and channel bars), yet it is in these two subenvironments that the greatest concentrations of cerithid shells are found. The shells occur in cross-bedded coquinas (Fig. 78A), sorted and deposited by storm waves or tidal currents. In each case the source of the shells is from exhumed and eroding pond sediments, which outcrop on the foreshore seaward of the beach ridge (Fig. 23A), and at the lower levels of eroding channel banks respectively. Cerithid shells rarely occur randomly in channel or beach-ridge cores.

In beach-ridge sediments, and to a minor extent in marsh and levee sediments, whole and broken shells are concentrated into coarse laminae or layers, representing shell pavements deposited during particularly violent storm overbank floodings.

The shell of *Geloina*, the only common pond bivalve, is distributed and redistributed over a similar range and in a similar fashion to the cerithid gastropods. The average size of the shell, however, bears an interesting relationship to the cohesiveness of the mud in which it lives. *Geloina* is a small suspension-feeding bivalve which lives buried in an upright position a few millimeters beneath the surface, with its siphons protruding out through a slit-like opening. In pond sediments, which near the surface are soft and poorly cohesive because of the deposit-feeding infauna, *Geloina* is usually quite common (around $20/m^2$), but it never grows beyond a maximum size of 12 mm. In the levee marsh zone, however, where the surface sediment is distinctly more cohesive, *Geloina* is rare, but sometimes grows to a length of 20 mm, despite the fact that the time it can spend feeding is considerably shorter, and it has to withstand greater desiccation and salinity fluctuations at this level.

The probable cause of this peculiar size distribution in *Geloina*, is that in the soft pond sediments, a large shell would sink into the mud and be unable to survive (body weight increases as the 3/2 power of surface area [Stanley 1970]). Also the siphons of bivalves clog easier in muddy sediments. Thus it appears that in this area, *Geloina* only reaches its mature size near the limits of its physiologically controlled distribution. This example bears out the sort of life habit exclu-

Fig. 78A. Channel coquina. Slab of channel sediment with cross-bedded cerithid shells and subparallel *Peneroplis* tests (*white flecks*) in pelleted muddy matrix.

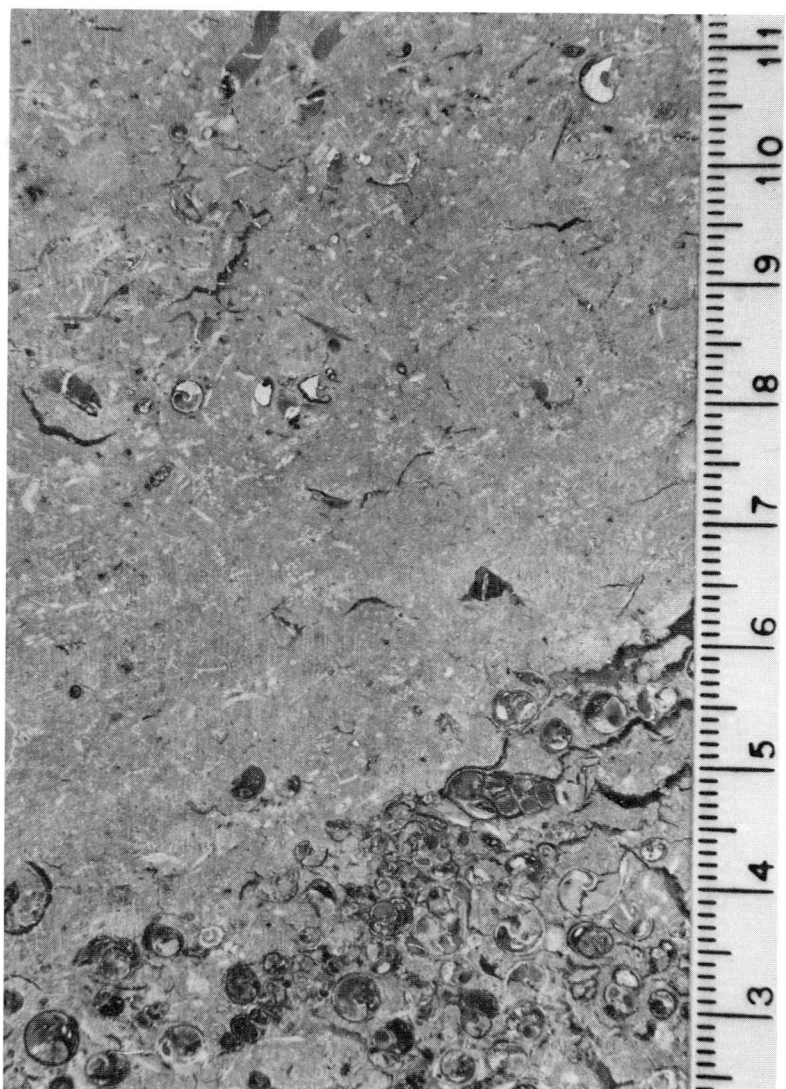

sion of suspension-feeders and deposit-feeders that is discussed by Rhoads and Young (1970).

Generalities. Conclusions about the skeletal record tend to be rather negative. All aspects are difficult to interpret. For instance:

1) The aragonite needles which constitute the bulk of the sediment, are certainly in part at least a skeletal record of codiacean algae, but, despite considerable research effort in the past, the proportion contributed by the algae is still debatable.

2) Small forams and ostracods are so readily transported from offshore with the pelleted aragonitic sediment, that their tests are distributed throughout the tidal flat sediments, including those of the supratidal, in approximately equal abundance.

3) An example is given of the production of a pellet-like particle by the foram *Peneroplis*, which may be skeletal or fecal or even a bit of both (!).

4) Studies on cerithid gastropods illustrate difficulties in explaining their live distribution. After death, their shells may be (a) preserved in their life environment, (b) preserved in their life environment, but concentrated as a lag deposit,

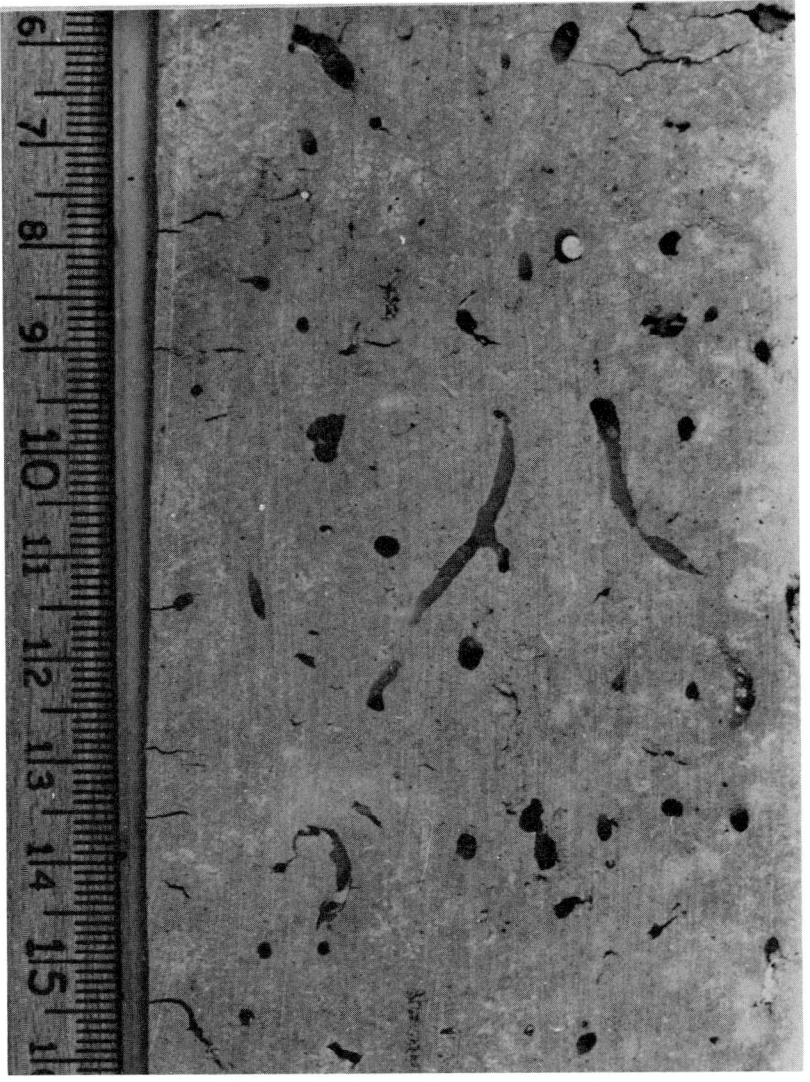

Fig. 78B. Burrows, mostly of *Dasybranchus* sp., a polychaete, preserved in cohesive muds of an intertidal channel bar.

or (c) eroded after initial burial and concentrated into a secondary coquina. Other permutations are possible.

4. The Trace Record: Modification of Sediments by Organisms

A large proportion of marine animals use the sediment as a source of food or shelter. On tidal flats this is true for the majority of animals, probably for the following two reasons:

1) The amount of organic detritus deposited on the sediment surface is greater than in other more turbulent environments.

2) The micro-environment within the tidal flat sediment is far more stable in terms of such parameters as temperature and salinity, than in the sediment surface or the water or air above (Seilacher 1964; Sanders et al. 1965).

Plants, too, interact with the sediment by abundantly colonizing its surface (cyanophytes), rooting in it (angiosperms), or making holdfasts (green algae).

With this dependence of most tidal flat animals and plants on the sediment, it should come as no surprise that the sediment is greatly modified by biogenic processes. Shells may disintegrate after being bored by certain cyanophytes or broken by predators. Filamentous cyanophytes living in and on surface sediments cause the entrapment and binding of sediment particles and sometimes the precipitation of new material. Sediments are bioturbated mainly by deposit-feeding animals, who modify its texture by making fecal pellets and by tunneling.

Bored Shells. Many shells in the sediment of both the tidal flats and nearshore show signs of boring by endolithic algae (Fig. 79A). Swinchatt (1969) has described these borings from depths down to 60 m, but here an upper limit can be added for the occurrence of these algae, at about the 50% exposure level. In fact the boring is frequently so intense in the shells of pond cerithids that they are given a distinctly corroded appearance, and finally may disintegrate completely, leaving only the columella intact (Fig.79B).

Shell Breakage. Broken shell material in sediments does not necessarily indicate a highly turbulent environment of deposition (Ginsburg 1957). In fact, the association of broken shells and fine sediment, as in this area, should definitely indicate some other cause of breakage than simple mechanical action. Apart from endolithic algae, which usually cause shells to disintegrate rather than break, predatory crabs, such as *Panopeus*, are probably the chief cause of molluscan shell breakage in subtidal sediments. Among the birds, the sandpipers and plovers might be able to break open shells with their beaks, though they do not seem to feed specifically on the gastropods. Flamingoes, however, are known for their epicurean delight in cerithids (Allen 1956), which they probably crack open at the aperture with their toothed tongues.

Bonefish and mullet are well known as bottom feeders by Bahamian fishermen, who sight their schools by looking for patches of stirred-up white water ("whitings"). The fish suck up sediment and masticate it between the grinding surfaces on their tongue and upper palate. In so doing they no doubt cause the breakage of foraminiferal tests and small pelecypod shells, though they avoid large shells in their diet.

Algal Mats. Filamentous algae, mostly cyanophytes, occur abundantly in almost all surface sediments of the tidal flat and nearshore environments and develop into strongly cohesive, and in some cases very conspicuous, mats, especially at higher tidal levels (above 65% Exposure Index). Beneath these mats, distinctively laminated sediments occur which in large measure owe their particular morphologies and internal structures to the mat type present at the time of their formation.

These high-level, algal-laminated sediments are described and discussed in detail by Hardie and Ginsburg in Chapter 5 (this volume). Briefly their preservation can be summarized as follows:

1. Filament mats of *Schizothrix* are "preserved" as smooth, thin sediment laminae (Chapter 5). Occasionally the mats are cemented at the surface into paper-thin crusts.

2. The coarse filament alga, *Scytonema*, can be preserved in a variety of modes. In certain environments the filaments induce the local precipitation of Mg-calcite as tiny, indistinct rhombs which surround the filaments, preserving them as imperfect molds (Monty 1967). Compaction can crush these rhombs into a mush with few filament molds preserved, or, alternatively, further precipitation can cause the cementation of the rhombs into a hard crust within which filament molds are well preserved (Chapter 5).

Influx of pelleted sediment in various quantities and at different intervals can cover or partly cover the irregular mat surface, producing wavy sediment laminae of varying thickness alternating with organic- and precipitate-rich mat laminae. Where the mats have not induced precipitation of Mg-calcite rhombs they decay and disappear as they become buried.

At lower levels, in the ponds, the sediment surface is colonized by *Schizothrix* (s.l.), the same alga though probably not the same ecophene as that which colonizes the levee crest. In the ponds it does not develop into a cohesive mat, because of the activities of cerithid gastropods and other grazing animals (Garrett 1970). However, in areas where the gastropods are absent, a soft, rubbery,

Fig. 79A.
Endolithic algal borings in *Geloina* shell. Thin section. Scale: × 50.

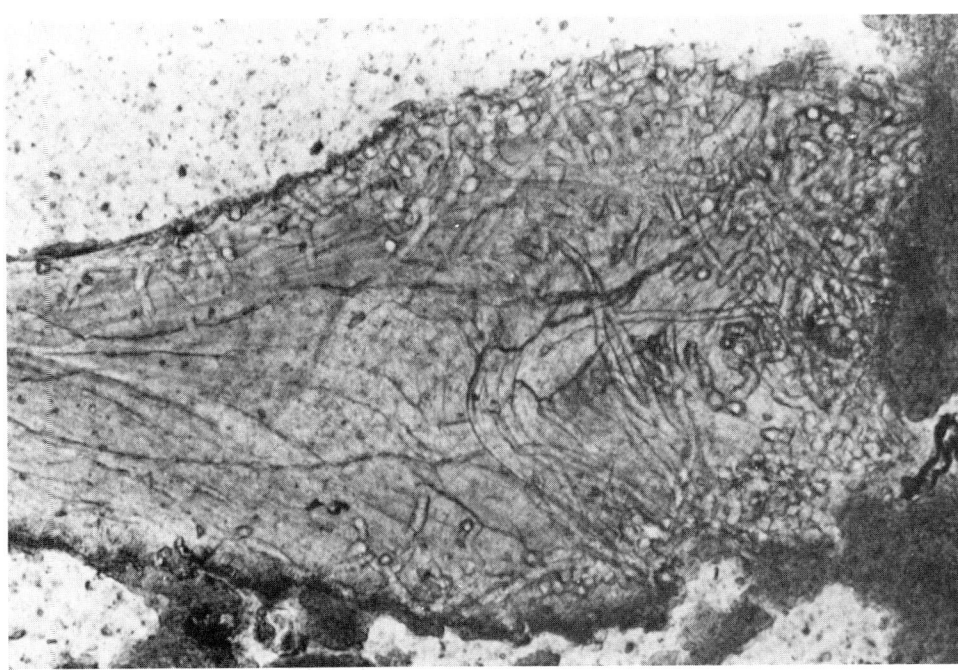

Fig. 79B.
Stages in the disintegration of cerithid shells by endolithic algal borings. Only the columella remains of the far right shell. Scale in mm.

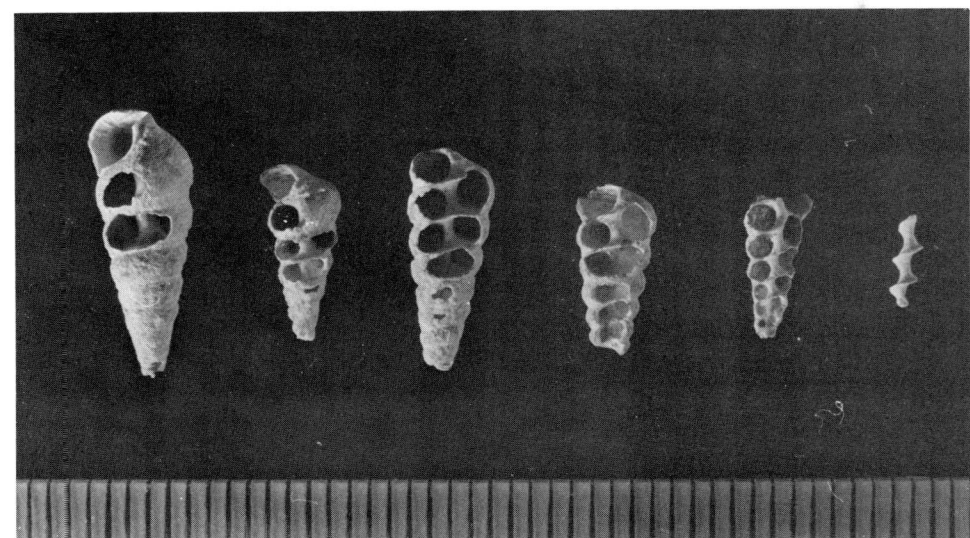

pink-colored mat develops which binds the surface sediment together. This mat is easily broken, most notably by polychaetes which burrow below it and through it, and because it contains air bubbles, pieces of the mat, or rather small algal-bound clots of sediment, are dislodged and transported by wind-waves around the ponds. These clots are unlikely to be recognizably preserved because of the later activities of burrowing animals.

On the crests of channel bars and beach mounds (Figs. 28 & 22 resp.) a smooth, cohesive, rubbery *Schizothrix* (s.l.) mat develops, which binds sediment in thin laminae when flooded by sediment-laden water. These laminae are sometimes preserved beneath the surface, but are more usually destroyed by the bioturbation of burrowing polychaetes. These mound-like features therefore provide peculiar examples of "stromatolitic" structures (i.e., structures built by the sediment-binding properties of cyanophytes at the surface), which, however, lack the characteristic laminations because they are destroyed by subsurface bioturbation.

Table 11. Pelleting rate of *Cerithidea costata* and *Batillaria minima* on wet supratidal mat

Cerithidea costata			Batillaria minima		
Time of observation (mins.)	No. pellets excreted	Comments	Time of observation (mins.)	No. pellets excreted	Comments
8	4		4	5	⎫
9	7		5	3	⎪
10	15		2	3	⎬ just taken
10	16		6	1	⎪ from a longish
5	33	moved 12 mm	6	1	⎪ period of
2	8	moved 32 mm	7	2	⎭ feeding
			7	8	
44	83		9	19	⎫
			7	18	⎭
			7	4	moved 48 mm hardly feeding
Average: 2 pellets/minute			60	64	
			Average: 1 pellet/minute		

Neumann, Gebelein, and Scoffin (1970) describe the occurrence and geological significance of subtidal mats similar to those occurring offshore from this area. Such mats, though in many cases quite cohesive at the surface, are unlikely to be preserved due to the grazing activities of such epifauna as bonefish and mullet and mat-dwellers such as ostracods, cumaceans, syllid polychaetes, and nematodes, and to the bioturbation of the burrowing infauna, chiefly polychaetes and shrimps.

Bioturbation. a) *In the ponds:* a special study was made of bioturbation processes in the ponds, the two major grazing gastropods *Cerithidea* and *Batillaria* and the burrowing polychaete *Marphysa* being singled out for quantitative studies of bioturbation rates.

Cerithidea and *Batillaria*, which flourish in the intertidal zone between Exposure Indexes 25 to 65%, feed by grazing on surface sediments, but only when flooded and for a few hours afterward, while there is sufficient water on the mud to keep their bodies wet.

About a dozen cerithids of both species were collected from a pond margin at about 40% exposure level, and placed on a piece of wet supratidal mat (previously ungrazed and therefore full of algal filaments). Their behavior was watched and the number of pellets excreted in short intervals of time noted (Table 11).

The gastropods fed on the mat, simultaneously excreting pellets behind them, but at greatly varying rates. Assuming that the rate of feeding can be extrapolated to the natural pond margins for the length of the tidal cycle that is spent feeding, the rate of grazing (sediment reworking) can be calculated as follows:

Pellet size, assuming an ellipsoid shape
 $= 4/3\, ab^2 = 4/3 \times 0.43 \times 0.0144 = 0.025$ mm^3
Pelleting rate $= 100$/hour (experimental average)
Cerithid abundance $= 1000/$m^2 (usual variation: 500 to 1500/m^2)
Period spent feeding $= 50\%$ of time (an underestimate, assuming a position at the 50% exposure level)
Volume worked per month by cerithid gastropods:
 $= 0.026 \times 100 \times 1000 \times 50/100 \times 24 \times 30$ mm^3/m^2/month
 $= 0.94 \times 10^6$ mm^3/m^2/month
i.e., *a continuous layer approximately 1 mm thick per month*, over an area covering the vertical interval between the 25 and 65% exposure levels, where the gastropods are most abundant and active.

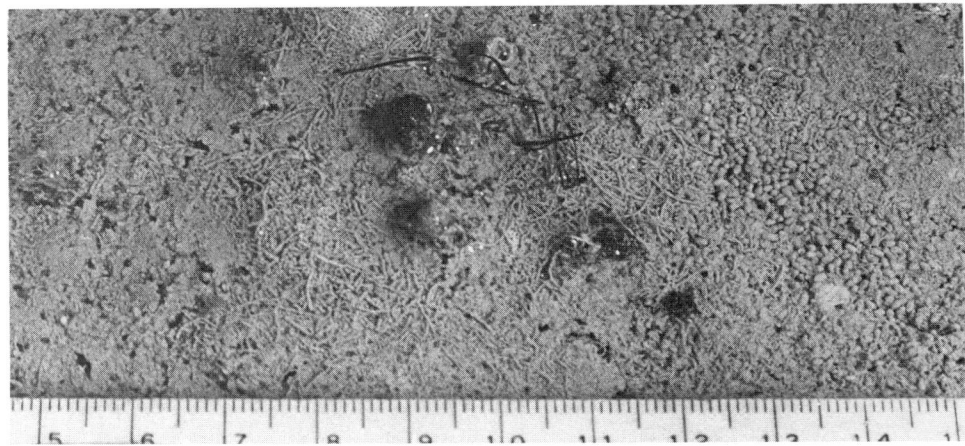

Fig. 80A. Pellet-strewn surface of subtidal pond. The long strings of small bead-shaped pellets are defecated by nereid polychaetes living just below the surface—the strings break easily, leaving fine, sand-sized particles. The larger pellets to the right are defecated by *Marphysa*—(the small brush-like plants are dwarf specimens of *Penicillus* sp., which is rare and seasonal in the ponds).

Table 12. Pelleting rate of *Marphysa sanguinea*

Pellet "Volcano" number	Number of pellets	Equivalent unpelleted	Total excreta*
Channel bar, 20–50% exposure level January			
1	120	60	180
2	60	120	180
3	170	45	215
4	70	35	105
5	52	50	100
6	0	150	150
Pond edge, 45% exposure level January			
1	90	135	225
2	50	75	125
3	40	60	100
4	0	80	80
Pond, 15% exposure level January			
1	120	60	180
2	80	40	120
3	120	0	120
4	70	70	140
5	40	120	160
6	150	0	150
7	20	0	20
8	20	0	20
9	0	0	0
10	10	0	10
11	170	0	170

Summary	
Subenvironment	Pellets/day (average)
Channel bar	155
Pond edge	130
Pond center	100

* All figures expressed as pellets (or pellet equivalents) defecated per day.

Because the surface sediment of the ponds is usually a mass of fecal pellets (Fig. 80A), and it is these on which the gastropods are grazing, it is clear that there must be sufficient colonization and regrowth of cyanophytes, bacteria, etc., since the time of last excretion, for them to be sufficiently nutritious for reingestion by the gastropods.

Bonefish are also surface-grazing animals: they suck up mouthfuls of sediment, leaving shallow pits 1 cm deep. Some of the sediment is dropped from the mouth, the rest being ingested and excreted as loose fecal material. They are not common in the ponds, and their bioturbating effects are believed to be minor.

Within the surface sediments, the meiofauna, chiefly consisting of nematodes, but also including a few ostracods and other small crustaceans, are responsible for small-scale (i.e., very local) movement of sediment particles, though their contribution to the degree of bioturbation is unknown. *Geloina*, a small infaunal suspension-feeding bivalve, and *Syncera* and *Tornatina*, two miniscule sediment-dwelling gastropods, must also be responsible for some small degree of

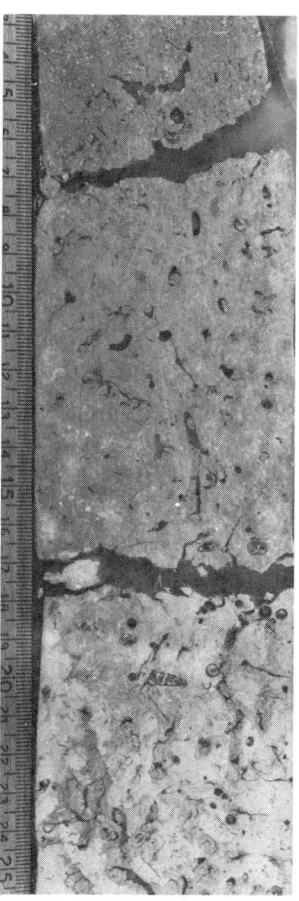

Fig. 80B.
Slab of pond core, showing relatively homogeneous nature of sediment. *Near top*, pelleted nature is obvious, there are many *Peneroplis* tests (*white*), and burrows are generally not preserved in the soft mud. Between 8 and 17 cm, pellets are only obvious in polychaete burrows. Below 18 cm, cerithid gastropods are present and polychaete burrows are preserved as a mottled texture.

Apseudes, a small tanaid shrimp, abundant in some ponds, continually makes and remakes U-burrows to depths of 2 cm. A patch of red-pigmented mud laid on the surface was completely churned into the sediment after three months, so that only a dull pinkish hue was discernible throughout the top 2 cm. Thus, *Apseudes*, where present, is a very active bioturbator.

Several worms burrow deeper into the sediment. One of them, the eunicid polychaete *Marphysa*, burrows deeply, feeding on buried sediment and dead mangrove roots, while depositing most of its fecal material at the surface in volcano-shaped mounds (Fig. 20A). Its bioturbation rate was studied by pouring red-pigmented mud over several fecal mounds and returning a day later to count the number of new fecal pellets and to estimate the volume of unpelleted fecal sludge covering the pigmented layer. The results of this field experiment are given in Table 12. It was found that *Marphysa* defecates an average of 100 to 155 pellets per day (in January; no measurements were made during summer). A burrowing rate for *Marphysa* at the pond location is calculated as follows:

$$
\begin{aligned}
&\text{Abundance of } \textit{Marphysa} \text{ (mounds)} &:& \quad 24/m^2 \\
&\text{Number of pellets per mound} &:& \quad 100/\text{day} \\
&\text{Size of pellets} &:& \quad 0.53 \text{ mm}^3 \\
&\textit{Volume of sediment reworked} &=& \; 24 \times 100 \times 0.53 \\
& & =& \; 1272 \text{ mm}^3/m^2/\text{day} \\
& & =& \; 464 \times 10^3 \text{ mm}^3/m^2/\text{year}
\end{aligned}
$$

equivalent to a layer of pellets (and unpelleted fecal matter) one-half millimeter thick over the whole pond surface in one year.

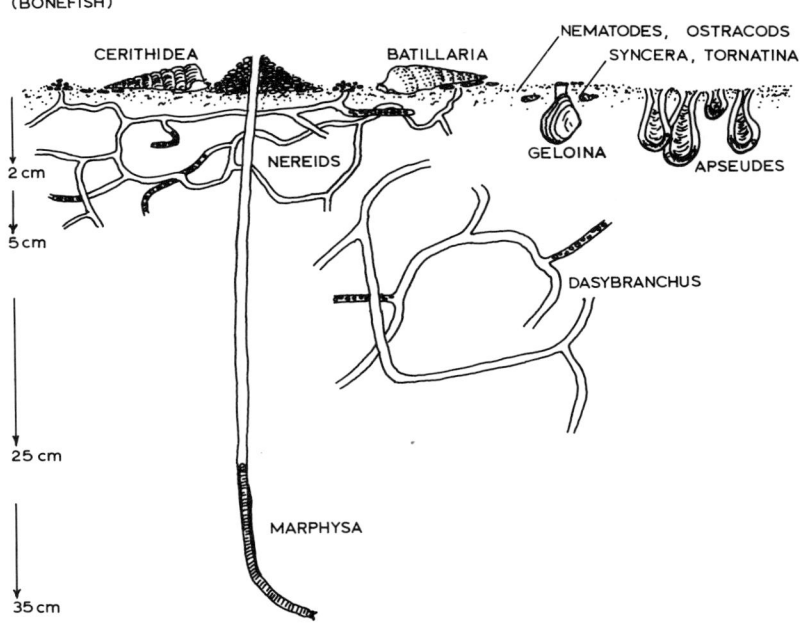

Fig. 81. Schematic cross-section through a pond, showing the chief bioturbating animals and the depths of their maximum effect. Note that the vertical scale is nonuniform, because of the concentration of bioturbators at or near the surface.

If the average burrow depth is taken to be 30 cm, and the burrowing rate is to be that calculated above, then *Marphysa* is theoretically capable of reworking all the pond sediments to the depth of its burrows in a period of from 600 to 700 years, though this capability may or may not be realized in practice.

The two species of nereid polychaetes identified from this area are common burrowers, especially in the upper 5 cm of pond sediments. At night they graze on the surface, and their habit is to deposit their fecal pellets both within their burrows and at the surface (Fig. 80A). Thus it is difficult to estimate their rate of bioturbation, though together with the tubifiscid oligochaetes, which burrow to similar depths, they are thought to be at least as active as *Marphysa*.

Dasybranchus, a slow-moving capitellid polychaete sometimes present in the ponds, burrows to depths of 30 cm, depositing all its fecal pellets within its burrow tubes. Its rate of bioturbation is probably between one-fourth and one-half that of *Marphysa*.

Pond sediments, then, are quickly turned over near the surface, the very surface being completely reworked in a period of one month by the cerithids alone. The top 2 to 5 cm, reworked mainly by *Apseudes* and the nereids, are probably completely reworked in three months to ten years, depending on the abundance and type of fauna. Deeper sediments, penetrated only by *Marphysa* and sometimes *Dasybranchus* may never be completely reworked, though much material is brought to the surface from below. Figure 81 illustrates pond bioturbation.

The sedimentary record of bioturbation in ponds is therefore that no laminations or other original sedimentary structures are preserved, instead the sediment is more or less homogeneous, though mottled, with pellets and burrows as the only preserved sedimentary structures (Fig. 80B). However, the deposition and preservation of storm layers (Hardie & Ginsburg, Chapter 5, this volume) serves to illustrate the limitations of bioturbating animals. During large storms, such as hurricanes, a layer of sediment up to 5 cm thick may be deposited locally in some ponds. Such events must kill off much of the pond fauna—the cerithid gastropods are frequently concentrated immediately beneath such storm layers—and those elements of the burrowing infauna that do survive are unable to rework the thick layer because it is usually of a different texture, becoming hard and compact if desiccated soon after deposition. Thus, in ponds with storm layers, reworking of the sediments below and above the layer has proceeded apace, but the layer itself is usually left relatively untouched (Fig. 59B), though its upper and lower boundaries may be obscured.

b) *In other subenvironments:* In the channels, the cerithids are absent except on the crests of some channel bars, though a similar suite of polychaetes is effective in bioturbating the sediment. These are aided especially by two burrowing shrimp, *Alpheus* and *Callianassa*, who throw out large amounts of excavated material at the surface. Though an overall rate of bioturbation was not estimated for channel sediments, two observations are pertinent. One is that the experimental rate of *Marphysa* pelleting was found to be approximately similar to that in the ponds. The second is that the coarser sediments of channel-bar deposits, which principally contain cerithid shells and *Peneroplis* tests, are usually cross-bedded in core sections and show little sign of extensive reworking (Fig. 78). It must be that the most active burrowers avoid coarse deposits such as these, or that the overall rate of resedimentation of channel-bar sediments exceeds the rate of bioturbation.

Bioturbation in nearshore sediments was not studied in any detail. However, surface-grazers, such as gastropods, are conspicuously absent, though schools of mullet and bonefish occasionally rework surface sediments by their sediment-grubbing activities. There is an abundant, near-surface meiofauna (Neumann, Gebelein, & Scoffin 1970), though the deeper-burrowing polychaetes are fewer than in pond sediments. *Callianassa* is the chief deep burrower. The record of bioturbation here is similar to that in the ponds—a general mottling, though with few polychaete burrows and soft pellets preserved because the sediment is sandier and less sticky and cohesive than in the ponds. No storm layers have been recognized in cores from offshore.

Levee and beach-ridge sediments are not bioturbated to nearly the same extent as are the sediments at lower tidal levels, the reason being that there are fewer animals present in the sediments, and those that are present are not particularly active in moving large quantities of sediment. An earthworm, *Pontodrilus*, though quite rare, is responsible for a considerable amount of pelleting and destruction of the laminated levee sediments (Fig. 36B). Similarly, a few species of ants burrow shallow tunnels into both beach-ridge and levee sediment, moving the sediment around in small rough mastication pellets. The only other animals living within levee and beach-ridge sediments are the land crab *Cardisoma*, the fiddler crab *Uca*, and the marsh crab *Sesarma*. Though these crabs construct sizable burrows with conspicuous talus heaps (cf., Shinn 1968), they are not, except where abundant, responsible for as much bioturbation as is the earthworm, because they build for shelter, not food, and do not continually work the sediment.

Inland marsh sediments are not bioturbated at all, there being no grazing or burrowing animals living in the environment.

c) *General points:*

1. Bioturbation is the most important sediment-modifying process at the lower tidal levels of this tidal flat complex: most primary sedimentary structures are obscured or eliminated, and because there are so many organisms disturbing the sediment, the result, usually a general mottling and pelleting of the sediment, is not ascribable to any particular bioturbating species.

2. Most bioturbating species are soft-bodied (e.g., worms), or at least poorly calcified (e.g., shrimps), and therefore stand little chance of being identified as fossils. Of the bioturbators on this tidal flat, only the molluscs are likely to be preserved as an identifiable skeletal record.

3. The total bioturbation rate is generally greater at or near the surface than at depth. This is due to the decrease in biomass with depth in the sediment and to the fact that most burrowing species burrow from the surface to a certain depth rather than mining a layer at a certain depth.

4. Because of (3), trails and shallow burrows are usually eliminated by the

activities of deeper burrowers, whose burrows are thus preferentially preserved.

5. In general, primary sedimentary structures are preserved when the sedimentation rate is greater than the bioturbation rate. In the ponds this occurs (rarely), when a storm layer is deposited which is of sufficiently different constituency to eliminate the usual bioturbators. The levee sediments illustrate the unusual case of a slow sedimentation rate of a few millimeters per year (Hardie and Ginsburg, Chapter 5, this volume) combined with a slow bioturbation rate caused by the scarcity of animals in that subenvironment: the primary laminations are still usually preserved, though sometimes only ghosts remain.

Pellets. Since Moore (1931) described fecal pellets as being common constituents of the Clyde Estuary muds, they have been recognized as common constituents of fine-grained marine sediments in many other areas (see Häntzschel et al. 1968), and many organisms are known to produce fecal pellets of sediment, though probably as many excrete an amorphous fecal sludge instead of or together with pellets. Not all "pellets" are fecal in origin however: some are balls of masticated but noningested sediment, others are rounded intraclasts, while others are micritized skeletal fragments.

At the lower tidal levels of this area (in ponds, channels, and nearshore) much of the surface sediment consists of freshly pelleted mud (Fig. 80A). The makers of the surface pellets are not difficult to find: they include *Batillaria*, which defecates 1 mm-long tapering rods (Kornicker & Purdy 1957); *Cerithidea*, which defecates 1 mm-long blunt-ended rods; *Marphysa*, which defecates 1 to 2 mm-long ellipsoid pellets (Fig. 80A) on the top of a characteristic pellet-mound which marks the end of its burrow; the nereid polychaetes, which defecate ½ mm-long ellipsoids in piles on the surface (Fig. 80A); and *Apseudes*, which produces tiny ill-defined pseudofeces.

Samples of sediment from several locations were wet-sieved in an attempt to obtain a quantitative estimate of the various shapes and sizes of pellets present. Unfortunately, however, the wet-sieving process, no matter how gently applied, broke up most of the larger pellets. Examination of the silt fractions of this sieved sediment did, however, reveal the presence of tiny clots of sediment, presumably pellets, in sizes down to 15 to 30 μ. These are possibly the fecal pellets of nematodes and the pseudofeces of *Apseudes*.

Thin-section study of pond sediments shows that near the surface the pellets are distinct because they are outlined by open pore space, but a few centimeters or more below the surface the rounded outlines are lost as the soft pellets are squeezed together by compaction (Fig. 75A). Then the whole sediment is a more or less nonporous mass of dirty brown or yellowish gray color, with some black spots of organic matter. Under crossed polarizers, a micro-granular or micro-acicular texture is apparent, the crystallites of which extinguish in clots or bundles. This clotted texture, though not directly attributable to particular pellets, should at least suggest that the sediment has been bioturbated and pelleted.

Pellets are preserved in a recognizable form under some circumstances. One of these is induration, a poorly understood process whereby a cement is precipitated within the pellet, making it hard enough to withstand compaction and/or transport. Indurated fecal pellets are reported from the sediments of large areas of the Bahama Banks (Illing 1954; Cloud 1962; Purdy 1963), and induration of cerithid fecal pellets in Bimini lagoon preserved them in a recognizable state some 22 cm below the surface (Kornicker & Purdy 1957). But induration of fecal material is not a process characteristic of these tidal flat sediments.

Some animals defecate pellets which are initially more cohesive than others; *Callianassa* pellets are partially indurated before excretion by a phosphatic compound (Weimer & Hoyt 1964, p. 763), while *Marphysa* pellets are probably more cohesive because of their high mucus content. Both *Callianassa* and *Marphysa* pellets survive the wet-sieving process that destroys most other pellets, and both

Fig. 82A. Beach-mound sediment with burrows mostly of the polychaete *Dasybranchus* (large burrows and black worm). Burrows which connect with the surface are lined with a tan-colored "oxidized" zone.

have been recognized in cores and thin sections (Shinn 1968, text Fig. 8: and this book, Fig. 75B).

Without induration or strong cohesion, pellets may still be preserved in a recognizable form if they are defecated within burrows in already compacted sediments. This case is common for polychaete burrows of the ponds and channels (Fig. 78B) and is especially characteristic of earthworm and ant burrows which are excavated in the compact and relatively dry levee and beach-ridge sediments (Fig. 36B). This again illustrates the generalization that structures (in this case pellets) made deep within the sediments stand more chance of being preserved than those made near the surface, because the rate of change in sediment texture (due to bioturbation, compaction, etc.) decreases with depth.

The special case of peneroplid pellets is discussed under the Skeletal Record above.

Burrows and Related Structures. The following burrow types are most characteristic of this tidal flat:

1. *Polychaete feeding burrows* (Fodinichnia [Seilacher 1964]), made by *Marphysa sanguinea* (family Eunicidae), *Dasybranchus lumbricoides?* (family Capitellidae), *Perinereis* cf. *anderssoni*, *?Ceratonereis tridentata* (family Nereidae), and others.

The above worms are all common in pond, channel-bank, and bar subenvironments below Exposure Index 40% (occasionally up to Exposure Index 60%).

Their burrows are cylindrical holes of more or less constant diameter, 0.4 to 3.0 mm, with a subvertical or random orientation, straight or sinuous, unbranched or branching with Y-shaped junctions. There is no ornament or special lining to the tubes, though the adjacent sediment may be oxidized to a paler or a reddish color (Figs. 78B, 82).

The method of burrowing is described above in the section on Bioturbation. The burrows of several polychaetes from several families are grouped together

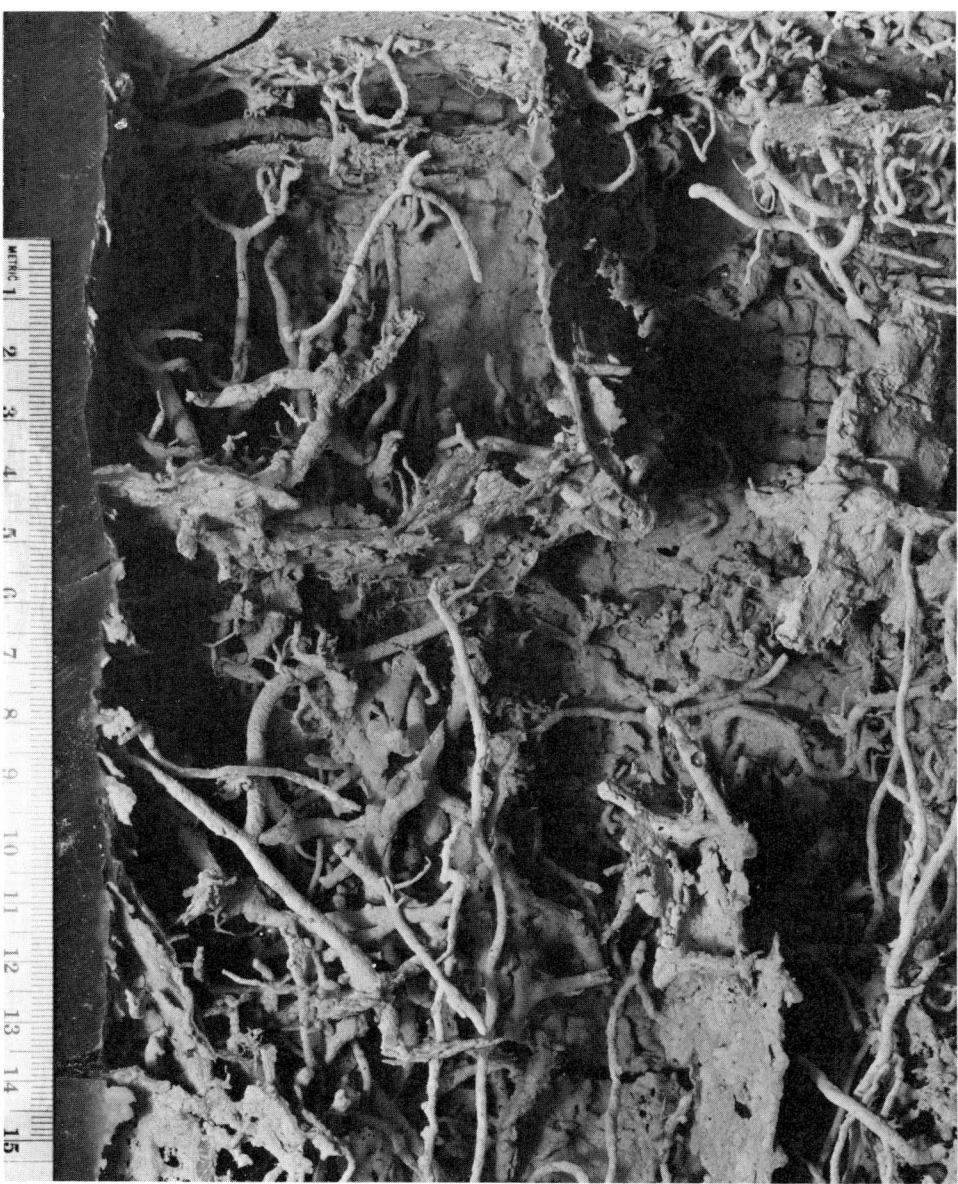

Fig. 82B. Polychaete burrows displayed as resin casts in beach-mound sediment. (The wet sediment was impregnated with polyester resin under pressure, then washed free of sediment to expose the resin-filled burrows.) Burrows display random orientation, branching, and uniform diameter. Most burrows are of *Dasybranchus*, though nereid burrows are common near the top. (Techniques and photo courtesy of Dr. R. N. Ginsburg.)

here because of the difficulty of distinguishing between burrows (vacated by their makers) in the field, and especially in sediment-core sections.

Most burrows are preserved as open tubes, though some are loosely backfilled with fecal pellets (Fig. 81).

Similar burrows are common wherever deposit-feeding errant polychaetes are common, in environments from shallow water to the deep sea. However, preservation of their burrows is favored by a stiff sediment texture such as characterizes these tidal flat muds, at least below 5 cm.

2. *U-shaped shrimp burrows* (Domichnia, ?Fodinichnia). Made by *Apseudes* sp., a tanaid shrimp, which usually occurs in colonies of several hundred individuals per square meter in some ponds and on accreting channel banks in soft mud: Exposure Index 0 to 45%.

The burrows are cylindrical tubes, approximately 1 mm in diameter, extending 0.5 to 2.5 cm below the sediment surface and curved into a U-shape with slightly contorted vertical shafts, with a *spreite* 2 to 3 mm wide (Fig. 83A).

Fig. 83A.
Apseudes burrows in pond sediment. These tiny filter-feeding shrimp (one in burrow just above 2.8 on scale) build smooth-walled, contorted, U-shaped tubes which are rarely preserved in cores because of the softness of the sediment.

Fig. 83B.
Panopeus burrow entrance in channel bank, showing flattened shape and coarse excavated material. The burrow extends into the sediment some 50 cm and has a similar flattened oval shape with unlined roughhewn walls. *Callinectes* builds similar burrows for shelter. Nearby burrow entrances are to an *Alpheus* burrow. Scale bar 30 cm.

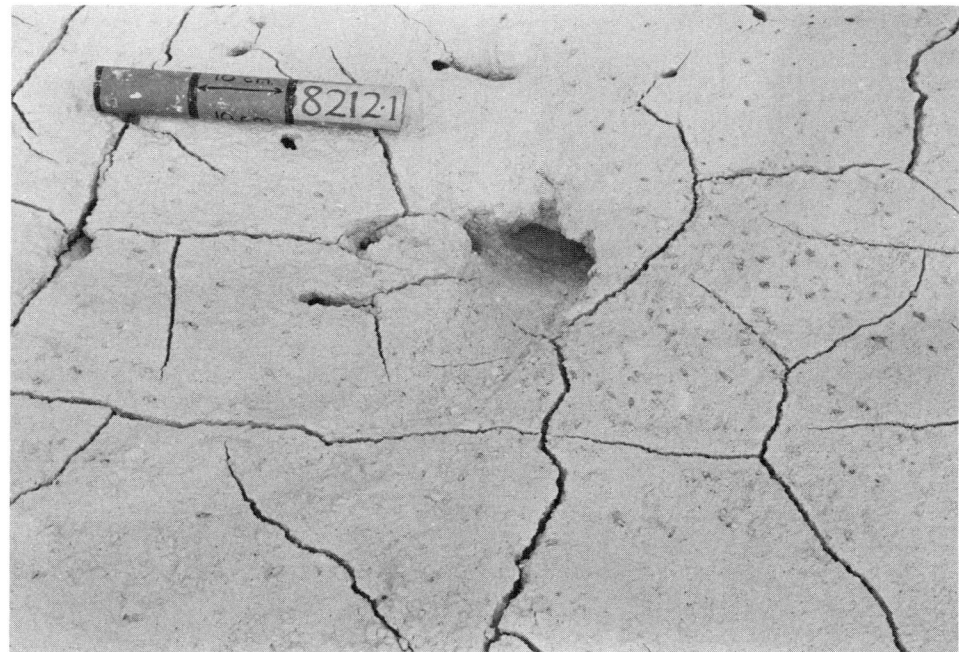

These burrows were not recognized in any core section, and their chances of preservation must be low because of the lack of cohesion of the mud in which they are constructed.

3. *Shrimp burrow complexes* (Domichnia and Fodinichnia). Two types are present on these tidal flats, those made by *Callianassa* sp., a thallassinoid shrimp, and those of *Alpheus heterochelis*, a caridid shrimp.

Callianassa burrows have been well described by Shinn (1968). They rarely occur on the tidal flats except in the channels, though they are common offshore. Where their burrows are preserved, the characteristic muddy or silty lining (Shinn 1968; Weimer & Hoyt 1964) is not in evidence. It may be that these

Fig. 84A.
Group of holes marking subsurface burrow of *Alpheus*, the snapping shrimp, in an abandoned channel. Excavated material surrounds the actively used burrow exits, but an old burrow has partly collapsed, forming a hollow in the background.

Fig. 84B.
Oblique view of resin cast of *Alpheus* burrows, showing interconnecting tunnels with smooth floors and rough-hewn roofs, which cut across polychaete burrows. Width of burrow complex: 1 m.

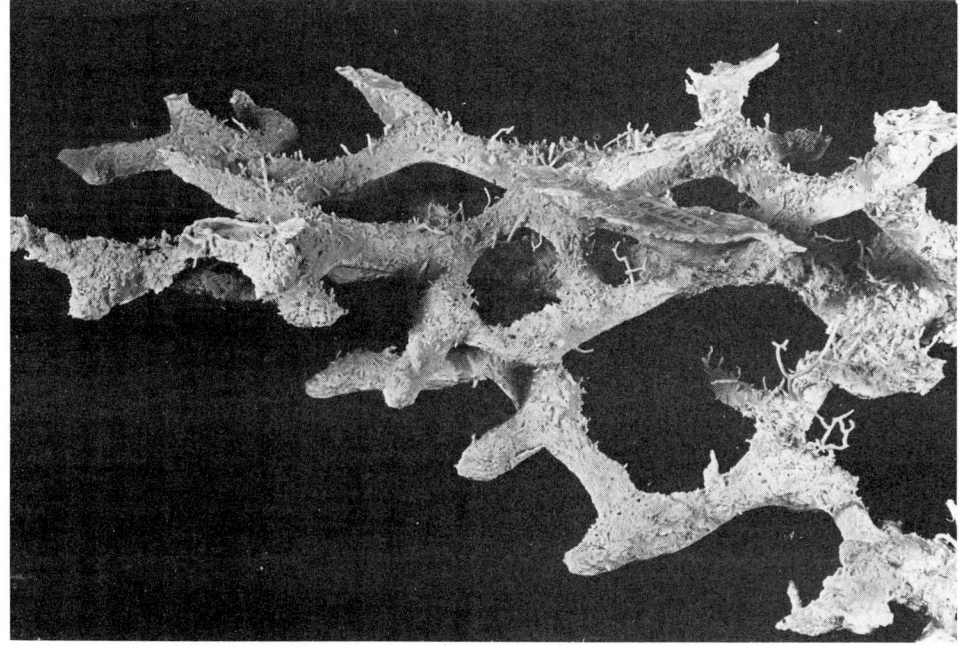

sediments are either sufficiently cohesive and smooth-textured to render a lining unnecessary, or are so similar in texture to the lining that the latter is rendered invisible.

Alpheus burrows, also described by Shinn (1968), are common on accreting channel bars, in abandoned channels, and in some ponds, especially near channel mouths. Burrow entrances (Fig. 84A) are usually between Exposure Indexes 10 and 50%, though at least some portion of the burrow network is constructed below low water so as to retain water for respiration.

The burrow consists of an anastomosing network of tunnels which vary in cross-sectional outline from circular to oval and in diameter from 18 to 50 mm. Most tunnels are subhorizontal, though those that connect with the surface and those that extend to lower levels may dip at angles of up to 45°. Many of the tunnels are dead-ends, but where two or more tunnels meet there is a slightly enlarged "room."

Shinn (1968) demonstrated how *Alpheus* burrowed in aquaria, using its claws to dig and its swimmerets to wash away loosened material. Plastic casts (Fig. 84B) of burrows made in Andros tidal flat sediments show features that demonstrate that *Alpheus* uses similar techniques in natural situations. The burrow roofs and walls show claw marks clearly, while the floors are relatively smooth, without claw marks. Sediment cores show the floors to be underlain by faintly layered sediment.

Questions arise as to why *Alpheus* should construct such a large burrow network, and why the network has a general horizontal orientation. The following observations suggest an answer: (1) no crustacean fecal pellets have been noted in association with *Alpheus* burrows—*Alpheus* is therefore not a deposit-feeder or a suspension-feeder; (2) *Alpheus* burrows are most common in areas of high polychaete density; and (3) most polychaete burrows are subvertical and are therefore intersected by *Alpheus* burrows. It is thus suggested that *Alpheus* is a carnivore which feeds on burrowing polychaetes. Its mode of life is then similar to that of the familiar terrestrial mole.

Because *Alpheus* burrows collapse easily, and no back-filled burrows were detected in our cores, only the layered burrow floors are likely to be preserved.

4. *Earthworm feeding burrows* (Fodinichnia). Made by *Pontodrilus bermudensis* (an oligochaete), which lives beneath surfaces of Exposure Index 90 to 98%, at depths of 2 to 6 cm, in levee and beach-ridge sediments.

The burrows consist of cylindrical or, less commonly, cavernous holes, the cylindrical holes being approximately 2 mm in diameter. Most burrows are subhorizontal and parallel to sedimentary laminations which are "mined" by Pontodrilus (Fig. 85A).

Burrows are well preserved in core sections either as empty holes or, more usually, loosely packed with rough ellipsoidal fecal pellets (Fig. 36B).

5. *Ant dwelling burrows* (Domichnia). Made by two species of terrestrial ants (unidentified), which excavate networks of tubular passageways (Fig. 85B) beneath surfaces of Exposure Index 97 to 100% in levee (rarely) and beach-ridge sediments.

The passageways are usually lined or backfilled with tiny (100 μ) mastication pellets, excess pellets being piled at the surface around the burrow entrance. The ants probably do not feed on the sediment, but use their burrows as a base of operations while foraging on the surface and as a shelter for the breeding pair and their young.

6. *Crab shelter burrows* (Domichnia). At low tidal levels, the swimming crab *Callinectes sapidus* (Portunidae) and mud crab *Panopeus herbsti* (Xanthidae) excavate low-angle burrows, some 10 to 15 cm wide and 5 cm tall, by picking at the walls with their claws and removing sediment to a talus heap at the burrow entrance (Fig. 83B). These burrows are used as resting places during moults and between scavenging expeditions. None have been recognized in cores, and it is probable that such a large unlined burrow would collapse on burial and compaction, leaving only a layered talus floor preserved.

On levees and beach ridges, fiddler crabs *Uca* sp. (Ocypodidae), marsh crabs *Sesarma* spp. (Ocypodidae), and the land crab *Cardisoma guanhumi* (Gecarcinidae) all excavate similar burrows. *Uca* and *Sesarma* generally occur in colonies on the levee backslopes or channel banks at Exposure Index 65 to 90%, while *Cardisoma* colonizes elevated marsh and beach-ridge areas at Exposure Index 95 to 100%.

The burrows (cf. Shinn 1968, and Fig. 86) are cylindrical holes with a rough claw-sculptured wall. The diameter (10 to 30 mm for *Uca* and *Sesarma*, 50 to 150

Fig. 85A.
Earthworm (*Pontrodrilus*) exposed within its burrow a few centimeters below the levee surface. Its ovoid fecal pellets, which surround it, lie in one plane, showing that it has "mined" a lamination. Scale in mm.

Fig. 85B.
Tunnels of ant burrow in levee sediments. Notice that the burrows are lined with tiny mastication pellets. Scale in mm.

mm for *Cardisoma*) is variable, even along one burrow. Orientation is also variable, though burrows extending downward from a horizontal surface are usually subvertical, with a slight helical twist, while burrows extending into channel banks are subhorizontal. All burrows extend down to tap the water table at least at high tide, usually at a depth of 20 to 35 cm.

All three crabs are surface scavengers, using their burrows only for rest and protection. The sediment is dug out with the claws and masticated into round lumps which are deposited on the surface near the burrow entrances. The burrows, being excavated in compact dry sediment, stand a good chance of being preserved, either empty or filled with poorly laminated sediment or leaves washed in by rain or fetched in by the crabs (Fig. 87A).

Fig. 86A. Territory of a fiddler crab (*Uca*). Around the burrow entrance are piles of masticated balls of excavated material, and scratch marks where the crab has fed on the surface algal mat. Small cylindrical fecal pellets are scattered around. (There is a small pellet mound of *Marphysa* immediately to the left of the crab burrow.)

Fig. 86B. Resin cast of *Uca* and *Sesarma* burrows, showing subvertical shaft and claw-sculptured walls.

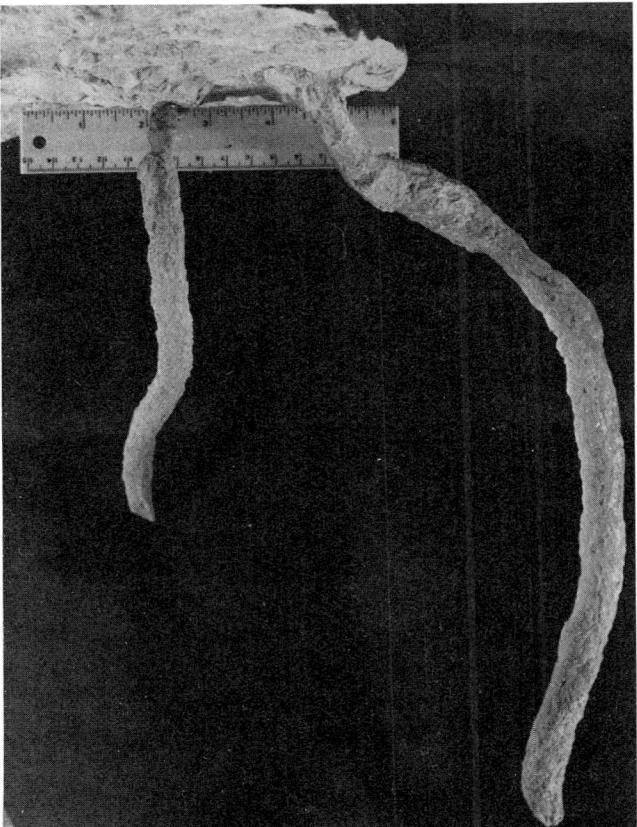

Other burrow types include:

a) emergency burial pits (Cubichnia) made by young *Callinectes*, which when in apparent danger dig rapidly backward until the whole body is buried up to the eyes. Such pits collapse when the crab reemerges and thus stand a poor chance of preservation.

b) Domichnia of the bivalve *Geloina*, a suspension feeder, which lives in a vertical position just beneath the surface. Preservation of *Geloina* in life position is possible only if its orientation is not disturbed after its death by the activities of other burrowers.

Fig. 87A. Vertical core section of levee-marsh sediments with *Uca* and *Sesarma* burrow. The burrow, which has a slight helical twist, is partially infilled near its base with laminated detritus. The sediment is poorly laminated, except near the top, and shows many root holes, some with color haloes. Scale in mm.

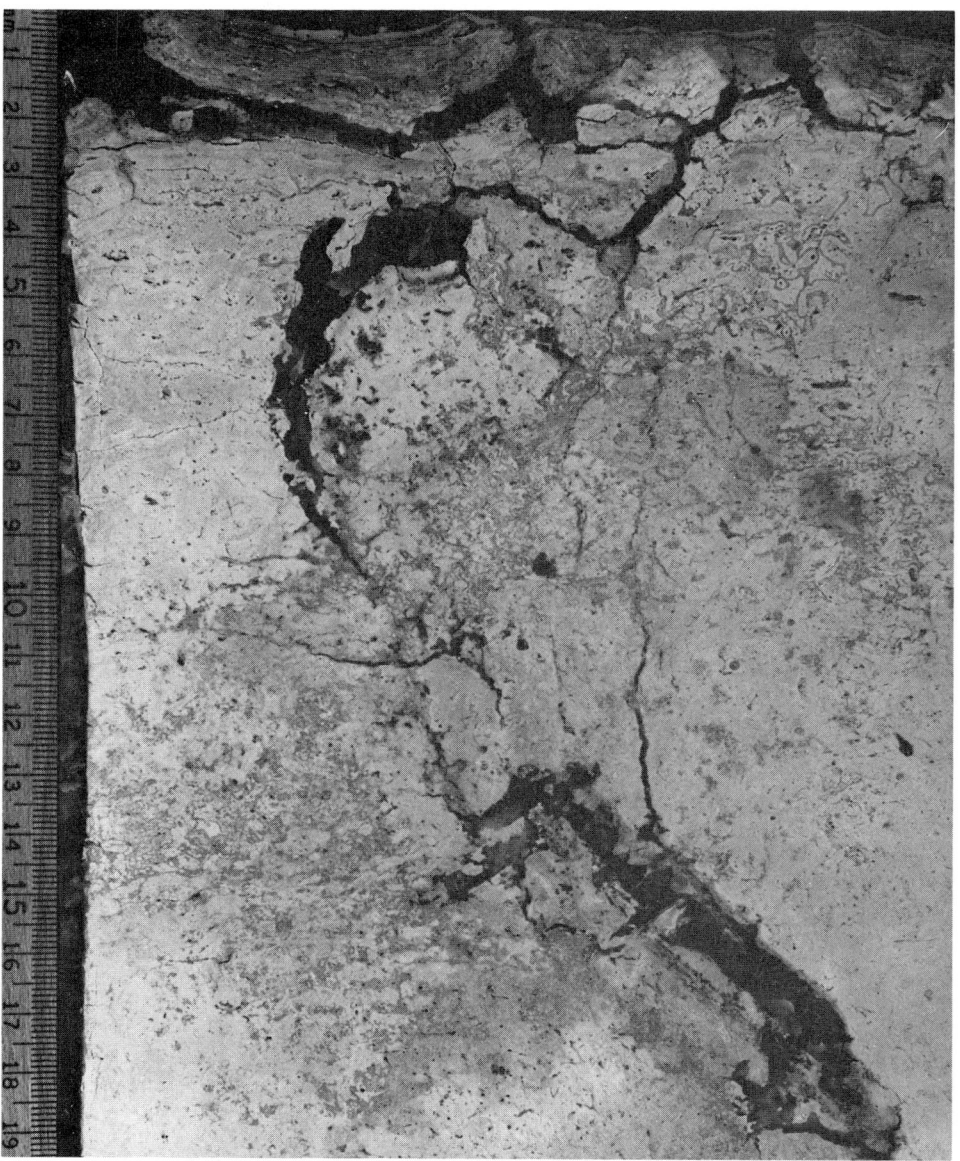

The following surface traces are also present:

1. *Gastropod grazing trails* (Pascichnia), made by the cerithids *Cerithidea* and *Batillaria* which graze apparently randomly over the surface, though only rarely turning so much as to cut back across the same track.

2. *Polychaete trails* (?Pascichnia, ?Repichnia). The nereid polychaetes emerge from their burrows at night and crawl over the surface for distances of a few centimeters, only to disappear down another burrow. Such feeding and/or locomotion trails, together with the gastropod grazing trails, are extremely unlikely to be preserved, because further surface and subsurface bioturbation continually destroys them.

3. *Fish feeding pits* (Pascichnia). The bonefish *Elops*, when it feeds, sucks up a mouthful of sediment, leaving a shallow pit 1 cm deep and some 2 to 3 cm in diameter (Fig. 87B). These pits could be preserved if made in cohesive sediment and if filled with sediment of a different texture, but in most instances their record will be erased by later burrowing.

4. *Bird footprints* (Repichnia). Flocks of several different types of birds, mostly migrants, stop over to feed in the ponds, leaving foot and beak marks,

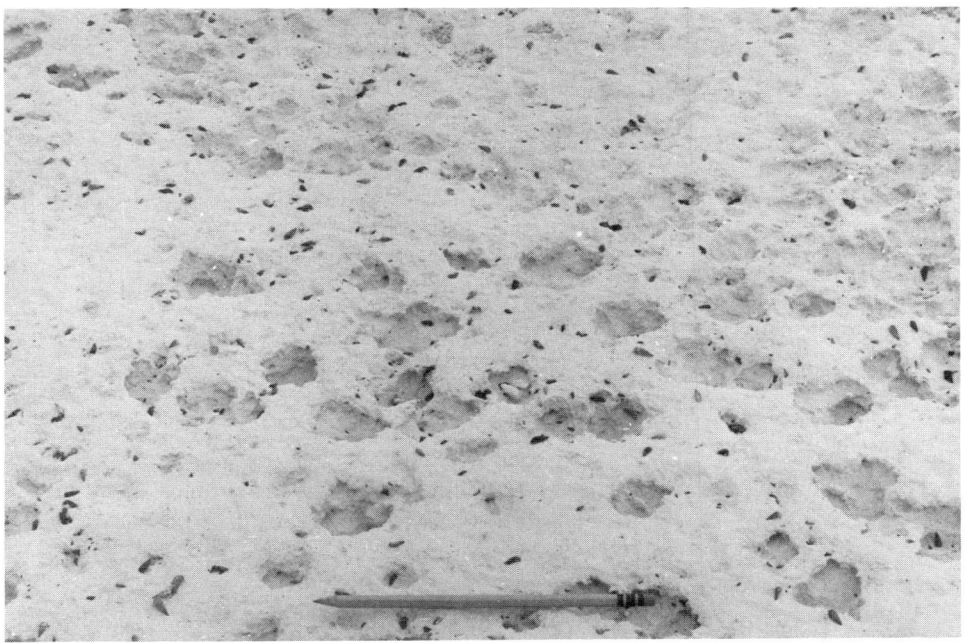

Fig. 87B. Bonefish feeding pits on the beach. These fish suck up mouthfuls of surface sediment which they grind between a specially adapted armored tongue and upper palate.

which however are soon erased by the grazing trails of gastropods. The levee, on which most resident birds live, has too hard a surface for footprints to be formed, except immediately after deposition of a muddy storm layer (Fig. 12B).

Flamingoes, not now present on this section of the west Andros shore, make relatively gigantic sedimentary structures during both feeding and nest-building activities (Allen 1956).

5. *Crab grazing scratches* (Pascichnia). Both *Uca* and *Sesarma* sometimes graze on surface algal mats around their burrow entrances, leaving distinctive scratch marks (Fig. 86A). Of all the surface traces these stand perhaps the best chance of preservation because levee sediments are less likely to be burrowed, and because it is possible to split the sediments along the lamination planes.

Burrows—General Points. Conditions for the preservation of burrows are generally similar to those under which pellets are preserved, namely:

1) If the burrow has an indurated or specially cohesive wall. The only burrow in this category is that of *Callianassa*, whose wall is generally constructed of finer and more cohesive material than the surrounding sediment (cf. Shinn 1968, Plate 110, Figs. 4 & 5), and is impregnated with a phosphatic compound (Weimer & Hoyt 1964).

2) If the burrow is constructed in cohesive, well-compacted sediment. The levee burrows of ants, oligochaetes, and crabs definitely fall into this category and can survive the compaction of at least a man's weight. Less definitely preservable are the burrows of *Marphysa* and *Dasybranchus* which anastomose in the deeper sediments of ponds, channel bars, and beach mounds. Though they are well preserved even in roughly handled cores (Fig. 78B), it is possible that further compaction by overburden pressure would squeeze some of them out of existence. Even less likely to be preserved are the burrows of *Alpheus*, *Callinectes*, and *Panopeus*, which though usually constructed in cohesive sediment are very likely to collapse because of their relatively large diameter. It is not surprising then that none of these crustacean burrow holes were recognized in core sections, though the laminated floors of *Alpheus* burrows (Shinn 1968, text Figs. 4B & 5) are sometimes faintly visible. *Apseudes* constructs its burrow in very soupy surface muds, and therefore all traces are lost on compaction.

3) If the burrow becomes infilled with distinctly different sediments than those surrounding the burrow. This type of preservation is possible for all types of burrows, but because infilling is likely to take some time, either during occupation or after vacation of the burrow by its occupant, it is not usually a factor in the preservation of burrows other than those in the previous two categories. Thus, *Callianassa* burrows are frequently filled with coarse sediment (Shinn 1968, Plate 110, Fig. 5); *Cardisoma* and *Uca* burrows are partially or completely filled with surface detritus (Shinn 1968, text Fig. 12); ant and earthworm burrows are usually filled respectively with mastication and fecal pellets (Fig. 36B); and the burrows of *Marphysa* and *Dasybranchus* are sometimes filled with fecal pellets (Fig. 78B).

It is a paradox that in the zones where burrows are least important (on the levees and beach ridges) they are most likely to be preserved *in toto*, whereas many of the abundant burrows in pond and channel sediments are lost to the record because later burrows destroy them, or because the bioturbation process decreases sediment cohesion, which in turn favors the erasure of burrows by compaction. The same applies to surface traces, most of which are made in the most bioturbated zones.

The chief types of traces present clearly do not conform to any simple pattern, including as they do all the ethological types (fodinichnia, domichnia, pascichnia, cubichnia, and repichnia) in patterns that penetrate deeply or only shallowly into the substrate. This complex situation contrasts somewhat with the conclusions drawn by Seilacher (1964) and Rhoads (1967) that intertidal burrowers tend to produce deep vertical burrows for shelter, and the reason may be that both authors were considering sandy substrates where frequent movement of the substrate has two important consequences: (1) it acts as an extra stress factor which would tend to select against shallow burrowing and surficial benthos, and (2) it does not favor the accumulation of detritus and the growth of infloral algae, which are the chief food sources of surface and near-surface deposit feeders.

It is also significant that the assemblage of trace-fossils preserved need not reflect anything like the total amount of sediment-modifying activity that takes place in any given environment. What are preferentially preserved are those activities which modified the sediment last, i.e., burrowing destroys surface traces, and deep burrowers destroy the record of shallow burrowers.

Plant roots. The role of vegetation (especially mangroves and grasses) as sediment-trappers (cf. Scoffin 1970) is not of major importance on these tidal flats, because of the generally sparse cover these plants afford the sediment surface. Phrased another way, the physiography and primary sedimentation would probably differ only slightly if all plants, with the exception of the algae, were absent from the tidal flats.

Beneath the surface, however, plant roots form sedimentary structures which are the major record of the existence of plants. However, only the red mangrove (*Rhizophora mangle*) colonizes wide tracts of the tidal flats *and* has a distinctive and extensive root system (Davis 1940; Hoffmeister & Multer 1965).

The *Rhizophora* leaf crown is supported above sediment level by aerial prop roots which curve out from the crown, entering the sediment at many points. Within the sediment the roots branch frequently and at all angles downward to depths of 1 to 2 m and produce a plethora of tiny root hairs each one only a few hundred microns in diameter. Thus sediment cores with mangrove roots show several cross, longitudinal, and oblique sections of large root stocks (up to 5 cm in diameter) and an abundance of root hairs at all angles from vertical to horizontal, but mostly straight not curved.

After death, the inner tissues of the root (vascular tissues and the spongy aerenchyma) decay, leaving several concentric layers of dark woody periderm around a cylindrical space which partly or completely fills with sediment. The method of filling is not always clear, but where the filling consists of loosely packed ellipsoidal pellets it is difficult to avoid the conclusion that burrowing polychaetes (*Marphysa* in particular) are responsible for the filling and may be responsible for hollowing out the inner tissues. The root hairs decay leaving a mass of tiny tubules.

It is usually easy to differentiate a *Rhizophora* root stock from a crustacean burrow of about the same diameter by the presence of a periderm around the sediment infilling. Root hairs can (with difficulty) be distinguished from small polychaete burrows by their straightness and their tendency to cluster around and branch off from the root stocks. In general, roots may be distinguished from animal burrows by their tendency to decrease markedly in diameter at a branching node.

In levee sediments, roots are less conspicuous. Grass roots are the commonest, with their horizontal runners and downward-branching nodal roots. The runners cut through laminations without disturbing them at all, suggesting that they make space for themselves by dissolving the immediately adjacent sediment with an acid secretion.

At the highest levels on the beach ridge, where plant growth is most diverse and dense and periods of sedimentation are few, the full effect of plants on sediments can be seen. As the plants grow, their roots lever up the sediment above them. Dead and rotting leaves and wood provide food for most of the soil animals and for such burrowers as the land crab *Cardisoma* (Shinn 1968, p. 887), and the inclusion of this organic material in the soil gives it a light brown color. In fact, the beach-ridge sediment becomes a sort of earthy "top soil," with no original laminations remaining.

5. Summary and Conclusions

1. Three major communities are recognized, based on the spatial association of included organisms. They correspond closely with the main physiographic divisions outlined by Hardie and Garrett (Chapter 4). They are:

a) The Nearshore Community, an entirely subtidal, semirestricted marine community;

b) the Pond Community, extending from 0 to 65% Exposure Index, which is a low-diversity, restricted marine community, including blue-green algae, foraminifera, desposit-feeding polychaetes, molluscs (especially grazing cerithid gastropods), and various crustaceans;

c) the Levee Community, extending from 75 to 100% Exposure Index, which is a low-diversity, salt-tolerant terrestrial community. It includes mangroves, certain grasses and other halophytic angiosperms, but few animals (the earthworm, crabs, and insects).

Gradational subcommunities colonize large areas, including the algal marsh (a levee-pond gradation with very low diversity), the channels (a pond-nearshore gradation) and the beach ridge (a levee-true terrestrial gradation).

2. Skeletal material derived from codiacean algae, foraminifera, ostracods, and molluscs provides a skeletal record in the following ways:

a) Clay-size needles of aragonite constituting the bulk of the sediment, were at least in part derived from the offshore codiacean algae *Penicillus* and *Rhipocephalus*. Yet it is at present impossible to distinguish their skeletal spicules from inorganically precipitated aragonite needles.

b) Tests of small foraminifera and ostracods are present in approximately equal numbers in sediments from all environments, including supratidal environments. Thus it appears that they were mostly carried on from offshore with the rest of the tidal flat sediment.

c) *Peneroplis*, a large foram, is notable for producing numerous pellet-like particles which may be at least in part skeletal.

d) Two cerithid gastropods and one bivalve are the most abundant molluscs. All three live in the ponds, and may be preserved there in random distribution or as storm-winnowed lag deposits. Erosion of pond sediments on the beach or along channel banks leads to concentration of their shells into cross-bedded coquinas.

3. Some *in situ* activities of organisms are preserved as a trace record as follows:

a) Algal borings in shells preserve casts of the blue-green algal filaments or cause disintegration of the shell.

b) Broken shells in fine sediments probably indicate the presence of heavily armed predators and scavengers, such as crabs.

c) Algal mats, which are present on virtually every sediment surface, are grazed and burrowed at low tidal levels. Only at high tidal levels in levees and marsh sediments are they preserved. *Schizothrix*-type mats are preserved as thin, smooth sediment laminae perforated by very fine filament molds. *Scytonema*-type mats may be preserved as wavy alternations of sediment and mat laminae and/or as filament molds outlined by precipitated Mg-calcite rhombs, which under certain circumstances cement into a lithified crust.

d) Bioturbation is a process characteristic of the lower tidal levels (ponds, channels, and nearshore) where animals are most abundant. In the ponds, gastropods, polychaetes, and shrimps are mainly responsible for a rapid turnover rate of once per month for surface sediments, and once per 700 years for sediments at 30 cm depth. Such rapid rates destroy almost all primary sedimentary structures. In levee sediments, primary laminations are partially destroyed by the burrowing and pelleting of the earthworm and ants. Because there are no animals inhabiting the inland algal marsh, the layered sediments remain undisturbed.

e) Pellets, produced mostly by deposit-feeding animals such as gastropods and annelids, are very abundant in surface sediments of the ponds, channels, and offshore. Unless they are specially cohesive or become indurated, however, they usually squash together and become indistinguishable on compaction.

f) Burrows include a diverse assemblage of vertical and random tubes of deposit-feeding annelids, and vertical shaft, U-shaped or galleried tunnels of scavenging or suspension-feeding crustaceans. The probability of their preservation depends on such factors as the presence or absence of a specially strengthened burrow wall, the cohesiveness of the burrowed sediment, the diameter of the burrow, and the nature of infilling processes. Most surface traces are destroyed by later burrowing.

g) Plant roots generally produce an insignificant trace record, though red mangrove roots are very prominent, especially in pond and marsh sediments. Their large roots fill with sediment after death and may resemble infilled crustacean burrows, while their root hairs resemble tiny straight polychaete burrows.

4. Ancient analogs of the Andros tidal flat deposits would be characterized by three distinct "biofacies" (the overall record of the three communities):

1) Offshore—bioturbated, pelleted sandy mud, with burrows and diverse skeletal record;

2) Pond—bioturbated pelleted or massive muds, with many polychaete burrows and a low-diversity skeletal fauna;

3) Levee—algal laminated sandy and muddy sediments, with conspicuous burrows (if present) and transported skeletal material.

Marsh, channel, and beach-ridge sediments are also distinctive.

7 ALGAL STRUCTURES IN CEMENTED CRUSTS AND THEIR ENVIRONMENTAL SIGNIFICANCE

Lawrence A. Hardie

1. Introduction

One of the most interesting features of modern carbonate tidal flats is the development of hard, cemented "crusts" on or near the surface of soft, unlithified Holocene lime sediments. These hard surface layers, *generally only a few centimeters thick*, occur just above the normal high water level in Andros Island, Bahamas (Shinn et al. 1965, 1969), the Florida Keys (Shinn 1968), Bonaire, Netherlands Antilles (Deffeyes et al. 1965; Lucia 1968), and over the whole tidal range in the Qatar Peninsula, Persian Gulf (Taylor & Illing 1969; Shinn 1969), and Shark Bay, Western Australia (Brian Logan, personal communication). Most early interest centered on the crusts because many of them were found to contain protodolomite or very high magnesian calcite. Current interest in these crusts, particularly the subtidal crusts, has turned toward what we stand to learn from them about the processes of lithification and early diagenesis. Of equal significance, however, is that these crusts provide an opportunity to see just what features diagnostic of the depositional environment are "frozen in" to the sedimentary record before burial. In this latter regard, the cemented crusts of the tidal flats of northwest Andros Island, Bahamas, were found to be particularly revealing. In many of these crusts are preserved delicate filament molds and mat structures of the blue-green alga *Scytonema* that thrives only in a very small elevation range (about 15 cm) in the high water tidal zone and in freshwater supratidal marshes (Chapter 4; Black 1933; Monty 1967, 1972). This preservation is a rather remarkable phenomenon when one considers that despite presumed ubiquity of blue-green algae in the seas for over two billion years (that is, the record of stromatolites), fossilized blue-green algae are in fact relatively rare in the geologic record; Johnson (1961, p. 194), for example, recognizes only *Girvanella*, *Ottonosia*, and *Samphospongia* as fossil genera of filamentous blue-green algae (although some workers believe that *Girvanella* is a green alga). The modern *Scytonema* crusts of the Andros Island tidal flats are all the more remarkable because they are forming in "specialized" environmental niches and so not only represent a record of the soft parts of algae but also a very sensitive environment indicator and water-level gauge. Of special interest are the crusts formed by direct precipitation of calcite around *Scytonema* filaments, what will be referred to as *algal tufa*. On Andros Island these algal tufas are found only in freshwater-dominated marshes.

In Chapter 5 Hardie and Ginsburg described the main petrographic features of the crusts (lamination with "palisade" structure). The present chapter describes the distribution of the crusts and summarizes the algal structures and the cement textures they carry. Because it is mainly the filament and mat forms of *Scytonema* that are preserved in the crusts, the factors influencing the distribution of *Scytonema*-dominated mats are examined. Finally, the origin and envi-

Table 13. Basic features of the cemented crusts of Three Creeks and Pumpion Cay areas

Crust type	Thickness	Characteristic structures	Mineralogy
I. *Crusts with internal structure*			
A–i) pure algal tufa with fibrous palisade structure	1–3 cm	laminated mushroom and mound-shaped heads made of calcite-encrusted radiating *Scytonema* filament molds; very porous (vertical fenestrae between filament molds)	Mg-calcite (12–13 mole % $MgCO_3$)
A–ii) sediment-rich algal tufa with fibrous palisade structure	1–3 cm	laminated knobs, mounds and sheets made of unlithified peloid laminae alternating with laminae of radiating calcite-encrusted *Scytonema* filament molds smothered by peloids; very porous (mainly vertical fenestrae between molds).	peloidal sediment made of aragonite needle mud (silt and sand-sized peloids); filament molds made of Mg-calcite (12–13 mole % $MgCO_3$)
A–iii) sediment-rich fibrous palisade-structured crusts without algal tufa	1–3 cm	laminated knobs and mounds made of lithified peloidal sediment with upright tubular fenestrae (filament molds); sheet cracks and other fenestrae partially filled with sparry aragonite cement.	peloidal sediment made of aragonite (silt and sand-sized peloids); Mg-calcite micrite (11–20 mole % $MgCO_3$) mud not distinguishable from cement; very high Mg-calcite micrite cement (35–43 mole % $MgCO_3$); sparry aragonite cement.
B) crusts with columnar palisade structure	1–5 cm	flat sheets and domal knobs with stout vertical palisade structure made of sediment filled columns set in a laminated peloidal sediment; in 3 dimensions the palisade structure is actually a continuous wall joined in a rough honeycomb pattern; sheet-cracks and other fenestrae partially filled with sparry aragonite cement.	as for IA–iii
C) compound crusts	3–8 cm	type IA–iii encrusting type IB crusts	as for IA–iii and IB
II. *Crusts with no internal structure*			
A) paper-thin sheet crusts	1 mm	flat films and sheets of mud	aragonite
B) crusts of bioturbated sediment	1/2–2 cm	flat sheets of churned peloidal mud with scattered gastropod (including land-snails) shells	as for IA–iii

ronmental significance of the crusts are discussed. It is hoped that this information will help to throw some light on the depositional environments of *fossil* blue-green algae as well as the processes responsible for lithifying them.

2. Types of Crusts and Their Distribution

Hardie and Ginsburg in Chapter 5 found it petrographically convenient to place the internally layered cemented crusts into two subtypes of lamination with "palisade" structure: (i) those with fibrous structure, (ii) those with columnar structure. To bring out the full environmental significance of the crusts, it is best here to use a more elaborate subdivision of crust types we found on the northwest Andros tidal flats, as follows:

I. Crusts with internal structure.
   A. Crusts with fibrous palisade structure.
      i) pure algal tufa crusts and "heads."
      ii) sediment-rich algal tufa crusts.
      iii) sediment-rich crusts without algal tufa.
   B. Crusts with columnar palisade structure.
   C. Compound crusts of type A–iii encrusting type B.
II. Crusts with no internal structure.
   A. Paper-thin sheet crusts.
   B. Crusts of bioturbated sediment.

The petrographic features of most of these crust types have been described in

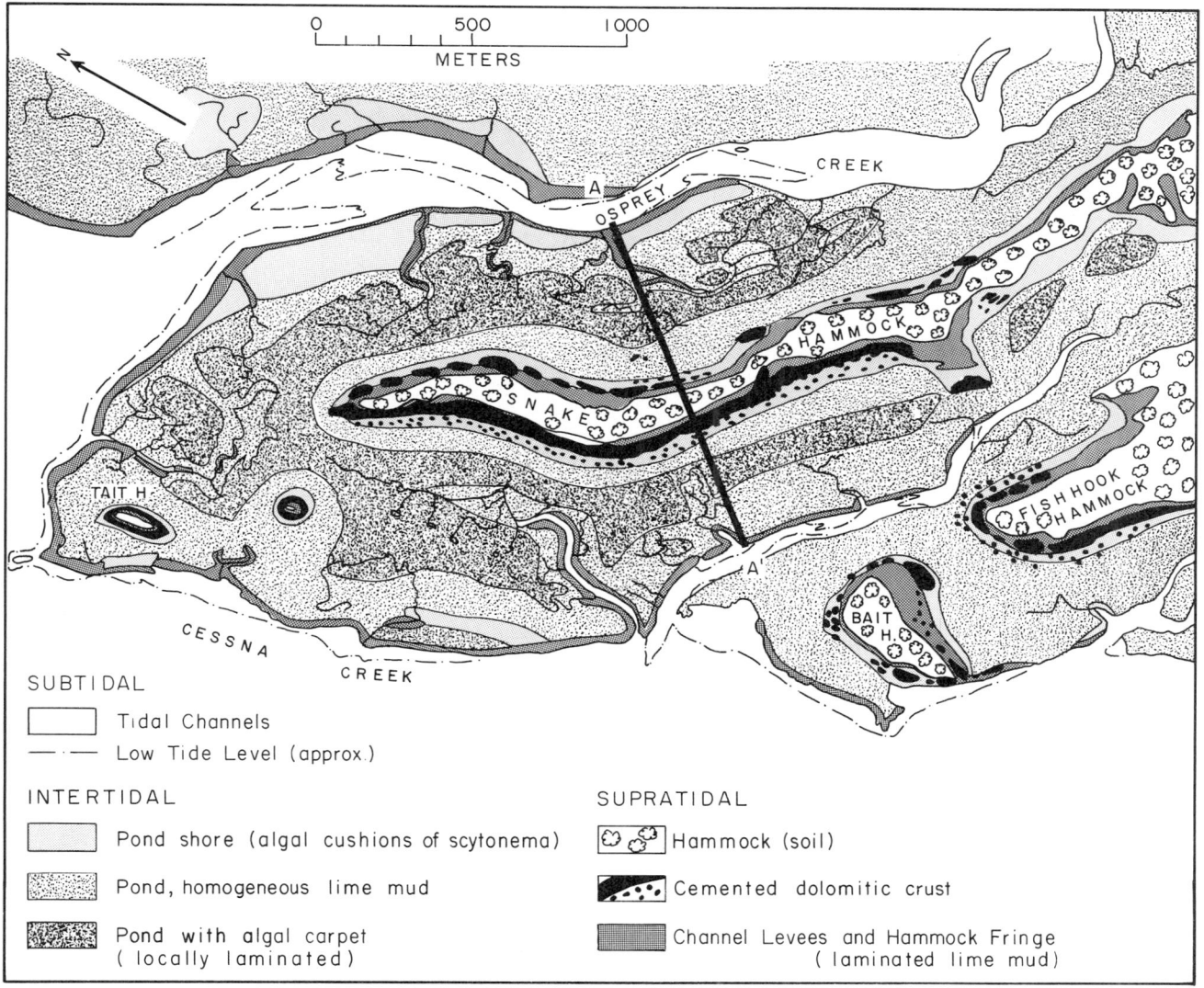

Fig. 88. Map of Pumpion Cay area near Williams Island (Fig. 1), showing the distribution of surface-cemented crusts with respect to the depositional subenvironments. A cross-section along line A–A' is shown in Fig. 8.

Chapter 5, but for convenience the major characteristic structures of the crusts are briefly summarized in Table 13. The special algal features we are focussing on in this paper will be further described below.

The distribution of surface crusts has been carefully mapped in two small areas of northwest Andros Island, one at Three Creeks (Figs. 6B & 10, see Fig. 1 for location) and one at Pumpion Cay near Williams Island (Fig. 88, see Fig. 1 for location). These crusts were found to cover 1.5% and 3% respectively of the tidal flat surface above mean tide level. In both these areas cemented crusts were also found in the subsurface as thin (up to 8 cms), hard, discrete layers buried within soft, unlithified lime sediment.

As the maps Figures 6B and 10 show, the *surface crusts* at Three Creeks occur almost exclusively on the back sides of levees, on the back sides of beach ridges, and on small isolated high crowns along the western (seaward) edge of the inland algal marsh. The surface crusts are specifically absent from the crests of levees and beach ridges and from the beach, offshore, pond, channel, and channel-bar surfaces. In the Pumpion Cay area near Williams Island, the surface crusts are found as fringes around "palm hammocks" (Fig. 88 & Shinn et al. 1965). At Three

Creeks, far back in the inland marsh near the pinelands, crust fringes occur around rare, small, tree-covered mounds (Pleistocene bedrock highs only a few tens of meters across) very much like the crusts around the palm hammocks in the Pumpion Cay area. Both at Three Creeks and Pumpion Cay the surface crusts, with rare exceptions, occur within the upper part of the high *Scytonema* algal marsh zone fringing the ponds (Figs. 6B & 88) slightly above the mean high water mark. This is true even of the inland algal marsh: across the wide expanse of this marsh, surface crust is found only on the few slightly higher drier parts, where *Scytonema* grows in discontinuous low knobby mats rather than in the thick, lush mats typical of most of the inland marsh.

The distribution of individual types of surface crust, particularly the algal tufa crusts, is quite selective. Algal tufa crusts (types IA–i and IA–ii) are found only in the freshwater algal marsh: type IA–ii occurs along the western edge of the seaward high crowns, where the surface is directly exposed to rising sediment-charged seawater that floods the adjacent ponds and channels during very severe onshore storms and hurricanes; type IA–i occurs on the eastern (landward) sides of the seaward high crowns and in high spots in the marsh interior where sediment is deposited only during hurricanes. Characteristically associated with these algal tufa crusts are the paper-thin aragonite crusts of type IIA. These paper-thin crusts occur as hard but fragile sheets no more than 1 mm thick, as brittle as thin ice, stretching between the isolated *Scytonema* tufa mounds and heads (Fig. 33). They are easily peeled from the underlying soft, unlithified, peloidal sediment layer. Not present in the inland marsh are surface crusts of types IA–iii, IB, IC, and IIB, instead, these crusts are typical of the seawater-dominated channeled belts of the Three Creeks and Pumpion Cay areas. Types IB and C are by far the most abundant in these areas and commonly carry very high magnesian calcite, which in some samples shows rudimentary ordering like that of protodolomite (Shinn et al. 1965, pp. 1218–20). Crusts of type IIB were found in only one place at Three Creeks, along the edge of a pond behind Point Simon (see Fig. 2 for location) as thin plates studded with gastropod shells (*Batillaria* and *Cerithidea* and many tests of the land-snail *Cerion*). Such crusts are much more common at Pumpion Cay, but are still far subordinate to the type I crusts. They appear to represent exposed pond edge sediment. In a few places, structured crusts of types IB and C have a weakly lithified basal zone made of bioturbated sediment. In these samples it is clear that the zone of cementation has included a little of the underlying churned sediment with the thin surface cap of layered sediment. Paper-thin crusts of type IIA do occur in the channeled-belt areas, but can only be recognized in thin section where they are seen as dense film-like caps on thin sediment laminae, especially in levee backslope and upper high algal marsh boundary zone sediment (see, for example, Fig. 42A and descriptions in Chapter 5). In the field the presence of these paper crusts is presumably masked by *Schizothrix* surface films and by the coherence of the dry surface sediment.

Subsurface crusts are common, even abundant in places, at both Three Creeks and Pumpion Cay. In the channeled belt at Three Creeks, crusts of types IB and IC are found beneath levees and beach ridges and beneath the sediment of the upper pond. For example, at locality 8274 (Fig. 2) trenching in the beach terrace showed the following sequence from top to bottom: 20 cm soft peloidal mud—5 cm crust—7 cm peloidal mud—2 cm crust—30 cm peloidal mud—0.5 cm crust—30 + cm peloidal mud. Erosion of the beach cliff at this locality has exposed the upper crust bed and has left a lag gravel of large flat plates of crust on the beach (Chapter 5). Similarly, along the south bank of Palmasola Creek (see Fig. 2 for location) erosion has exposed a bed of type IC crust, rich in very high Mg-calcite and protodolomite, beneath 20 cm of soft, laminated levee-crest sediment. Slow cannibalistic migration of the channel has added large protodolomite-bearing crust plates (flat pebbles) to the channel and has allowed laminated levee sedi-

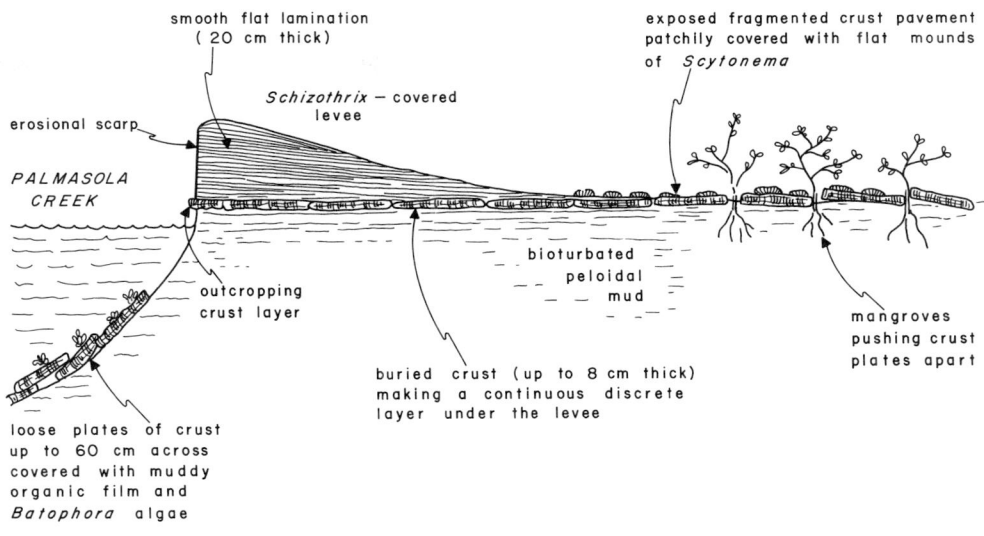

Fig. 89.
Schematic cross-section of channel-levee complex on the south bank of Palmasola Creek (Fig. 2), showing three important aspects of the distribution of proto-dolomite-bearing crusts: (1) exposed crust pavement behind the levee; upgrowing mangroves and their roots have fragmented the crust into plates and pushed them askew, exposing the soft unlithified underlying peloidal mud; (2) buried crust layer, continuous with exposed pavement, covered by smooth, flat, laminated sediment of levee crest: (3) large loose plates of crust eroded out of the channel bank and slumped down into the channel as the channel migrates by lateral erosion.

ment containing sand and silt-sized protodolomite crust intraclasts to prograde over the crust bed (Fig. 89).

In the inland algal marsh, buried paper-thin crusts (type IIA) are a characteristic feature of all sediment cores we collected. They occur as thin caps on top of hurricane layers (Fig. 52B), and because of their fragility and brittleness they are the main source of the flat intraclast flakes dispersed in the inland marsh sediment. At the seaward edge of the marsh (locality 8055, Fig. 2), coring and trenching revealed at least five hard buried crusts of types IB and C in 1.7 meters of sediment. One other relevant aspect of the inland marsh is that as much as 60% of the sediment column beneath the marsh consists of lightly to heavily calcified *Scytonema* mats (Chapter 5). While these friable, spongy algal tufa layers are not hard, indurated crusts they do represent a well-preserved record of algal filament structures and are therefore included here.

In the Pumpion Cay region, subsurface crusts of types IB, IC, and IIB are found in almost all areas surrounding the palm hammocks well into the ponds and even cropping out in the low beach cliff. In several places at Pumpion Cay, as many as three "horizons" of crust were found in 1 m of Holocene sediment, for example, at the eastern edge of Snake Hammock (Fig. 88) along the pond shore, crusts separated by soft peloidal mud were located at 10, 30, and 65 cm below the surface. It is in the Pumpion Cay area that Shinn et al. (1965) showed that the crusts traced beneath pond sediment increase in age away from the palm hammocks.

The significance of the buried crusts will be discussed below.

3. Algal Structures in the Crusts

It is the mat and filament structures of *Scytonema* that are preserved in the cemented crusts (both the surface and buried crusts). Hardie and Ginsburg described these features in Chapter 5, but it is worthwhile here to reemphasize the main points and, where necessary, to elaborate on their descriptions.

In the field the most striking feature of all the crusts of type I is their similarity in surface morphology to living *Scytonema* mats. At the upper boundary of the high algal marshes of the channeled belt and on the high crowns of the inland algal marsh, *Scytonema* grows as patchy discontinuous mats consisting of low, flat topped mounds and sheets and as small mushroom-shaped knobs (Figs. 17B &

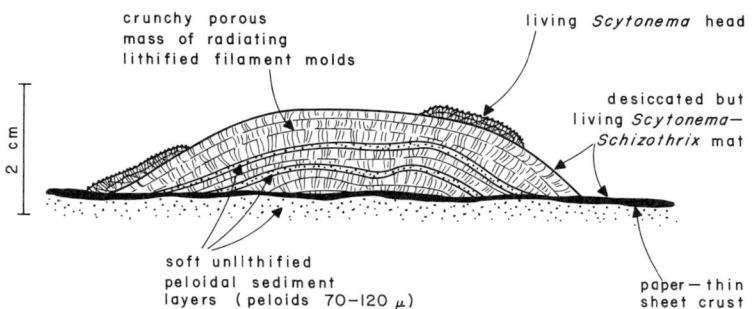

Fig. 90. Schematic cross-sectional sketch to show the typical spatial relationships of cemented algal tufa layers, living *Scytonema* mounds, and peloidal sediment layers in crusts of the upper high marsh zone of the channeled belt.

33). Commonly, these mounds and knobs are located directly over widely spaced polygonal cracks in the underlying sediment substratum (Fig. 33B). It is precisely this mound and knob morphology that characterizes the surface of type I crusts, as comparison of Figure 17B with Fig. 47A shows so beautifully. These figures give a good idea of the typical shapes and sizes of the mounds and knobs in both the *Scytonema* mats and the crusts. Not only is there a morphological similarity between the crust and algal mat surfaces, but very commonly the mounds and knobs of the crusts are preferred sites for new growth of living *Scytonema* patches. The sketch in Figure 90 shows schematically the typical structure of the mounds and knobs of the crusts and the relations among the crusts, living *Scytonema* and *Schizothrix* mats, and unlithified sediment.

Internally it is the fibrous palisade structure (Figs. 46B & 49A) that preserves the most delicate record of *Scytonema* (Chapter 5). In thin section the fibrous structures of type IA–i and IA–ii crusts are seen to be crystalline carbonate envelopes (up to 30 μ in wall thickness and composed of 12 to 13 mole % $MgCO_3$ magnesian calcite) around thick (up to 40 μ) upright *Scytonema* filaments, or more commonly are simply empty crystalline tubular envelopes that are perfect molds of upright *Scytonema* filaments (Figs. 48 & 49B; see also Monty 1967, plates 5 & 6). No trichome or cell structure of the filaments are preserved, the preservation is simply entombment, with calcification occurring on the mucilage sheath (as observation of very lightly calcified *Scytonema* from the inland marsh showed). This algal tufa structure is restricted to crusts and buried *Scytonema* mats from the inland freshwater-dominated marsh. The crusts range from small mushroom-shaped heads of pure algal tufa made of radiating, upright, filament molds (Figs. 46B & 48) to flat, laminated mounds made of filament tufa smothered by unlithified detrital peloids and shell fragments (Figs. 49A & B). The buried lightly calcified *Scytonema*-mat layers of the inland marsh sediment column consist mainly of a flattened and twisted mass of Mg-calcite tubular filament mold tufa with isolated upright heads preserved in some layers; they have been described in Chapter 5 under "thin bedding and thick lamination" (see particularly, Figs. 52 & 53). They represent a most important record of algal filament structures.

Filament molds that are preserved in lithified detrital sediment and not encased in precipitated calcite tufa envelopes are typical of the type IA–iii and IC crusts of the channeled belt. In these crusts the filament molds are 30 to 50 μ wide and up to 1.5 mm long and make a tangled twisting mass of essentially upright tubular fenestrae in a tightly packed peloidal mud matrix (Fig. 50B).

The characteristic internal feature of crusts of type IB is the stout columnar palisade structure (Fig. 46A) that is organized into a remarkable honeycomb pattern (Fig. 47B). As described in Chapter 5, these columns appear to be sediment-filled "cracks" that most likely represent casts of tufts of *Scytonema*. The features of these honeycombed palisades that support a *Scytonema* precursor, rather than some kind of desiccation-crack, origin are as follows (Chapter 5): (1) columns are very closely spaced (usually < 5 mm), of irregular cross-sectional shape, are divided by twisting fenestrae and not uncommonly coalescent; (2) living *Scytonema* mats have a decided honeycomb pattern to the filament tufts (Fig. 51A), and where covered with sediment the tufts in cross-section are remarkably similar in morphology to the columns (Fig. 51B); (3) the columnar structure is common in *Scytonema*-type domal mounds in crusts of type IB (Figs. 46A & 47A).

4. Cementation Textures in the Crusts

There are three basic types of cements that lithify the crusts:

1) *Scytonema* filament envelopes. In the algal tufa crusts (types IA–i and IA–ii), blocky micrite crystals (1 to 3 μ) of calcite (bulk X-ray analysis gives 12 to 13 mole % $MgCO_3$ range among crusts examined) envelop the filaments and meld together where neighboring filaments approach or touch one another (Fig. 48). This produces a rigid but lightly cemented, very porous crust that is easily crushed between the fingers. In places the outermost crystals of the envelopes approach a microspar (5 to 7 μ) in texture. Under the binocular microscope, on freshly broken surfaces the inside walls of filament molds as well as the walls of some of the cavities between molds appear extremely smooth, almost like dissolution surfaces. It is significant that in the sediment-inundated tufas the peloid sediment laminae (Figs. 49A & 90) are either not cemented or at most very lightly cemented; these peloid layers can be easily washed out with a weak jet of water. It is only the filaments and filament bundles that appear to be firmly calcified by micrite envelopes.

2) Interstitial micrite cement. In crusts of types IA–iii, IB, IC, and IIB the silt and sand peloid sediment appears to be held together by a dense pervasive micrite cement. Under the microscope this cement is difficult to tell from detrital mud and deformed soft peloids, but the overall hardness of the crust is persuasive evidence of cementation by micrite. Also, X-ray analysis shows large amounts (up to 60%) of high magnesian calcite (usually two phases, one about 16 to 20 mole % and one 35 to 43 mole % $MgCO_3$) not present in unlithified sediment with the same peloidal framework texture. Preliminary electron microprobe traverses and Mg-Ca maps show that the magnesium in these crusts is mainly confined to the matrix between peloids and to the outermost edges of peloids.

3) Sparry cavity linings. Relatively large fenestrae (millimeter-scale animal burrows, filament and mat molds, primary air pockets, etc.) and vuggy sheet-cracks in crusts of types IA–iii, IB, IC, and IIB are partly filled with acicular to bladed aragonite crystals up to 50 μ long and 20 μ wide, growing into the cavities from the walls (Fig. 91; see also Shinn et al. 1965, Fig. 7B). In cases where vertical cavities pinch to form a narrow neck, this sparry aragonite cement fills the constriction as a meniscus cement. In some long, horizontal sheet-cracks, the cement forms a thin, fine-grained floor-coating and is absent from the roof (Fig. 92). These types of cement are typically vadose cements.

An enigma are the paper-crusts of type IIA. Under the microscope they are seen as a fine micrite, but unlike the micrite cements of (2) above, X-ray analysis shows only aragonite.

In looking at the micrite cements of all the Andros crusts the limitations of the petrographic microscope become very obvious. So the brief descriptions above must be regarded as preliminary. Currently SEM and microprobe studies of these crusts are under way and hopefully will yield more penetrating and reliable observations.

5. Factors Influencing the Distribution of Scytonema-dominated Algal Mats

Scytonema-dominated algal mats are very selectively distributed on the northwest Andros Island tidal flats, and so their preservation in the cemented crusts produces a remarkable "built-in" environment indicator and water-level gauge. For the environmental significance of the structured crusts to be used to its fullest, it is necessary to know what controls the distribution of *Scytonema* mats.

Black (1933) first pointed out the zonation of algae on Andros Island's interior

Fig. 91. S.E.M. photo of aragonite wall cement partly filling vug in cemented crust from Three Creeks. Note the radial growth and the acicular as well as *bladed* habit of the aragonite. 1500 × magnification; the euhedral crystal in upper left corner is about 30 μ long.

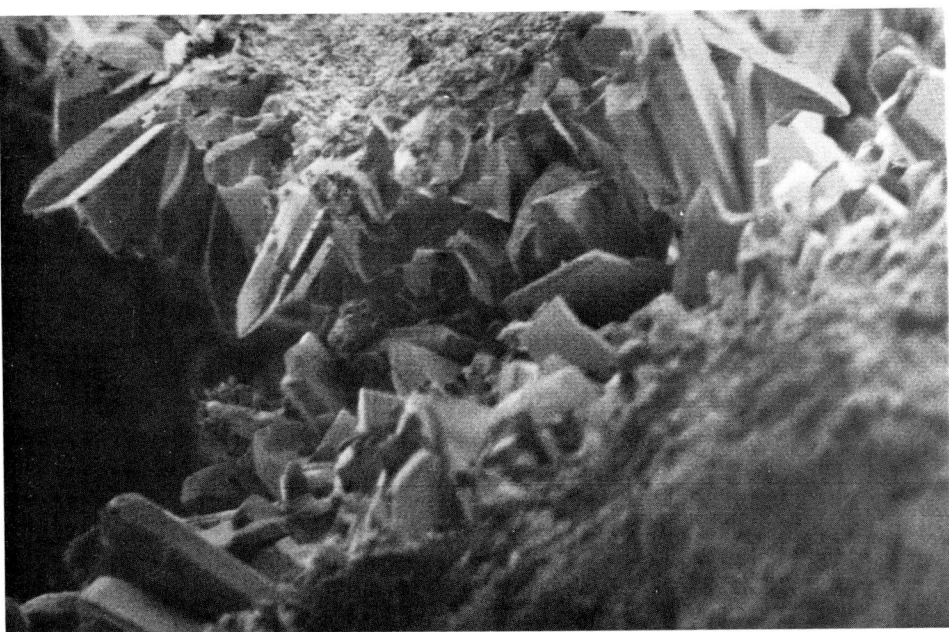

Fig. 92. Thin section photomicrograph of cemented crust from Three Creeks, showing aragonite wall-cement lining floors of flat elongate vugs (*bottom half of photo*). Note that cement is not present on the ceilings of the cavities. *Upper half of photo* shows *Scytonema* filament molds pictured in Fig. 50B. Scale bar is 500 μ.

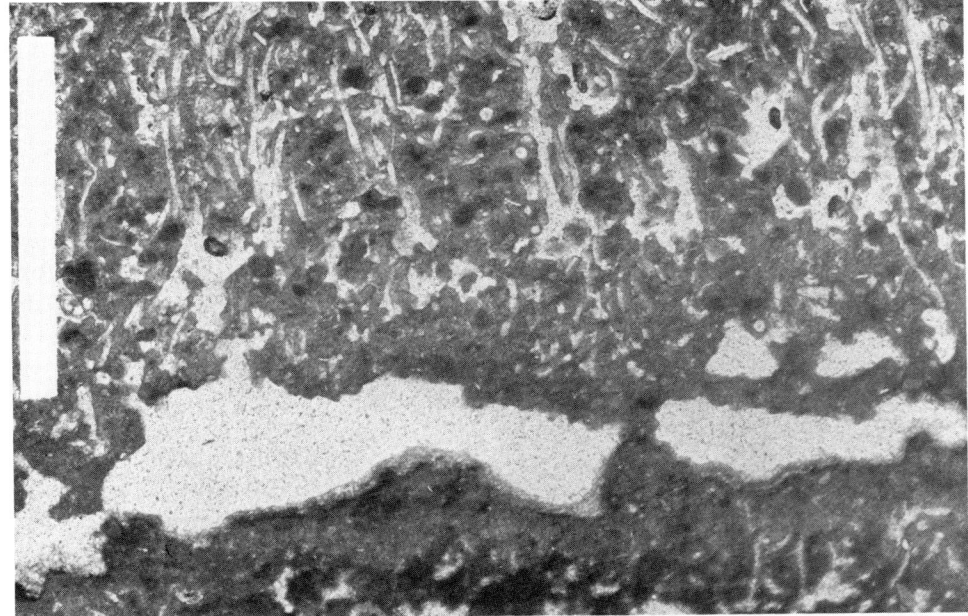

mudflats. He recognized the following distribution: subtidal—*Udotea, Penicillus, Halimeda*; upper intertidal—*Phormidium, Symploca*, and unicellular forms; supratidal (above mean high water)—*Scytonema, Plectonema, Schizothrix*, and unicellular forms; on limestone outcrops (freshwater)—*Scytonema* alone. Black noted that "with *Scytonema* . . . salinity has a strong controlling influence. Very little *Scytonema* was found in places liable to be washed by undiluted seawater. Where the salt content of the water is appreciably less than one part per thousand, flourishing colonies of pure *Scytonema* were found" (p. 178). Black recognized two species of *Scytonema*, *S. androsense* and *S. crustaceum*, but found no preferential specie distribution, rather they always coexisted. Monty (1967, 1972), on the other hand, working on the east side of Andros, found *S. crustaceum* characteristic of intertidal settings and *S. myochrous* restricted to freshwater environments. However, as Monty points out, (1967, p. 67), the difference between *S. crustaceum* and *S. myochrous* is simply one of filament thickness, the thicker *S. myochrous* being thought to be better protected against long desiccation periods. At Three Creeks I could find no other differences than filament thickness between *Scytonema* at the intertidal pond edges (typically 15 to 20 μ in filament diameter) and *Scytonema* from the interior of the inland marsh (typically 30 to 40 μ in filament diameter): the "string of beads" trichome, the plump heterocysts, the false branching, the deep yellowish-brown pigmentation (in transmitted light), and the strongly diverging lamination of the sheaths are identical. Perhaps, along with workers such as Jaag (1945; quoted in Monty 1967, p. 66), these two differently sized *Scytonemas* should not be regarded as different species but simply as different environmental varieties of the same specie, e.g., *S. myochrous* var. *crustaceum* and perhaps even this distinction is academic, because at Three Creeks in the same mat, filament size and mat morphology changes within a few meters as Exposure Index changes, as noted below. Therefore, in the present work no special distinction is made and the genus name *Scytonema* only will be used.

At Three Creeks, *Scytonema*-dominated mats are restricted to the inland algal marsh and to narrow, sharply defined zones (that occupy a vertical elevation range of only about 15 cms) fringing the ponds in the channeled belt (Chapter 4, and Figs. 6B. 10, & 21). Not only that, but it is possible to subdivide these *Scytonema* zones into two separate subenvironments, the high algal marsh and low algal marsh (Figs. 10 & 21), based on mat morphologies and their relation to elevation (Figs. 9A & B). As described in Chapter 4, in the channeled belt the low algal marsh is characterized by a lush meadow of "pincushions" of *Scytonema* (Fig. 18A & B), while the high algal marsh is covered by a continuous flat to wrinkled tufted *Scytonema* carpet (Fig. 16A). At the upper edge of the high marsh zone, where it grades into the *Schizothrix*-dominated levee backslope subenvironment, the *Scytonema* mat is patchy and discontinuous, consisting of the low mounds and knobs so typical of the structured crusts (Fig. 17B). The lower boundary of the low algal marsh against the pond is very sharply marked by the abrupt dying out of the "pincushions" and is always found at the same absolute elevation (within $\pm$ 2 cm) for any given pond (and even from one pond to another), corresponding to an Exposure Index of 65% (Chapter 3), just above mean tide level (Fig. 9). The upper boundary of this low marsh with the high marsh is gradational but still rather well defined and quite consistent at an elevation ($\pm$ 3 cm) just about at mean high water, corresponding to an Exposure Index of about 85% (Fig. 9). It is the upper boundary of the high marsh zone that shows the most variation and least consistency in absolute elevation (and hence Exposure Index) in a single levee-pond system: this boundary when traced laterally may vary by as much as 10 to 15 cm in elevation (as measured with a Dumpy level) giving a variation of $\pm$ 4% about the mean Exposure Index of 95%.

The reasons for this restricted distribution of *Scytonema* at Three Creeks appear to be related mainly to the degree of submergence in seawater and to the frequency and extent of burial by sedimentation. Cropping by invertebrates was seriously considered to account for the sharp lower boundary, but in the field the gastropods *Batillaria* and *Cerithidea* were observed to graze on the *Anacystis-Schizothrix* scums at the pond edge and not on the large filamented cushions of *Scytonema* (manipulative experiments were not done to check this out and are clearly needed before we can entirely rule out cropping).

Doty (1946) has pointed out the relation between tide levels and zoning of marine algae in the rocky intertidal of the Pacific crust of Oregon and California. The significant factor here was that zones immediately above a particular tide level such as lowest higher low water (LHLW), were found to be subject to two- or three-fold increases in maximum single exposure periods. Such large increases

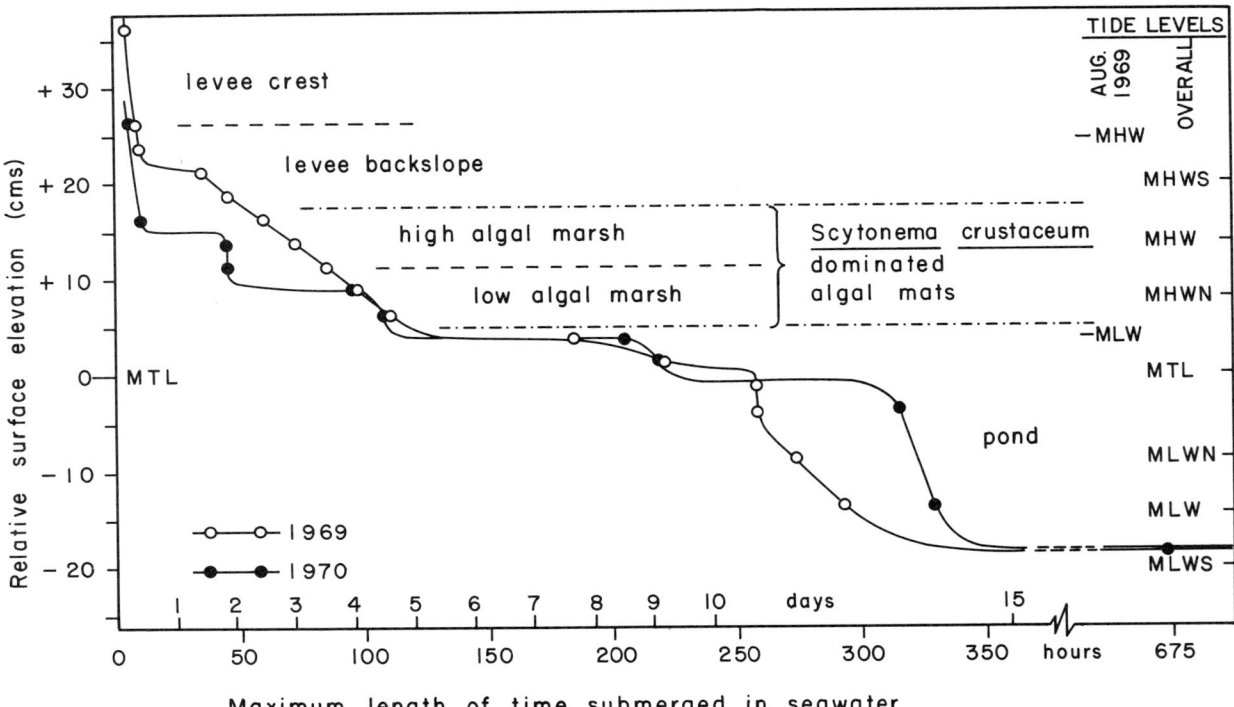

Fig. 93. Plot of relative surface elevation of a levee-pond complex against maximum length of time the surface was submerged in seawater. The data points, taken from continuous tide records, represent the *single* longest period of uninterrupted submergence each surface elevation was subjected to in the two periods June 1969–May 1970 and June 1970–May 1971. The figure also shows the relative elevations of the different subenvironments that make up this complex. At the right, the overall mean tide levels for the whole study period 1968 to 1971 are compared with the seasonal mean tide levels for August 1969, which was a typically rainy month.

in duration of exposure would be enough to kill the thalli of marine algae which thrive below LHLW. In this way, Doty accounts for the abrupt restrictions in vertical distribution of the intertidal algae. I would like to suggest that the same kind of thinking can be applied to the distribution of *Scytonema* on the Andros tidal flats, but *in reverse*. It is a matter of record that *Scytonema* has a wide tolerance to freshwater and to desiccation (e.g., Fig. 18B) and, in fact, is probably best considered as a terrestrial alga. But it is also clear that *Scytonema* endures submergence in seawater and even produces lush growths in the upper intertidal zones. So in applying Doty's kind of approach we would postulate that *Scytonema* has only a *limited tolerance* to seawater, and if submerged continuously for more than a critical length of time, the thalli are killed or their growth is seriously curtailed, so that aggressive colonization becomes impossible at the elevation levels frequently subjected to such harmfully long submergences. This can be tested by looking at the relations between critical tide factors (Doty 1946) and distribution of *Scytonema*-dominated mats. These relations are shown for a levee-pond environment at Three Creeks in Figure 93, in which surface elevation is plotted against maximum length of time the surface was submerged in seawater during a single event. Two curves are shown, one for the period June 1969–May 1970 (a rather wet twelve-month period = 188 cm rain) and one for the period June 1970–May 1971 (a dry twelve-month period = 72 cm rain), prepared from continuous tide records in the pond (Fig. 2). A number of steps or plateaus are sharply marked in these curves, but no truly consistent relation between the plateaus and annual mean tide factors such as MHW can be made. However, what is most remarkable is that on both curves the lower boundary of the low algal marsh, the lowest elevation at which *Scytonema*-dominated mats occur in the pond where the water levels were monitored, coincides with the large well-defined plateau between MHWN and MTL where submergence time *abruptly* doubles. It is very tempting to interpret this to mean that *Scytonema* can tolerate no more than 4 to 5 days of continuous submergence in seawater and that any new colonization even a centimeter or two below this critical elevation will be subject to at least one lethal (certainly harmful) dose (> 9 to 10 days) of seawater each year. This would certainly account for the sharpness and consistent elevation over wide areas of the lower boundary of the low algal marsh zone in the channeled belt. Quantitative laboratory studies on cultured *Scytonema* might be helpful in rigorously testing this idea.

The origin of the submergence plateaus in Figure 93 is related to seasonal summertime floodings rather than to the overall annual mean tide factors such as Doty (1946) found for the rocky intertidal of the Pacific coast. This is really not unexpected because, as was pointed out in Chapter 4, on a shallow tidal flat like Three Creeks the daily and seasonal weather play an overwhelming role in determining water levels. For example, with an offshore wind the spring tide high water can be lower than normal neap tide low water! So the *maximum* submergence times are found in the mid-summer months when tides reach higher than normal because of increased rainfall and seasonal low pressure as the Bermuda High moves north. At these times sustained onshore winds will increase water levels by as much as 20 cms and keep them at these levels for days at a time. The large central plateau that coincides with the lower limit of *Scytonema*-dominated mats, then, is a seasonal flood level and seems to represent the low-water mark of the normal summertime floods (Fig. 93, see MLW for August 1969).

The upper boundary of the high algal marsh of the channeled belt has a quite different significance. The greatest deterrent to *Scytonema* growth on the dry upper levels of levees and beach ridges appears to be burial by sediment. As explained in some detail in Chapter 5, severe onshore storms, as frequent as three times a year, stir up lime mud and peloids on the Great Bahama Bank and flood the Andros Island tidal flats with sediment-laden seawater. The sediment is dropped preferentially on the beach ridge and levee crests and backslopes, building these features up lamina by lamina at an average rate of about 1.5 mm/year. During these violent storms nothing but a blush of sediment reaches the algal marsh zones behind the levees and beach ridges, and no sediment whatsoever reaches the inland algal marsh interior (Table 5). It is only during infrequent hurricanes, which cause widespread flooding of the entire tidal flats about once every 4 to 5 years on the average, that these *Scytonema* subenvironments become buried in sediment. Black (1933, p. 178) noted that the Scytonemataceae are very slow growing and growth measurements made in the field in the present study confirm this. Using hematite pigment layers as time-markers, *Scytonema* was found to increase filament length by about 0.5 mm in one year. At this rate, *Scytonema* growth cannot keep up with the sedimentation rate of 1.5 mm/year on the levee and beach-ridge crests and backslopes. On the other hand, the highly motile Oscillatoriaceae such as *Schizothrix* and *Phormidium* can quickly colonize a newly deposited sediment lamina. Ginsburg et al. (1954, p. 27) found in a laboratory culture box that *Phormidium* can "grow" through as much as 4 mm of sediment and colonize as much as 50% of the surface in twenty-four hours. It is well-known that Oscillatoriaceae have the capacity to move rapidly (rates up to 5 μ/sec.; Burkholder 1934), and as Fritsch (1952, p. 801) explains, such movements "are always accompanied by secretion of mucilage and take place only when some part of the thread is in contact with a substratum." Thus when Oscillatoriaceae become covered with sediment their response is to quickly move up through the sediment, traveling within an envelope of mucilage. This mucilage envelope "is fixed to the substratum and does not move with the trichome (but) remains behind as the latter advances" (Fritsch 1952, p. 801). The Scytonemataceae, on the other hand, do not have this capacity to move rapidly.

Therefore, on the levee and beach-ridge crests and backslopes where sedimentation is relatively frequent (once every few months), it is not surprising that *Schizothrix* mats dominate and *Scytonema* filaments are minor. The reverse is true on the pond shores and in the inland marsh, where sediment is deposited only once every few years; here the large filaments of *Scytonema* easily outgrow the tiny *Schizothrix* threads over the long run. The frequent sedimentation on the levee and beach-ridge crests will not only account for the absence of *Scytonema*-dominated mats from these subenvironments but will also explain the variability in elevation of the upper boundary of the high marsh zone. This follows from the facts that not all parts of the levee or beach ridge receive sediment during a storm (this depends upon the sheltering effects of local topography, etc.) and that each storm spreads its sediment load over a different limit (this depends on the intensity and duration of the storm, see Table 5). Therefore, the position of the levee backslope—high algal marsh boundary varies somewhat in both space and time. After some storms, sediment will reach into the high marsh, burying the *Scytonema* and allowing *Schizothrix* to colonize the surface, while at times when sedimentation on the backslope is infrequent the *Scytonema* mat of the high marsh might encroach onto the levee backslope zone. The record of this variability is clearly seen in the laminated sediment beneath the lower backslope and upper high marsh zones (see Chapter 5 under "Disrupted flat lamination" and "Crinkled fenestral lamination"; see also Fig. 40).

In summary, then, colonization by *Scytonema* in the seawater-dominated channeled belt is difficult on the levee and beach-ridge crests, levee and beach-ridge backslopes, the beach terrace and channel banks, and intertidal channel bars because of frequent sedimentation, and is impossible in the ponds, channels, beach, and offshore because of seawater "poisoning." This leaves only a narrow zone on the protected back sides of levees and beach ridges conducive to healthy *Scytonema* growth. The inland algal marsh, on the other hand, is protected from both frequent sediment and seawater inundation by its supratidal elevation and inland position, so *Scytonema* thrives over a vast area. Finally, it should be pointed out once again that within any *Scytonema*-dominated zone the morphology of the mat changes with changes in frequency of wetting. This is seen in the progressive decrease in mat thickness and continuity as Exposure Index increases, both in the channeled belt (compare the lush "pincushions" of the low marsh, the thinner, flat continuous mat of the high marsh, and the patchy growth of the high marsh upper boundary) and in the inland marsh. This relationship holds despite the abundance of rainfall, so that in addition to frequency of wetting a critical factor must be *duration* of an individual wetting. This is beautifully demonstrated in the inland marsh: as Figure 33A shows, where *Scytonema* is growing in shallow depressions that collect and hold rainwater for long periods, the mat forms large lush "pincushions," but on the immediately adjacent, higher, drier areas the mats are patchy and thinner. It is in these higher, drier areas of patchy *Scytonema* mats of both the channeled belt and inland marsh that the cemented crusts are found.

6. Origin of the Crusts: Some Preliminary Guesses

This is a difficult and complex problem that we will not be able to answer satisfactorily until we have far more precise petrographic, geochemical, and hydrologic data on the crusts and their micro-environments as well as more fundamental data on the kinetics of nucleation, growth, and dissolution of alkaline earth carbonate minerals at surface temperatures and pressures. Here I can only try to define the problems a little more sharply by framing the constraints set by the field data and to make some very preliminary guesses at the geochemical processes involved.

Three basic questions come to mind immediately: (1) What causes cementation to occur only in certain subenvironments and not in others? (2) Why is the precipitated cement aragonite in some cases, low magnesian calcite in others, and high magnesian calcite or even proto-dolomite in still others? (3) Why does algal tufa form only in the freshwater-dominated inland marsh and not in the seawater-dominated channeled belt?

In partial answer to the first question, we know empirically from the field relations that the necessary settings for the development of the crusts are: (i) a normally dry supratidal environment; (ii) a stable surface free from sediment influx and erosive scour for long periods of time (several years at a stretch). At the present stage in the evolution of the northwest Andros Island tidal flats only a small part (< 10%) of the complex meets these requirements, but if infilling of the ponds and abandoned channels continue then a vast area will become covered by the surface crusts (see discussion in Chapter 5). One further relevant point is that it is no accident that *Scytonema* is intimately associated with the crusts, because, as discussed in the preceding section, *Scytonema* also requires an environment free from sedimentation for long periods. The actual geochemical processes of cementation within these particular crust environments are difficult to resolve with any certainty. At the outset a clear distinction must be made between the algal tufas of the inland marsh and the cemented sediment crusts of the channeled belt. The algal tufas appear to have formed by preferential precipitation around individual algal filaments, whereas the other types of crusts are lithified by micritic, intergranular void cement and sparry vug-filling cement. For this reason I have dealt with them separately below.

Algal tufas. Calcareous tufa (usually almost pure calcite to magnesian calcite, with only a few percent $MgCO_3$) precipitated around blue-green algal filaments has been reported by many workers from terrestrial settings both fresh and saline (Bradley 1929; Eardley 1938; Irion & Müller 1968, especially Fig. 5; and the many works discussed by Fritsch 1952, pp. 868–69). However, of special interest here is the well-known tendency of *Scytonema* species to calcify, even epiphyte types growing on the walls of greenhouses (Schönleber 1936; Thunmark 1926; Elenkin 1936; among others). In such *Scytonema* tufa, the calcite crystals (< 10 μ, usually about 2 to 3 μ) are formed in and on the enveloping mucilaginous sheath of the filaments and not in the cell interior or walls, so that calcification is not a primary physiological process. However, Elenkin (1936) believes that *S. Julianum*, for example, has "developed through heredity the ability to extract calcium from the substrate which is present in minimal quantities, with which it encapsulates itself." Schönleber (1936), also working with *S. Julianum*, believed that the encrusting "calcspar crystals derive from the guttation water," but she found many examples where heavy calcite sheaths accumulate without the extrusion of water ("guttation drops") from the filaments. Schönleber looked at natural as well as cultured communities of *S. Julianum* and found that magnesian calcite encrustations ($MgCO_3$ content not measured) form both in apical and intercalary zones and that the thickness of the calcite crusts is a function of the age of the filaments, as is the size of the individual crystals (3 $\mu \rightarrow$ 7 μ). Most significantly, she found that encrustation only occurred on those parts of living *Scytonema* filaments subaerially exposed above the culture fluid, *the submerged parts of the filaments did not calcify*. My own field observations on Andros support this: tufa encasing upright, in place, living *Scytonema* filaments is found only on the highest, driest parts of the inland marsh, where the algae seldom if ever are submerged in water for more than a few hours at a time. In those lower parts of the inland marsh that remain moist or wet even through the dry winter months, the lush green carpet of living *Scytonema* is not calcified. However, an additional crucial observation is that beneath this thick living mat, in the decaying and squashed mass of yellow-tan thalli of old buried *Scytonema* mats can be seen the growths of scattered calcite crystals in and on the mucilage of *dead* filaments. This latter mode of post-mortem calcification appears to be responsible for the bulk of the inland marsh tufa beds, because discrete, domal stromatolite structures made of erect calcified filaments are far subordinate to tangled masses of crushed and deformed tufa and tufa fragments in the inland marsh sediment record (Fig. 52). This probably means that two different mechanisms of tufa formation are operating in the Andros *Scytonema* inland marsh. In both cases, the precipitation must be in some way an algally influenced process, because, as noted above, sediment layers interbedded with the tufa are not normally cemented. For the discrete stromatolitic tufa heads (crust types IA–i and IA–ii) two possible mechanisms are: (1) evaporation-induced precipitation from fluids exuded by dry subaerially exposed *Scytonema* filaments (the "guttation water" of Schönleber 1936); (2) evaporation and/or photosynthesis-induced precipitation from groundwater drawn up as capillary films around the *Scytonema* filaments—in other words, the filaments act simply as wicks. As far as I can ascertain, nothing is known about the ionic composition of algal "guttation water" and how it may be influenced by the composition and concentration of the rainwater, groundwater, or seawater absorbed by the algae, so the applicability of this mechanism to the origin of the tufa cannot be properly judged. Until we have more information, my prejudice is for the "wick" mechanism because the composition of the Mg-calcite of the tufa heads seems to be controlled by the water chemistry: at Three Creeks the tufas at the seaward edge of the inland marsh are of 12 to 13 mole % $MgCO_3$ calcite, while in the fresher interior lakes of eastern Andros, Monty (1972) reports 1.5 to 8 mole % $MgCO_3$ calcite, depending on the amount of contamination with seawater. With this mechanism, the Mg content of the calcite should be determined by the Mg/Ca activity ratio and temperature of the parent water (Füchtbauer & Hardie 1976; Katz 1973). Seawater at 25°C should yield a calcite with 13 ± 1 mole % $MgCO_3$, while fresher waters with lower Mg/Ca ratios should produce calcite with much lower Mg contents (Füchtbauer & Hardie 1976). The particular role of the algae would be "catalytic," providing the "evaporating-wick" micro-hydrologic setting, a stable gelatinous nucleation substratum, and a localized pH increase around the filaments (daytime pH's as high as 9.5 were measured within moist photosynthesizing *Scytonema* cushions). The calcite precipitation would presumably occur in tiny increments as the subaerially exposed *Scytonema* heads dried out after flooding by heavy summer rainstorms or infrequent seawater inundations. For the buried calcified *Scytonema* mats, the post-mortem nucleation-promoting mechanism is difficult even to guess at. One attractive possibility is the idea presented by Gebelein and Hoffman (1973). They found in the laboratory that Mg was strongly absorbed onto the mucilage of dead algal filaments and that this Mg-enriched organic mass would, with addition of $(NH_4)_2CO_3$ to force supersaturation, precipitate calcite with 17 to 20 mole % $MgCO_3$, no matter what the Mg/Ca ratio of the coexisting aqueous solution. Their purely inorganic reference solution produced only 10 mole % $MgCO_3$ calcite. This idea would certainly be consistent with the observation that calcite precipitation in the buried mats of the inland marsh occurs in and on the mucilage of dead filaments only. It would also be consistent with the high $MgCO_3$ content (up to 16 mole %) found in the calcite of these calcified mats. However, the model does not solve the basic problem of what process actually triggers nucleation of the Mg-calcite. It is evaporative concentration of vadose waters? Bacterial uptake of CO_2? Evaporative concentration would be most intense at the surface in the living mats, rather than below the surface, so some kind of internal process associated with organic decay would seem most logical. Bacterial and fungal breakdown of cellulose and other plant matter usually ends in liberation of CO_2, which decreases interstitial water pH, and carbonate dissolution rather than precipitation would be promoted. Perhaps in the Andros marsh, heterotrophic bacteria that utilize CO_2 to convert, for example, pyruvic acid to oxaloacetic acid are responsible for *raising* pH during the algal decay process. The roles of bacteria in modifying the geochemical micro-regime in carbonate environments certainly needs looking into in earnest. As a concluding aside, one puzzling aspect of Gebelein and Hoffman's study concerns the validity of the numbers they present. Their reference solutions were prepared with a wide range of Mg/Ca mole ratio and should have produced a correspondingly wide range of Mg-calcite compositions controlled by the distribution coefficient (Füchtbauer & Hardie 1976; Katz 1973; Winland 1969), but instead Gebelein and Hoffman report calcite with a uniform 10 mole % $MgCO_3$ value for all their reference experiments. This inconsistency needs resolving, because the basic idea they present is a very viable one for producing calcite with a high Mg/Ca ratio in algally dominated depositional environments.

Nontufa crusts. As pointed out above, the crusts of the seawater-dominated channeled belt (types IA–iii, IB, IC, IIB) carry none of the calcified filament features of the tufas of the inland marsh, but instead are cemented by void-filling micrite and microspar. The algal structures preserved in these crusts appear as molds in a cemented detrital sediment matrix and not as autochthonous calcite filament envelopes. Inorganic precipitation from supersaturated interstitial vadose waters seems the most logical cementation mechanism for these nontufa crusts. Within this framework two separate hydrologic processes operating in the same crust appear necessary, one to account for the intergranular Mg-calcite micrite cement and one to account for the vug-filling aragonite sparry cement. I offer the following speculations. The micrite cement that lines the intergranular pores (both between peloids and within the porous peloids themselves) may well represent precipitation from films of water drawn up from the shallow seawater water table by evaporation-driven capillary draw. Such a caliche-type mechanism is suggested by the observation that the crusts are hardest (well cemented, broken only with a hammer blow) at the very surface and become progressively less hard downward. In fact, the boundary between the underlying coherent but uncemented sediment and the cemented crust is not always apparent; the bottoms of loose plates of crust lifted free from the sediment are very uneven and can be easily scratched with a fingernail. Support for a progressive upward cementation comes also from unpublished work of David Hepp (Johns Hopkins University), who showed that for analogous crusts on Sugarloaf Key, Florida Keys, the Mg-calcite content increased upward in concert with increased hardness. The very high $MgCO_3$ content of some of the micrite (35 to 43 mole %) of the Andros crusts would be an expected consequence of progressive evaporation of these seawater films, because the Mg^{++} distribution coefficient of calcite is considerably less than unity and favors the enrichment of Mg in the mother liquor. Continued precipitation of calcite (the initial calcite should be about 13 mole % $MgCO_3$ for a seawater parent solution at 25°C) would therefore force the Mg/Ca mole ratio in the concentrated seawater to increase, which in turn would lead to higher and higher Mg/Ca ratios in the precipitated calcite. Ultimately, a proto-dolomite-like phase could be produced (see Füchtbauer & Hardie 1976, for a more elaborate discussion). An evaporation-induced precipitation mechanism rather than an aragonite → very high Mg-calcite "dolomitization" process is supported by the preliminary electron microscope and microprobe evidence that the Mg-calcite appears to be lining voids between the grains and filling borings in shells. Shinn et al. (1969, p. 1218) also noted that electron micrographs "show concentrations of rhombohedral dolomite (*sic*) crystals lining very small vugs" which "suggests that the dolomite (*sic*) is a primary precipitate." Kinetic difficulties with low temperature conversion of aragonite to "proto-dolomite" contrasted with the ease with which very high Mg-calcite will precipitate from supersaturated solutions gives further support to the proposed mechanism (see further details in Füchtbauer & Hardie 1976). With this mechanism the very porous nature of the peloidal host sediment, coupled with the high porosity of the constituent soft peloids themselves, would easily allow cement to become more voluminous than the original framework material! For the large vugs lined with aragonite spar, the presence of flat, floor-lining cement points to an origin from surface waters that made their way *down* the network of large fenestral pores and sheet-cracks. These downward-percolating waters may have been rainwater (contaminated more or less by seawater or seawater salts) which initially enlarged the pores by dissolution of the aragonite peloidal sediment walls and then reprecipitated aragonite when the subsequent drying out caused the trapped capillary droplets and films in the surface zone to evaporate. At times the source waters may also have been undiluted seawater washed over the levee and beach-ridge backslopes by onshore storms. That a cementation mechanism of this kind can operate in the rainy Andros Island setting is shown by the "case hardening" that was found to take place when loose plates of crust were left exposed on dry levee surfaces. The porous, lightly cemented bottom surface of the crust plates, easily scratched when first exposed, developed a thin, hard rind within a year to two. This has also happened naturally where mangrove roots have pushed crust plates up and exposed the bottom surface.

The reason why the vug-filling aragonite cement is not pervasive but is restricted to the larger pores and cracks is probably related to the fluid forces involved. Downward-percolating rainwater or other floodwaters will be driven by a very small hydraulic head and so will find ready access only through the larger vugs and sheet-cracks, where surface tension forces are weakest. On the other hand "evaporative pumping" can apparently develop upward capillary forces strong enough to overcome surface tension forces of even the smallest intergranular pores (Hsu & Siegenthaler, 1969), allowing precipitation of a pervasive micrite cement. This leads us directly to the problem of why some void fillings are aragonite and some Mg-calcite, the essence of the second question we posed at the beginning of this discussion. At 25°C magnesian calcite is metastable with respect to aragonite in solutions with Mg/Ca activity ratios greater than 2 (the free energy difference progressively increases with increasing Mg/Ca ratio in the solution), while aragonite is metastable with respect to very low Mg-calcite below this ratio (Füchtbauer & Hardie 1976). So rainwater that has dissolved aragonite or very low magnesian calcite would have a very low Mg/Ca mole ratio and should, on evaporation or CO_2-degassing, produce very low Mg-calcite and not aragonite. On the other hand, seawater or rain-diluted seawater should produce aragonite and not Mg-calcite, but, of course, a large number of Holocene marine cements in addition to the Andros crusts are Mg-calcite or Mg-calcite + aragonite. In some preliminary experiments Füchtbauer and Hardie (1976) found that in aqueous chloride solutions with Mg/Ca mole ratios between 2 and 20 mixtures of Mg-calcite and aragonite always precipitated no matter what the "salinity" (ionic strengths ranged from 0.06 to 6.0 for the Mg-Ca-Na-Cl-CO_3 experimental solutions). Obviously, for mono-mineralic cements neither Mg/Ca ratio nor salinity can be the crucial nucleation and crystal-growth-controlling factors. This has convinced me that we must look to the influence of physical factors, such as temperature, rate at which critical supersaturation is reached and exceeded, presence or absence of seeds, presence or absence of turbulence, and so on. Work is proceeding in this direction, but is not complete at the time of this writing. However, it does look as though slow approach to supersaturation might favor the preferential nucleation of the more stable aragonite over Mg-calcite in solutions with Mg/Ca ratios greater than 2. With this in mind, perhaps aragonite forms in the open vug system of the Andros crusts, because here, where large droplets must be evaporated, supersaturation could be approached slowly. On the other hand, rapid supersaturation would be expected in the intergranular void system, where *thin films* would evaporate rapidly and so favor nucleation of Mg-calcite. This whole problem, of course, leads us into the third question of why algal tufa forms in the inland marsh and not in the channeled belt. This is another difficult problem, directly related to the first two. On the face of it, the answer would appear to lie in the salinity differences between the freshwater-dominated inland marsh and the seawater-dominated channeled belt, the algal tufa apparently requiring a freshwater regime to form. So to be consistent with the origin proposed above for the tufa crusts, that is, the calcite precipitation mechanism is basically inorganic and only triggered by the algae, I suggest the following possible answer. In the channeled belt crust environments the cementation is pervasive (including, without preference, both the algal filaments and their sediment matrices), because seawater is very close to critical supersaturation with respect to Mg-calcite, and so a little evaporation would be as effective as photosynthesis in promoting nucleation. On the other hand, in the inland marsh-crust environments, where undersaturated brackish waters prevail, maybe it is the algal photosynthesis that provides the extra push needed to reach nucleation supersaturation, so producing preferential cementation around the filaments. Or perhaps tufa *does* form in the channeled belt, but the filament envelopes are easily masked by void-filling micrite cement that co-precipitates with the tufa or precipitates at times between tufa-forming episodes. Of course, the question of the origin of the tufa itself is by no means solved, and we might still consider the possibility that algal physiological processes are quite different in freshwater than in seawater.

The absence from the channeled belt of subsurface production of post-mortem tufa such as we find in the inland marsh could be explained by the fact that buried *Scytonema* mats are quickly bioturbated and the organic matter scavenged by infaunal burrowers. It is also likely that bacterial reduction is much more vigorous in this seawater setting, because little organic matter survives burial here: gray color and the faint smell of H_2S are characteristic of even supratidal sediments in the channeled belt! In other words, there is little chance that lightly calcified *Scytonema* mats, even if formed, would survive in recognizable forms in the channeled belt environment.

7. Environmental Significance of the Algal Crusts

The particular environmental importance of the Andros Island crusts lies in the fact that they beautifully preserve a record of macroscopic filamentous blue-green algae that live in a very narrow elevation range (15 cms) in the high water tidal zone and in freshwater-dominated supratidal marshes. As shown above, the algal crusts of the channeled belt are sensitive tide level gauges marking a narrow zone just above the normal high water line, but below the level of the levee and beach-ridge crests. The tufa crusts of the inland marsh, on the other hand, are freshwater regime indicators.

The immediate application of this would be in resolving the Holocene stratigraphic record on Andros, particularly relative to sea-level changes, using the buried crusts that are so common in many parts of the Andros tidal flats from south of Williams Island north to Three Creeks. These buried crusts are especially valuable because almost all the uncemented sediment that gets buried below the water table in the channeled belt is soon homogenized by infaunal burrowers and its environmental record wiped out. It is the presence of upright *Scytonema* stromatolitic heads and honeycomb mat structures in these subsurface crusts that is the most convincing proof of the subaerial, supratidal origin of the crusts. Such buried algal crusts must, therefore, record lower stands of sea level (and/or subsidence) during the last 5,000 years or so. Shinn et al. (1965) showed that a crust layer increased in C_{14} age as it dipped progressively deeper below pond sediment away from a palm hammock in the Pumpion Cay area. They suggested that the crust faithfully recorded the latest progressive Holocene sea-level rise. Examination of the buried crusts from this area showed that they carry *Scytonema* structures and so are indeed supratidal zone indicators and not crusts formed by some subsurface process. However, we found more than one "horizon" of buried algal crust, both at Pumpion Cay and Three Creeks (see above), so a complex sea-level history in the last 5,000 years is indicated; for example, at least five such crusts were found at different levels below the inland marsh at its

Fig. 94A. Bedding-plane (*top*) and vertical (*bottom*) views of laminated algal mounds covering mud-cracks in tidal flat carbonate sequence of the Ordovician St. Paul's Group of Western Maryland. Scale in mm. Compare with Figs. 33B and 47A.

seawardmost edge in the Three Creeks area! Placed in the context of a full stratigraphic study, these buried algal crusts could reveal a most precise story on the Holocene sea-level patterns that would help refine our understanding of Holocene sedimentation in particular and sediment accumulation mechanisms in general.

Ancient analogues of these Andros-type algal crusts in the rock record should, of course, carry the same environmental message. I know of two examples where mat structures like those in the Andros crusts have been preserved in ancient carbonates. The first example comes from the Ordovician St. Paul's Group of the central Appalachians in Western Maryland, U.S.A. (Matter 1967). On a few well-preserved bedding surfaces can be found sheet- and ribbon-like, low, flat mounds "growing" over mud-cracks, as Figure 94A shows so beautifully. These structures are amazingly similar in surface morphology to the *Scytonema* patch mats of the high algal marshes, especially in the inland marsh, where *Scytonema* growths establish themselves preferentially over mud-cracks up which moisture leaks (Figure 33B). The second example comes from the Precambrian Pethei Group of the Great Slave Lake region, N.W.T., Canada (Hoffman 1968). Figure 94B shows a strong, columnar palisade structure in a thin bed from this Precambrian

Fig. 94B. Vertical view of Precambrian laminated dolomite of Pethei Group of Great Slave Lake, Canada, area, showing palisade structure very much like the palisade-structured crusts of Andros Island pictured in Figs. 46A and 50A. Pencil for scale. Sample courtesy Paul Hoffman.

carbonate unit. The similarity to the Andros palisade-structured crusts of type IB (Figure 46A) is suggestive of a similar origin.

Finally, the algal tufas carry, I believe, a special significance: they make exceptionally fine analogues for ancient fossil algae, like *Girvanella*. This is especially true of the buried, calcified *Scytonema* mats of the inland marsh, as comparison of Figures 50B & 53B with illustrations of Cambrian *Girvanella* given by Ahr (1971) and Johnson (1961), for example, shows. If the analogy is carried all the way through, then *Girvanella* and similar fossil blue-greens that show carbonate filament envelopes may tell of a freshwater-dominated environment like the Andros inland marsh, and, of course, a climate and weather setting like that of the Bahamas.

8 DISTINCTIVE FEATURES OF A RAINY, LOW-ENERGY, TROPICAL CARBONATE TIDAL FLAT: A SUMMARY

Lawrence A. Hardie

In the foregoing chapters we have described and discussed in detail bedding and associated sedimentary and organo-sedimentary structures that are distinctive of the Three Creeks tidal flat deposit. Tables 14 and 15 bring together in summary

Table 14. Catalog of major sedimentary and organo-sedimentary structures in the Three Creeks tidal flat sediments

Structure	Type	Subtype	Illustrations
1) Bedding	a) millimeter lamination (< 3 mm thick)	i) smooth flat lamination	Figs. 36, 37, 38
		ii) disrupted flat lamination	Figs. 40, 41, 42
		iii) crinkled fenestral lamination	Figs. 43, 44
		iv) wavy fenestral lamination	Fig. 45
		v) lamination with "palisade" structure	Figs. 46, 47, 48, 49, 50
	b) thin beds (1–60 cm) and thick lamination 3–10 mm)	i) algal tufa-peloidal mud interbeds	Figs. 52, 53 (see also Shinn et al. 1969, Figs. 21, 22)
		ii) disrupted fenestral bedding	Fig. 54
		iii) flat-pebble gravel	Figs. 26A, 55, 56, 57
		iv) round-pebble gravel	Fig. 24B
	c) thin to thick cross-bedding (1–120 cm)	i) rippled skeletal-peloid sands	Figs. 24A, 26B
		ii) festooned skeletal sands	Shinn et al. (1969, Fig. 15)
	d) thick to very thick bedding (> 60 cm)	i) bioturbated peloidal mud	Figs. 58, 59, 80B, 75A
2) Bed forms	a) ripple marks	i) current ripples	Fig. 26B
		ii) oscillation ripples	Fig. 24A
	b) current lineation (parallel to current direction)	i) pervasive system	Fig. 12B
		ii) on backslopes of starved ripples	Fig. 26B
	c) depression fills	i) small (cms) flat lenses	Fig. 63
	d) lee-deposits	i) small (cms) elongate mounds behind bushes	Fig. 62C, 12A
	e) scours	i) cusp-shaped depressions in front of plant stems	

Table 14. Catalog of major sedimentary and organo-sedimentary structures in the Three Creeks tidal flat sediments (continued)

Structure	Type	Subtype	Illustrations
3) Proto-stromatolites and stromatolites	a) unlithified structures made by sediment trapping	i) small (< 10 cm) domal SH-structures	Figs. 11, 34, 35
		ii) LLH "raised disc" structures	Fig. 17A and Black (1933, Figs. 4, 8)
		iii) small unlaminated clots	Fig. 45
		iv) small (< 15 cm) LLH-S structures	Fig. 25A
	b) lithified structures made by sediment trapping	i) small palisade-structured domes, mounds and flat sheets	Figs. 46A, 47A, 50
	c) lithified structures made by direct precipitation (algal tufa)	i) small palisade-structured domes, knobs and mounds	Figs. 33, 46B, 48, 49
4) Desiccation structures	a) mud-cracks	i) 1–5 cm polygons	Figs. 14, 40, 42A
		ii) 5–15 cm polygons (wide, shallow cracks)	Fig. 62B
		iii) 20–30 cm polygons (deep prism cracks)	Figs. 20B, 28
	b) sheet-cracks	i) bedding plane cracks	Figs. 44A, 42A
	c) algal-mat deformation features	i) wrinkled, cracked, curled, and chipped mats	Fig. 27
		ii) raised hollow domes (made by air-pressure but cracked and detached by drying out)	Fig. 13B
5) Fenestral pores (excluding burrows)	a) tubular filament molds	i) vertical, horizontal and anastomosing	Figs. 39A, 42B, 50B
	b) irregular to amoeboidal voids	i) subhorizontal and subvertical	Figs. 43A, 44, 46, 49, 54
		ii) randomly oriented	Figs. 43A, 44, 54
6) Intraclast lenses	a) sand and granule dominated	i) open framework of flat chips	Figs. 15, 41, 52
	b) pebble dominated	i) 5–15 cm flat and curled discs	Figs. 26A, 55, 56, 57
7) Burrows	a) crab burrows	i) *Uca* and *Sesarma*	Figs. 86, 87A, 11
		ii) *Cardisoma*	
		iii) *Callinectes* and *Panopeus*	Fig. 83B
	b) shrimp burrows	i) *Alpheus*	Fig. 84
		ii) *Callianassa*	see Shinn (1968)
		iii) *Apseudes*	Fig. 83A
	c) worm burrows	i) polychaete	Figs. 78B, 82
		ii) oligochaete	Figs. 85A, 36B, 42A
	d) insect burrows	i) ants, etc.	Fig. 85B
8) Root-holes	a) trees	i) mangroves	Figs. 23A, 52, 53, 59
		ii) pine, etc.	
	b) shrubs	i) various halophytes	
	c) grasses	i) various halophytes	
		ii) sea-grasses	

Table 15. Structures and textures of sediments of the different subenvironments of the Three Creeks tidal flats, Northwest Andros Island

Subenvironment*	Sedimentary structures	Sediment texture
Levee-pond subenvironments		
Channel banks (40–90%)	Knobby, hemispherical proto-stromatolites (2–10 cm across), thinly laminated (0.1–0.7 mm). Crab (*Uca*) burrows. Prism cracks with medium sized polygons (5–15 cm).	Dense mud laminae with few interlayered peloid (50–150 μ) laminae. Scattered small tests of forams. Few small isolated algal filament molds (5–10 μ wide).
Levee crest (98–99.7%)	Very thin (0.1–1.0 mm) essentially flat lamination. Undulations and pinch-outs fairly common. Sparse grass, shrub, and black mangrove root-holes and ant burrows. Oligochaete worm burrows where sediment is moist.	Continuous dense mud laminae alternating with thicker discontinuous peloid (50–150 μ) laminae. Scattered small foram tests and gastropod shell debris. Few small isolated algal filament molds (5–10 μ wide).
Levee backslope (95–98%)	Very thin to thin lamination (0.3–2.0 mm), both flat and wrinkled, commonly fragmented. Small lenses of sand and granule intraclast chips. Small mud-cracks, closely spaced (~1 cm) and only one lamina deep. Sparse grass, shrub, and black mangrove root-holes. Oligochaete worm burrows where sediment moist.	Dense mud laminae alternating with peloid (50–150 μ) laminae. Wrinkled layers underlain by red-brown algal parting. Algal filament and filament molds permeate all layers.
High algal marsh: flat *Scytonema* zone (85–95%)	Thin (0.6–3.0 mm) wavy to wrinkled laminae with dark algal partings. In upper part of this zone the mat becomes discontinuous mounds which may be cemented into thin crusts with radiating filament or with palisade structure. Cracks with medium polygons (5–15 cm) where mat is dried out. Crab (*Uca* and *Sesarma*) burrows. Oligochaete worm burrows where sediment moist. Grass, shrub, and red and black mangrove root-holes.	Dense mud laminae and poorly sorted mud and peloid (50–150 μ) laminae. Sorted peloid layers uncommon. Red-brown algal partings separate most laminae. Filament molds (5–10 μ) and vertical tufts of filaments throughout. Fenestrae outstanding features. Scattered foram and gastropod debris. Crusts are cemented with Mg-calcite and/or aragonite. Some crusts rich in "protodolomite."
Low algal marsh: "pincushion" *Scytonema* zone (65–85%)	Thin lamination to thin beds (0.1–3 cm), wavy and fragmented. Pockets of intraclasts up to cobble size. Shallow mud-cracks (5–15 cm apart) separate algal pincushions at the surface. Crab (*Uca* and *Sesarma*) burrows common. Thick red mangrove roots prominent.	Mud (with scattered peloids) and peloid (50–150 μ) layers. Scattered whole and broken tests of forams and gastropods.
Ponds: mangrove pond (35–65%) open pond (0–35%)	Unlayered, although in a few places there are remnants of thin bedding. Worm burrows abundant. In mangrove pond, large prism cracks (20–30 cm across) and mass of hair-roots and major roots of red mangrove.	Clotted mud supporting scattered dense ovoid peloids (50–150 μ). Patches of open framework of mud peloids also 50–150 μ. Scattered whole gastropod and foram tests.
Shoreline subenvironments		
Offshore (0%)	Unlayered. Crustacean (mainly *Callianassa*) burrows and worm burrows. Seagrass roots in places.	Open framework of ovoid to spherical mud peloids (30–750 μ) with < 10% skeletal sand. Patches of clotted mud.
Beach (0–80%)	Unlayered. Crustacean (*Alpheus* and *Callianassa*) and worm burrows. In uppermost sandy zone thin laminae, ripple marks, and pockets of coquina, "roundstone" and "flat-pebble" gravel.	Pelleted mud (30–750 μ), skeletal sand and gravel. Mud and cemented crust intraclast gravel.
Beach terrace (Approx. 85–95%)	Very thin (0.3–1.0 mm) flat lamination. Pinch outs common. Small clots (draped by laminae) where algal tufts grew. Few scattered flat pebbles draped by small LLH protostromatolites. Very sparse grass roots.	Continuous mud laminae alternating with discontinuous peloid and skeletal grain (30–500 μ) laminae. Algal filament molds in mud layers, particularly in the isolated clots of sediment.
Beach-ridge washover crest (approx. 95–99.7%)	Very thin to thin (0.3–2.0 mm) flat to lenticular lamination. Pockets of flat pebbles and isolated sand ripples (up to 7 mm high). Sparse grass, shrub, and black mangrove root-holes. Oligochaete worm burrows where sediment is moist.	Continuous mud laminae, alternating with discontinuous peloid and skeletal grain (30–750 μ) laminae. Gastropod and foram debris commonly concentrated into thin lenticular shelly laminae. Flat pebbles of algally laminated sediment and cemented crust intraclasts in pockets. Algal filament molds in mud layers.
Beach-ridge hummock (> 99.9%)	Inclined lamination of skeletal sand and granules. Organic-rich "soil" cap. Grass, shrub, black mangrove and pine roots. Land crab and insect (mainly ant) burrows.	Very coarse skeletal sand and granules. Well sorted.

Table 15. Structures and textures of sediments of the different subenvironments of the Three Creeks tidal flats, Northwest Andros Island (continued)

Subenvironment*	Sedimentary structures	Sediment texture
Beach-ridge backslope	see levee backslope	
Inland algal marsh subenvironment		
High algal marsh: flat *Scytonema* zone (> 85%)	see levee high algal marsh	
Low algal marsh "pincushion" *Scytonema* zone (approx. 65–85%)	Thin lamination to thin beds (0.1–10 cm). Mainly continuous sediment layers interbedded with spongy algal and algal tufa layers. Layers of markedly variable thickness. Fragmented by wide shallow cracks in places, making in situ flat pebbles. Cracking of algal mat and draping by sediment makes "raised disc" structure. Large roots and hair-roots of red mangrove typically disrupt layering.	Sediment layers consist of dense mud (with scattered peloids) and well-sorted peloid (50–150 μ) laminae and beds. Algal layers are a complex of filaments and filament molds smothered by sediment. Filaments are commonly coated and welded by clear, fine granular carbonate to make an algal tufa. Scattered whole and broken shells of gastropods and forams.
Channel subenvironments		
Channel bars (0–25%)	Mainly unlayered with a few thin beds of skeletal gravel to make inclined bedding. Crustacean (*Alpheus* and *Callianassa*) and worm burrows common. Higher bars (reaching above LW) are normally very thinly laminated (0.1–0.6 mm) in the upper 10–50 cm; these bars also have large prism cracks (20–30 cm polygons) and red mangrove roots.	Peloids (50–150 μ) suspended in clotted mud. Scattered foram tests and gastropod debris. Pockets of sand-sized sorted mud clasts or peloids. Texture of laminated bar sediment is same as that of levee bank and crest sediment (see above).
Channel lags (0%)	Ripple cross-lamination.	Skeletal sand and gravel, mainly gastropod and foram tests. Intraclasts of pelleted mud and cemented crusts.

* Exposure Index (Ginsburg et al. 1970, and Chapter 3) in mean % of time exposed is given in parentheses for each subenvironment.

form all these features; Table 14 is simply a catalog of all the major sedimentary features we observed in the tidal flat sediments, while Table 15 briefly recapitulates the descriptions of the major structures and textures found in the surface sediment of each subenvironment. Figure 4 shows the exposure index range and elevation range for a number of important features, while Figures 65, 66, and 67 give some indication of the stratigraphic relations of major bedding types.

While this overall set of features does represent a diagnostic assemblage for recognizing ancient analogs of Three Creeks-type tidal flat deposits, it is particularly interesting to isolate those features that are a clear signature of the prevailing environmental elements, such as physiographic setting, climate, weather patterns, and wave and tidal energies. In this way significant comparisons can be made with the sedimentary record of other tidal flats deposits—comparisons that could lead to a clear understanding of the similarities and differences among tidal flat deposits, old and modern. As a start in this direction, I briefly present below my analysis of the relations between environmental elements and the sedimentary record of the Three Creeks tidal flat deposit: a summary review is given in Table 16.

The large-scale physiographic setting of the Bahama platform itself, as a massive, long-lived carbonate build-up completely isolated from the continental mainland, ensures an environment free from terrigenous sediment (except very minor atmospheric dust). As a result the entire sediment blanket of the Great Bahama Bank is a "pure" carbonate accumulation, generated *in situ*, for the most part, from the skeletal remains of calcareous algae and marine invertebrates. The broad, shallow, open-shelf structure of the platform also ensures that

Table 16. The environmental elements and their specific sedimentary record at Three Creeks

Environmental element		Sedimentary record
Isolation from continental mainland		"pure" carbonate sediment (no terrigeneous material)
Low tidal energies and slow circulation, Low wave energies		1) predominance of mud 2) lack of extensive point bar sequences; instead channel "fills" 3) lack of rapid-burrowing filter-feeders; instead stable substrate surface-grazing and burrowing deposit-feeders 4) lack of extensive cross-bedded and rippled sands 5) lack of club-shaped stromatolites
Rainy climate		1) lack of evaporites 2) fleshy freshwater algal layers and algal tufa in marsh sediments 3) extensive burrowing even where circulation is very restricted 4) extensive rooting by higher plants
Weather patterns	winter storms	deposition of mm-lamination on levees, beach terraces and beach-ridge washovers
	late summer hurricanes	1) deposition of thin beds and thick laminae in ponds, offshore and algal marshes 2) deposition of flat pebble gravels on beach ridges
	winter dry spells	1) wrinkled algal mats 2) algal polygons 3) surface cemented crusts
	summer rainy spells	see under rainy climate

the tides will be of low amplitude and low energy. The islands on the eastern edge of the platform shelter the bulk of the Bank from the trade winds, creating a lee-shadow shelf lagoon with restricted low velocity circulation (Traverse & Ginsburg 1966).

The low tidal energies and slow circulation at Three Creeks are reflected in: (1) the predominance of mud (either as individual particles or aggregated into peloids) in the tidal flat and nearshore sediments; (2) the lack of laterally continuous point bar sequences in the channeled belt sediments; instead, the sediment record is mostly of channel fills and channel-axis bars; (3) the absence of rapid-burrowing, filter-feeding marine invertebrates which normally are adapted for life in shifting substrates (Stanley 1970), as for example the cockles of the Dutch Wadden Sea (Verwey 1952); instead, because of the stable substrate and long periods between sediment movement, all sediments below normal high water are *completely* bioturbated by a low diversity, mud-burrowing and mud-ingesting fauna; (4) the absence of abundant current-produced structures, such as large-scale cross-bedding interlayered with small-scale cross-bedding, "herringbone" bedding, "flaser-bedding," etc. All of these features are strikingly different, for example, from the North Sea tidal flats, where the tidal energies are relatively high, sand is the major sediment, and daily movement of sand by strong tidal currents is the rule.

The rainy climate ensures that tidal water salinities remain relatively low (5 to 40%) the year round, even in areas where circulation is quite restricted. The "low" salinities leave their mark in the sediments by (1) the absence of evaporite minerals, (2) the extensive growth of the fleshy, freshwater alga *Scytonema* which produces *algal tufa* layers on burial, (3) extensive bioturbation of marine and terrestrial invertebrates, particularly in subtidal and intertidal environments, (4) an extensive system of rootholes in the sediment, produced by abundant growth of higher plants.

The significant weather pattern consists of rainy summers, dry moderate winters punctuated by onshore storms ("northers"), and infrequent late summer hurricanes. The summer rains encourage the luxuriant growth of the freshwater alga *Scytonema* in the inland marsh and pond marshes. Burial of these thick algal mats under sediment deposited catastrophically by hurricanes produces the characteristic thin-bedded, pelleted mud/algal tufa "hurricane layer" structure of the marsh sediment. Winter dry spells cause wrinkling and polygonal cracking of the marsh algal mats, producing distinctive wavy algal lamination and Black's type-C "raised disc" proto-stromatolites in the high algal marsh sediments. Slow evaporation of capillary groundwater during these dry winter periods, coupled with algal photosynthesis or decay, produces cemented surface crusts and encrusted algal "heads" in the marshes. Especially noteworthy are the winter "northers" which leave their particularly distinctive signature in the fine millimeter-lamination that caps the levees, beach terraces, beach-ridge washovers, and intertidal channel bars.

Finally, the combination of physiography, climate, and weather patterns ensures that the Andros tidal flats are not affected by significant wave action, except during infrequent severe onshore storms. Therefore, large, columnar, club-shaped, indurated stromatolites so spectacularly developed on wave-pounded headlands in restricted Hamelin Pool, Shark Bay, are characteristically absent from the Andros Island sediments.

9 SOME MISCELLANEOUS IMPLICATIONS AND SPECULATIONS

Lawrence A. Hardie and Peter Garrett

1. Storm Deposition: A Widespread Phenomenon on Tidal Flats

As we have shown above, onshore storms are the *exclusive* mechanism of deposition on the Three Creeks tidal flats. The particular conditions of low tidal energies and surface sediment binding by subtidal algal scums combine to prohibit sediment transport by the normal diurnal tide currents or even disturbance of the sediment surface by ordinary wind-waves. Only storm-induced bottom currents are capable of stirring sediment into suspension and carrying it onto the tidal flats.

In Florida Bay, like Three Creeks, layering is exclusively storm deposited (Ginsburg et al. 1954; Pray 1968). In the North Sea, Reineck (1970, pp. 36–37) reports that the lamination of the supratidal salt marshes of the Jado Bay tidal flats are deposits of storm "tides." At Scammons lagoon, Ojo de Liebre, Baja California, Phleger (1969) notes that supratidal "ponds" are replenished entirely by hurricane floods; obviously, the lower tidal flats are also subject to such storm deposition. On the tidal flats of the Colorado River delta, flooding of the supratidal "high flats" (Thompson, 1968) occurs only when high onshore winds coincide with spring high tides (personal communications from local inhabitants). In Shark Bay, Western Australia, Davies (1970) reports hurricane layers in "ponded" tidal channels. In the Persian Gulf the "shamals" are well known for flooding the sabhkas with seawater (Illing et al. 1965). In Laguna Madre, Texas, where the tidal range is virtually zero, flooding of the flats occurs only when strong onshore storm winds are blowing (Alan Scott, personal communication).

It is clear, then, that storm deposition is a widespread phenomenon. More than that, we think storm deposition is a characteristic feature of tidal flats of all kinds. After all, supratidal areas require, *a priori*, abnormally high water levels for flooding, and onshore storms are the primary cause of catastrophic inundation of the high ground of tidal flats. In fact, on littoral flats in areas of very low or negligible tidal range, storm flooding is the *only* mechanism of deposition. It is a peculiar and somewhat paradoxical feature of low-energy tidal flats that layering only reflects abnormal catastrophic high-energy events! We suggest that more attention be given to the storm model for tidal flat deposition.

2. Distinctive Carbonate Tidal Flat Communities

The three communities described in Chapter 6 characterize the physiographic divisions after which they are named: nearshore, pond, and levee. But these physiographic divisions, which may be only locally developed on this sort of tidal flat, in turn approximate to certain limits of water level, namely, subtidal, intertidal, and supratidal, into which most workers have heretofore divided the spectrum of tidal flat surfaces. Thus we must consider whether there are distinctive similarities between these tidal flat communities and others elsewhere, and whether the similarities can be extended back in time for comparison with ancient communities.

The chances seem good because the one parameter of water level is dominant in controlling the distribution of organisms, there being two major discontinuities, low and high water, to separate our three communities. But other discontinuities can accentuate the differences between the communities, or provide additional subdivisions. For instance, the presence of a barrier island chain has the effect of making the tidal flat waters distinctive from those offshore with a greater range of salinity and temperature, and thus accentuating the difference between the pond and nearshore communities. An extensive subtidal lagoon behind the barrier islands, such as occurs on other coasts, would have another community distinct from nearshore and lagoonal intertidal communities. The boundary between pond and levee communities is the most distinct, because it is the boundary between the marine and terrestrial realms, the point of least biologic diversity, where a few land-tolerant marine organisms extend up to meet a few marine-tolerant terrestrial organisms. The upper boundary of the levee community might grade up with increasing diversity into a truly terrestrial community were it not for the fact that there is a spatial separation of the highest tidal flat environments (levees and beach ridges) from the truly terrestrial environments of mainland Andros, due to the mechanisms of tidal flat deposition. The terrestrial environment does not concern us here because it is nondepositional.

Many studies and syntheses of tidal flat biotas (Hedgpeth 1957, pp. 693f.; Ricketts & Calvin 1968; Teal & Teal 1969; to name a few) show that there are indeed similar communities, usually with the same taxonomic groups represented or not represented. The true tidal flat communities, pond and levee here, have a low group and species diversity, because few organisms can tolerate the rigors of such variable conditions. But populations are frequently large, because competition is low and food resources plentiful. Diversity increases offshore, where "normal" marine conditions with little variability are met.

As for ancient tidal flat communities, there have been few detailed studies, though Fischer (1964), and especially Walker and Laporte (1970), show that strikingly similar "congruent" communities can be recognized in carbonates at least back to the Middle Ordovician. Before that, as Walker and Laporte warn, such communities may not have evolved to the stable state they have apparently held since, though Lochman (1968) has described a facies-restricted Cambrian fauna of three or four trilobite species which consistently occur in sequence with algal laminated beds.

3. Speculations on What a Precambrian Bahama-Type Tidal Flat Deposit Would Look Like

Destruction of layering by grazing and burrowing animals is a diagnostic feature of the sediment below the high-water mark on all modern tidal flats, except hypersaline complexes like Hamelin Pool, Shark Bay. It would be very interesting to see what these sediments would have looked like in Precambrian (or even early Paleozoic) times, before the evolution, or radiation, of the kinds of life now typical of modern tidal flats. To get some idea we need to "remove" the overprint of bioturbation and so "see" the primary unmodified depositional structures and textures. For the Three Creeks deposit we must establish what kinds of layering we get in the ponds, channel bars, beaches, and offshore subenvironments. For the other subenvironments the features would be virtually unchanged, except for "removing" the roothole structures of the higher plants. In the ponds, hurricane flooding would be the major sediment supplier, and we would expect thin beds, sometimes finely laminated or cross-bedded, to alternate with thin, compressed, spongy mat layers with little sediment. The layering would look somewhat like the marsh layering, but without the thick, freshwater algal layers and tufas. The *Schizothrix*-type mat covering the pond surface probably wouldn't be able to prevent shallow mud-cracking of newly deposited hurricane layers. On the channel bars, laminae would be deposited by winter storms as they are being deposited today; deep prism-cracking probably would be typical (Fig. 28). On the beach it is likely that both winter storm lamination and hurricane layers would be deposited; the surface probably would be covered by a *Schizothrix* mat as it is today. In the offshore, which is the main source of sediment for the tidal flats, the sediment would be thin to thick-bedded wedges (Perkins & Enos 1968) almost surely deeply scoured by hurricane and storm bottom currents. A coherent mat (*Schizothrix* and other types) very likely would cover the offshore surface, and from it in certain offshore environments might develop some of the peculiar stromatolites (digitate forms, domal forms built from laterally-linked hemispheroids, and thrombolites) that are no longer to be found, though it is unlikely that substantial, subtidal, head-shaped stromatolites (Garrett 1970) would develop, because the daily wave and tidal currents on the Great Bahama Bank are very low and special holdfasts for mats would be unnecessary (cf. intertidal stromatolites of Shark Bay, Logan 1961).

In a broad view, then, progradation would leave a vertical sequence of a laminite cap over a complex intertidal-subtidal unit of thin-bedded pond, laminated channel bar, fining-upward channel fills and laminated and thin-bedded beach sediments, over subtidal thick to thin lenticular beds of the offshore. The essential picture would be an upward decrease in layer thickness and an increase in desiccation features. Cycles of this general scheme are found in the Precambrian of Canada (Paul Hoffman, personal communication), but what is particularly interesting is that it is a familiar theme in Cambro-Ordovician carbonates of the Appalachians. The abundance of layering and subordinate role of burrowing in these deposits contrasts strongly with younger rocks and modern sediments of shallow marginal marine environments where bioturbation is predominant. The stratigraphic record of increase in bioturbated subtidal sediments must surely coincide with the radiation of grazing and burrowing animals.

4. The Significance of Islands

It is no accident that tidal mudflats and subtidal muddy bottom sediment are accumulating only on the *west* side of exposed Pleistocene bedrock islands on the Bahama Banks. The islands act as very effective water circulation barriers and, because of the persistent easterly trade winds, well-defined "lee-shadows" of low surface currents and low wave energies develop on the western sides of these solid barriers. It is in such "lee-shadows" that lime mud most effectively accumulates. This was beautifully demonstrated by the work of Traverse and Ginsburg (1966, 1967). They showed that pine pollen, which is known from laboratory studies to have a hydrodynamic behavior like that of very fine mud, is preferentially accumulating along with mud in the lee-shadow of Andros Island (Traverse & Ginsburg 1967, Fig. 3) and is virtually absent from wave-agitated ooid sand shoals. This is also true of Eleuthera Island, where there are no pine trees nor any pine-covered islands to windward. So on Eleuthera the pollen must have been transported in (by the occasional westerly winds and/or by surface water currents) to finally settle with the lime mud in the lee of the island: sedimentation pattern and not particle source is the controlling factor.[1] The presence of the island barriers on the windward sides of the Bahama Banks, therefore, turns the banks into extensive leeward *platform-lagoons* of restricted circulation and low turbulence.

A second major point of significance of the islands concerns the spatial relations between the islands and the coral reef tracts. On the Bahama Banks the coral reefs are not only restricted to the windward (eastern) edges of the platforms, but along these windward edges they occur *only in front of islands* (Fig. 15 in Ginsburg 1972). This is true also of the coral reef distribution in the Persian Gulf: inspection of maps of the Abu Dhabi area, such as those of

1. An obvious implication of the transported nature of the lime-mud in the lee-shadows of the Bahama islands is that any geochemical tests for *in situ* chemical or biochemical precipitation of such mud that do not take this factor into account are, of course, invalid (Lowenstam & Epstein 1957; Cloud 1962; Broeker & Takahashi 1966).

Kendall and Skipwith (1969), shows that the reefs are situated only on the Gulf side of bedrock islands. Gaps in the reef tract coincide with tidal channels that cut through the islands to the restricted, muddy, leeward lagoons. Ooid sand shoals are building at the channel mouths in the gaps between the reefs. The reason for the preferential growth of reefs to the seaward of islands in the Bahamas and Persian Gulf would seem to lie in the protection such islands offer from mud-carrying and/or hypersaline tidal and storm currents. With this explanation, the reef build-ups must be younger than the islands and owe their distribution to the islands. The geological implication is fascinating—in the geological record a muddy lagoon or evaporite facies must be physically separated from a reef-tract facies by a solid island barrier.[2]

As a final point it would seem rewarding to turn the question of the islands around. What would the present sedimentary environments and Holocene record of the Bahama Banks be like if there were no islands? The entire platform would be exposed directly to constant easterly wind-wave and tidal current agitation and Atlantic Ocean storm-surge onslaughts. Mudflats could not develop, and only subtidal (and some intertidal) lag sand sheets and shoals (like those described so well by Ball 1967) could accumulate. The biological communities would likely be quite different and so affect the kinds of sediment produced. *Thallasia*-type seagrass communities rich in molluscan fauna (Jackson 1973; Davies 1970b; Read 1974) would perhaps dominate, and mud-producing green algal colonies would be minor. Whatever mud was produced by seagrass epibionts (encrusting forams and red algae), by green algae, and by comminution of large skeletal grains would be washed off the platform if not firmly bound by the seagrass root-mass. Even so, periodic large storms would stir up mud and silt from the seagrass stands and carry these fines off the platform. In these circumstances, where no protecting island barriers exist to ward off such muddy ebb and storm currents, it is unlikely that fringing reefs would be successful.

---

2. It is a point not often recognized that coral reefs are not barriers to anything but waves. Backreef organisms are little different from those of the reef itself, reflecting that backreef lagoons are not "restricted" environments. Reefs certainly are not water-circulation barriers and cannot be the cause of the restriction necessary to produce a hypersaline backreef lagoon in which evaporite minerals can precipitate.

10 A CONCLUDING NOTE: SENSITIVITY OF THE RECORD

Lawrence A. Hardie

Perhaps the most encouraging finding we made in this study was the remarkable ability of the tidal flat sediments to record small differences in environmental conditions. A simple illustration of this is given in Figure 4, which shows the ranges of exposure index and surface elevation of several types of sedimentary structures. The figure brings out the important point that very small surface elevation differences (measured in centimeters) across the tidal flat result in major differences in frequency of exposure and in sedimentary structures like layering, mud-cracks, and burrows. Of course, many complex and interdependent parameters are actually involved in producing the spectrum of sedimentary structures at Three Creeks—for example, frequency of sedimentation, proximity to sediment source, frequency and duration of flooding by seawater, frequency and duration of flooding by rainwater, topography, competition among organisms (and a whole host of biologic parameters), etc.—but, for one particular topographic feature like the channel levee, for example, elevation (or else frequency of subaerial exposure) is a conveniently simple "summary" parameter. However complex the operational parameter array actually is, the fact remains very clear that the end results—the sediments, with their specific structures, textures, and fossils—do reflect even very small changes in overall conditions from one physiographic subenvironment to another across the entire tidal flat. The implication is obvious and significant: highly refined environmental interpretations of ancient tidal flat deposits are within our grasp, *if only we have the proper calibrations.* If we had enough working models it should be possible to read with some precision the climate, weather patterns, tidal regime, and general hydrology, physiography, depositional mechanisms, water chemistry, organism activities, etc., of ancient tidal flat deposits. An attempt is made in Chapter 8 to show the possibilities.

With the sediment record being sensitive to so many variables, it is easy to see why no two tidal flat deposits, ancient or modern, are quite alike, a point stressed in the introduction to this book. It follows that it is unlikely that there exist in the geologic column any deposits exactly analogous to the modern tidal flat accumulations of the Bahamas, Shark Bay, the Persian Gulf, etc. If we try to force ancient deposits to fit one or other of the few studied modern deposits (as appears to be quite prevalent) then at best only general interpretations of the depositional environment can result. Certainly, as a first approximation such a method has value, but if we hope to go beyond broad comparisons and to exploit the wealth of environmental information stored in ancient tidal flat deposits then we must use an approach that stresses the individuality of each deposit. Our only recourse is to build "thought models" by piecing together whatever we need from an array of basic "building blocks." These building blocks are the sedimentary structures, fabrics, textures, produced by the spectrum of processes that operate on tidal flats of all kinds, rainy or arid, high energy or low energy, carbonate or noncarbonate, tropical or temperate, land-locked or surrounded by the ocean. With a full storehouse of such building blocks we should be able to construct any model we need, some pieces coming from one modern example, some from another.

The calibration of the building blocks must come mainly from studies of modern active tidal flats, where processes can be monitored. To be operational, the data must be in the form of *detailed and carefully illustrated descriptions of the sedimentary structures, fabrics, and textures and the processes that produced them*. For this approach to the study of modern deposits, the "parts" of the deposit are more important than the "whole," the "kind" of process more important than the "scale." Unfortunately, at the present time we have far too few of such data, as anyone who has tried to do any rock-by-rock comparative sedimentology can surely testify! The solution to the problem rests with the sedimentologists working in modern sediments.

In a wider perspective one can hope that this approach will be applied to all kinds of sedimentary environments, because the goal of making highly refined environmental interpretations (which requires a deep understanding of what it takes to make a distinctive lithology or subfacies) is not simply an academic sedimentologic exercise, but has important implications for other branches of geology. For example, in paleoecology it is more often necessary to know the details of the depositional environment before a proper understanding of the enclosed communities can be reached, than the other way around (see, for example, Makurath 1975). The same applies to studies of organic evolution, where it is important to recognize environment-determined changes in species distribution and succession. It is also true that, for the most part, the geochemistry of a particular deposit—contemporaneous, pene-contemporaneous, or late diagenetic chemical processes—cannot be properly evaluated without first understanding the primary depositional environment in some detail. This is certainly so for evaporite deposits (Hardie & Eugster 1971; Bosellini & Hardie 1973; Eugster & Hardie 1974) and, I believe, is also true for dolomites. The application of detailed information obtained from sedimentary rocks to the study of paleoclimatology and paleooceanography is quite obvious, but there are, I think, even wider implications: crucial tests of the theory of plate tectonics and models of crustal evolution (mainly continent-ocean history) might very well lie in the detailed environmental information stored in sedimentary deposits. Finally, the economic significance of refined facies maps and block diagrams has long been recognized by the petroleum industry, and, indeed, it was the geologists of the oil companies who were mainly responsible for the advances we have made in the last twenty years in understanding ancient and modern depositional environments.

REFERENCES

Ahnert, F. 1960. Estuarine meanders in the Chesapeake Bay area. *Geographical Review* 50:390–401.

Ahr, W. M. 1971. Paleoenvironment, algal structures and fossil algae in the upper Cambrian of central Texas. *J. Sediment. Petrol.* 41:205–16.

Aitken, J. D. 1967. Classification and environmental significance of cryptalgal limestones and dolomites with illustrations from the Cambrian and Ordovician of southwestern Alberta. *J. Sediment. Petrol.* 37:1163–78.

Allen, R. P. 1956. The Flamingos: their life history and survival. National Audubon Soc., *Res. Rept.*, No. 5.

Ball, M. M. 1967. Carbonate sand bodies of Florida and the Bahamas. *J. Sediment. Petrol.* 37:556–91.

Ball, M. M., Shinn, E. A., and Stockman, K. W. 1967. The geologic effects of hurricane Donna in south Florida. *J. Geol.* 75:583–97.

Bathurst, R. G. C. 1971. *Carbonate sediments and their diagenesis.* Developments in Sedimentology 12, Amsterdam: Elsevier Publ. Co.

Beerbower, J. R. 1960. *Search for the Past.* Englewood Cliffs, N.J.: Prentice-Hall.

Bernoulli, D., and Wagner, C. W. 1970. Subaerial diagenesis and fossil caliche deposits in the Calcare Massiccio formation (lower Jurassic central Appennines, Italy). *Neues Jahrb. Geol. Palaontol.*, Abh., 138:135–49.

Black, M. 1933. The algal sediments of Andros Island, Bahamas. *Phil. Trans. Roy. Soc. London, Ser.* B. 222: 165–92.

Bosellini, A. 1967. La tematica deposizionale della Dolomia Principale (Dolomiti e Prealpi Venete). *Boll. Soc. Geol. Ital.* 86:133–69.

Bosellini, A., and Hardie, L. A. 1973. Depositional theme of a marginal marine evaporite. *Sedimentology* 20:5–27.

Bradley, W. H. 1929. Algae reefs and oolites of the Green River Formation. *U.S. Geol. Surv.*, Prof. Paper 154:225–66.

Broeker, W. S., and Takahashi, T. 1966. Calcium carbonate precipitation on the Bahama Banks. *J. Geophys. Res.* 71:1575–1602.

Bucher, W. H. 1938. Key to papers published by an institute for the study of modern sediments in shallow seas. *J. Geol.* 46:726–55.

Burkholder, P. R. 1934. Movement in Cyanophyceae. *Quart. Rev. Biol.* 9:438–59.

Butler, G. P. 1970. Holocene gypsum and anhydrite of the Abu Dhabi sabkha, Trucial Coast: an alternative explanation of origin. *In:* J. L. Rau and L. F. Dellwig (editors), *Third Symposium on Salt.* Cleveland, Ohio: Northern Ohio Geol. Soc., pp. 120–52.

Cloud, P. E. 1962. Environment of calcium carbonate deposition west of Andros Island, Bahamas. *U.S. Geol. Surv.*, Prof. Paper 350:1–138.

Davies, G. R. 1970a. Algal-laminated sediments, Gladstone embayment, Shark Bay, Western Australia. *Am. Assoc. Petrol. Geologists*, Mem. 13:169–205.

———. 1970b. Carbonate-bank sedimentation, eastern Shark Bay, Western Australia. *Am. Assoc. Petrol. Geologists*, Mem. 13:85–168.

Davies, G. R., and Ludlam, S. D. 1971. A basinal model for Middle Devonian "laminites," Elk Point basin of Western Canada (abstract). *Geol. Soc. Am.*, Abs. with Prog., 1971 Ann. Meetings, Washington, D.C.

Davis, J. H. 1940. The ecology and geologic role of mangroves in Florida. *Papers Tortugas Lab.* 32:303–412.

Deffeyes, K. S., Lucia, F. J., and Weyl, P. K. 1965. Dolomitization of Recent and Plio-Pleistocene sediments by marine evaporite waters on Bonaire, Netherlands Antilles. *Soc. Econ. Paleontologists Mineralogists*, Spec. Publ. 13:71–88.

Doty, M. S. 1946. Critical tide factors that are correlated with the vertical distribution of marine algae and other organisms along the Pacific coast. *Ecology* 27:315–28.

Drouet, F. 1963. Ecophenes of *Schizothrix calcicola*. *Proc. Acad. Nat. Sciences Philadelphia* 115:261–81.

Dunn, G. E., and Miller, B. I. 1960. *Atlantic Hurricanes*. Baton Rouge: Louisiana State University Press.

Eardley, A. J. 1938. Sediments of Great Salt Lake, Utah. *Am. Assoc. Petrol. Geologists, Bull.* 22:1305–1411.

Elenkin, A. A. 1936. Sinezelenyye Vodorosli S.S.S.R. *Izdvo Akad. Nauk S.S.S.R.*

Eugster, H. P., and Hardie, L. A. 1974. Sedimentation in an ancient playa-lake complex: the Wilkins Peak Member of the Green River Formation of Wyoming. *Geol. Soc. Am., Bull.* 86:319–34.

Evans, G. 1965. Intertidal flat sediments and their environments of deposition in the Wash. *Quart. J. Geol. Soc. London* 121:209–45.

Fischer, A. G. 1964. The Lofer cyclothems of the Alpine Triassic. *Geol. Surv. Kansas, Bull.* 169:107–49.

Folk, R. L. 1962. Spectral subdivision of limestone types. *Am. Assoc. Petrol. Geologists, Mem.* 1:62–84.

Fritsch, F. E. 1952. *The Structure and Reproduction of the Algae*. Volume 2, Cambridge: University Press.

Füchtbauer, H., and Hardie, L. A. 1976. Experimentally determined homogeneous distribution coefficients for precipitated magnesian calcites: application to marine carbonate cements (abstract). *Geol. Soc. Am., Abs. with Prog.*, 1976 Ann. Meetings, Denver, Colo., p. 877.

Garrett, P. 1970. Phanerozoic stromatolites: noncompetitive ecological restriction by grazing and burrowing animals. *Science* 169:171–73.

———. 1971. *The sedimentary record of life on a modern tropical carbonate tidal flat, Andros Island, Bahamas*. Ph.D. thesis, The Johns Hopkins University, Baltimore, Maryland.

Gebelein, C. D. 1969. Distribution, morphology, and accretion rate of Recent subtidal algal stromatolites, Bermuda. *J. Sediment. Petrol.* 39:46–69.

Gebelein, C. D., and Hoffman, P. 1968. Intertidal stromatolites and associated facies from Cape Sable, Florida. *Geol. Soc. Am.*, Spec. Papers, 121:109 (abstract).

———. 1973. Algal origin of dolomite laminations in stromatolitic limestone. *J. Sediment. Petrol.* 43:603–13.

Ginsburg, R. N. 1955. Recent stromatolitic sediments from south Florida (abstract). *J. Paleo.* 29:723.

——— 1957. Early diagenesis and lithification of shallow-water carbonate sediments in South Florida. *Soc. Econ. Paleontologists Mineralogists*, Spec. Publ. 5:80–100.

——— 1960. Ancient analogues of Recent stromatolites. *Report XXI Int. Geol. Congress*, Part XXII, Int. Paleontol. Union, 26–35.

———. 1972. *South Florida carbonate sediments*. Univ. Miami, Comparative Sed. Lab.

———. (editor), 1975. *Tidal Deposits*. N.Y.: Springer-Verlag.

Ginsburg, R. N., Bernard, H. A., and Moody, R. A. 1966. The Shell method of impregnating cores of unconsolidated sediment. *J. Sediment. Petrol.* 36:1118–25.

Ginsburg, R. N., Bricker, O. P., Wanless, H. L., and Garrett, P. 1970. Exposure index and sedimentary structures of a Bahama tidal flat. *Geol. Soc. Am., Abs. with Prog.*, Annual Meetings, Milwaukee, Wisconsin, pp. 744–45.

Ginsburg, R. N., and Hardie, L. A. 1975. Tidal and storm deposits, northwestern Andros Island, Bahamas. *In:* R. N. Ginsburg (editor), *Tidal Deposits*. N.Y.: Springer-Verlag.

Ginsburg, R. N., Isham, L. B., Bein, S. J., and Kuperburg, J. 1954. Laminated algal sediments of south Florida and their recognition in the fossil record. Report 54–21, *Univ. Miami Marine Lab.*, unpublished.

Ginsburg, R. N., and Lloyd, R. M. 1956. A manual piston coring device for use in shallow water. *J. Sediment. Petrol.* 26:64–66.

Graf, W. H. 1971. *Hydraulics of Sediment Transport*. New York: McGraw-Hill Book Co.

Guy, H. P., Simons, D. B., and Richardson, E. V. 1966. Summary of alluvial channel data from flume experiments, 1956–1961. *U.S. Geol. Surv.*, Prof. Paper 462 I.

Häntzschel, W., El-Baz, R., and Amstutz, G. C. 1968. Coprolites, an annotated bibliography. *Geol. Soc. Am.*, Mem. 108.

REFERENCES

Hardie, L. A. 1969. Algal crusts from the Bahamas. *Am. Assoc. Petrol. Geologists*, Bull. 53:721 (abstract).

Hardie, L. A., and Eugster, H. P. 1971. The depositional environment of marine evaporites: a case for shallow clastic accumulation. *Sedimentology* 16:187–220.

Harms, J. C. 1969. Hydraulic significance of some sand ripples. *Geol. Soc. Am.*, Bull. 80:363–96.

Harris, D. L. 1959. An interim hurricane storm surge forecasting guide. *Nat. Hurricane Res. Project Report* 32, U.S. Weather Bureau, Washington, D.C.

———. 1963. Characteristics of the hurricane storm surge. *Tech. Paper* 48, U.S. Dept. Commerce, U.S. Weather Bureau, Washington, D.C.

Hedgpeth, J. W. 1957. Biological Aspects of Estuaries and Lagoons. *In:* Treatise on Marine Ecology and Paleoecology, 1, Ecology, *Geol. Soc. Am.*, Mem. 67:693–729.

Hjulström, F. 1935. Studies of the morphological activity of rivers as illustrated by the river Fyris. *Bull. Geol. Inst. Upsala* 25:221–527.

Hoffman, P. 1967. Algal stromatolites: use in stratigraphic correlation and paleocurrent determination. *Science* 157:1043–45.

———. 1968. Stratigraphy of the Lower Proterozoic Great Slave Supergroup, East Arm of Great Slave Lake, District of Mackenzie. *Geol. Surv. Canada*, Paper 68–42.

Hoffmeister, J. E., and Multer, H. G. 1965. Fossil mangrove reef of Biscayne Bay, Florida. *Geol. Soc. Am.*, Bull. 76:845–52.

Hsü, K. J., and Siegenthaler, Ch. 1969. Preliminary experiments on hydrodynamic movement induced by evaporation and their bearing on dolomite problem. *Sedimentology* 12:11–25.

Illing, L. V. 1954. Bahaman calcareous sands. *Am. Assoc. Petrol. Geologists* 38:1–95.

Illing, L. V., Wells, A. J., and Taylor, J. C. M. 1965. Penecontemporary dolomite in the Persian Gulf. *In:* L. C. Pray and R. C. Murray (editors), *Dolomitization and Limestone Diagenesis: A Symposium. Soc. Econ. Paleontologists Mineralogists*, Spec. Publ. 13:89–111.

Irion, G., and Müller, G. 1968. Mineralogy, petrology and chemical composition of some calcareous tufa from the Schwäbische Alb, Germany. *In:* G. Müller and G. M. Friedman (editors), *Recent Developments in Carbonate Sedimentology in Central Europe.* N.Y.: Springer-Verlag, pp. 157–71.

Jackson, J. B. C. 1973. The ecology of molluscs of *Thalassia* communities, Jamaica, West Indies. I. Distribution, environmental physiology, and ecology of common shallow-water species. *Bull. Marine Sciences* 23:313–50.

Johnson, J. H. 1961. Limestone-building algae and algal limestones. *Colo. School Mines*, Spec. Publ.

Johnson, R. G. 1964. The community approach to paleoecology. *In:* J. Imbrie and N. D. Newell (editors), *Approaches to Paleoecology.* New York: Wiley, pp. 107–34.

Katz, A. 1973. The interaction of magnesium with calcite during crystal growth at 25–90° C and one atmosphere. *Geochim. Cosmochim. Acta* 37:1563–86.

Kendall, C. G. St. C., and Skipwith, Sir P. A. d'E. 1968. Recent algal mats of a Persian Gulf Lagoon. *J. Sediment. Petrol.* 38:1040–58.

———. 1969. Geomorphology of a Recent shallow water carbonate province: Khor al Bazam, Trucial Coast, Southwestern Persian Gulf. *Geol. Soc. Am.*, Bull. 80:865–92.

Kornicker, L. S., and Purdy, E. G. 1957. A Bahamian faecal-pellet sediment. *J. Sediment. Petrol.* 27:126–28.

Laporte, L. F. 1967. Carbonate deposition near mean sea-level and resulting facies mosaic: Maniius formation (Lower Devonian) of New York State. *Am. Assoc. Petrol. Geologists*, Bull. 51:73–101.

Leopold, L. B., Wolman, M. G., and Miller, J. P. 1964. *Fluvial Processes in Geomorphology.* San Francisco: W. H. Freeman.

Lochman, C. 1968. *Crepicephalus* faunule from the Bonneterre Dolomite (Upper Cambrian) of Missouri. *J. Paleo.* 42:1153–62.

Logan, B. W. 1961. *Cryptozoon* and associate stromatolites from the Recent, Shark Bay, Western Australia. *J. Geol.* 69:517–33.

Logan, B. W., and Cebulski, D. E. 1970. Sedimentary environments of Shark Bay, Western Australia. *Am. Assoc. Petrol. Geologists*, Mem. 13:1–37.

Logan, B. W.; Davies, G. R.; Read, J. F., and Cebulski, D. E. 1970. Carbonate sedimentation and environments, Shark Bay, Western Australia. *Am. Assoc. Petrol. Geologists*, Mem. 13.

Logan, B. W., Harding, J. L., Ahr, W. M., Williams, J. D., and Snead, R. G. 1969. Carbonate sediments and reefs, Yucatan shelf, Mexcio. *Am. Assoc. Petrol. Geologists*, Mem. 11.

Logan, B. W., Hoffman, P. and Gebelein, C. D. 1974. Algal mats, cryptalgal fabrics, and structures, Hamelin Pool, Western Australia. *Am. Assoc. Petrol. Geologists*, Mem. 22: 140–94.

Logan, B. W., Rezak, R., and Ginsburg, R. N. 1964. Classification and environmental significance of algal stromatolites. *J. Geol.* 72:68–83.

Lowenstam, H. A., and Epstein, S. 1957. On the origin of sedimentary aragonite needles of the Great Bahama Bank. *J. Geol.* 65:364–75.

Lucia, F. J. 1968. Recent sediments and diagenesis of South Bonaire, Netherlands Antilles. *J. Sediment. Petrol.* 38:848–58.

———. 1972. Recognition of evaporite-carbonate shoreline sedimentation. *Soc. Econ. Paleontologists Mineralogists*, Spec. Publ. 16:160–91.

Makurath, J. H. 1975. Sedimentology and paleoecology of the Keyser Limestone (upper Silurian-lower Devonian) of the central Appalachians. Ph.D. dissertation, The Johns Hopkins University, Baltimore, Maryland.

Matter, A. 1967. Tidal flat deposits in the Ordovician of Western Maryland. *J. Sediment. Petrol.* 37:601–09.

Matthews, R. K. 1965. Genesis of Recent lime mud in Southern British Honduras. Ph.D. thesis, Rice University.

McGugan, A. 1967. Possible use of algal stromatolite rhythms in geochronology (abstract). *Geol. Soc. Am.*, Spec. Paper 115:145.

McKee, E. D., and Weir, G. W. 1953. Terminology for stratification and cross-stratification in sedimentary rocks. *Geol. Soc. Am.*, Bull. 64:381–90.

Monty, C. 1967. Distribution and structure of Recent stromatolitic algal mats, eastern Andros Island, Bahamas. *Ann. Soc. Geol. Belg.*, Bull. 90:55–100.

Monty, C. L. V. 1972. Recent algal stromatolitic deposits, Andros Island, Bahamas, Preliminary Report. *Geol. Rundschau* 62:742–83.

Moore, H. B. 1931. The muds of the Clyde Sea area, III. Chemical and physical conditions; rate of sedimentation; and fauna. *Marine Biol. Assoc. U.K. Jour.*, n.s. 17:325–58.

Moore, R. C. (ed.) 1964. *Treatise on Invertebrate Paleontology*, Part C, Protista 2 (Sarcodina). University Kansas Press.

Neumann, A. C., Gebelein, C. D., and Scoffin, T. P. 1970. The composition, structure and erodability of subtidal mats, Abaco, Bahamas. *J. Sediment. Petrol.* 40:274–97.

Newell, N. D., Imbrie, J., Purdy, E. G., and Thurber, D. L. 1959. Organism communities and bottom facies, Great Bahama Bank. *Bull. Am. Museum Nat. Hist.* 117 (4):181–228.

Newell, N.D., and Rigby, J. K. 1957. Geological studies on the Great Bahama Bank. *In:* R. J. Leblanc and J. G. Breeding (editors), *Regional Aspects of Carbonate Deposition. Soc. Econ. Paleontologists Mineralogists*, Spec. Publ. 5:15–72.

Perkins, R. D., and Enos, P. 1968. Hurricane Betsy in the Florida-Bahama area—geologic effects and comparison with hurricane Donna. *J. Geol.* 76:710–17.

Phleger, F. B. 1969. A modern evaporite deposit in Mexico. *Am. Assoc. Petrol. Geologists*, Bull. 53:824–29.

Pray, L. C. 1968. Hurricane Betsy (1965) and nearshore carbonate sediments of the Florida Keys. *Geol. Soc. Am.*, Spec. Paper 101:168–69 (abstract)

Purdy, E. G. 1963*a*. Recent calcium carbonate facies of the Great Bahama Bank, 1. Petrography and reaction groups. *J. Geol.* 71:334–55.

———. 1963*b*. Recent calcium carbonate facies of the Great Bahama Bank. 2. Sedimentary facies. *J. Geol.* 71:472–97.

———. 1968. Carbonate diagenesis: an environmental survey. *Geol. Romana.* 7:183–228.

Rabbi, E., and Ricci Lucchi, F. 1968. Stratigrafia e sedimentologia del Messiniano Forlwese (Dintorni di Predappio). *Giornale di Geologia, Bologna*, Ser. 2, 34:595–624.

Read, J. F. 1974. Carbonate bank and wave-built platform sedimentation, Edel Province, Shark Bay, Western Australia. *Am. Assoc. Petrol. Geologists*, Mem. 22:1–60.

Reineck, H. E. 1967, Layered sediments of tidal flats, beaches, and shelf bottoms of the North Sea. *In:* G. H. Lauff (editor), *Estuaries*, Publ. 83, Am. Assoc. Adv. Science, Washington, D.C., pp. 191–206.

———. 1970. *Das Watt, Ablagerungs und Lebensraum.* Frankfurt am Main: Waldemar Kramer, pp. 1–142.

———. 1972. Tidal flats. *In:* J. K. Rigby and H. R. Gould (editors), *Recognition of ancient sedimentary environments. Soc. Econ. Paleontologists Mineralogists*, Spec. Publ. 16:146–59.

Reineck, H. E., and Singh, I. B. 1972. Genesis of laminated sand and graded rhythmites in storm-sand layers of shelf mud. *Sedimentology* 18:123–28.

Reineck, H. E., and Wunderlich, F. 1968. Die Entstehung von Schichten und Schichtbänken im Watt. *Senckenbergiana maritima* 1:85–106.

Rhoads, D. C. 1967. Biogenic reworking of intertidal and subtidal sediments in Barnstable Harbor and Buzzards Bay, Massachusetts. *J. Geol.* 75:461–76.

Rhoads, D. C., and Young, D. K. 1970. The influence of deposit-feeding organisms on sediment stability and community tropic structure. *J. Marine Res.* 28:150–78.

Ricketts, E. F., and Calvin, J. 1968. *Between Pacific Tides,* California: Stanford University Press, 4th ed., revised by Joel W. Hedgpeth.

Roehl, P. O. 1967. Stony Mountain (Ordovician) and Interlake (Silurian) facies analogs of Recent low-energy marine and subaerial carbonates, Bahamas. *Am. Assoc. Petrol. Geologists*, Bull. 51:1979–2032.

Sanders, H. L., Mangelsdorf, P. C., and Hampson, G. R. 1965. Salinity and faunal distribution in the Pocasset River, Massachusetts. *Limnol. Oceanog.* 10:216–229.

Schäfer, W. 1962. *Aktuo-Paläontologie nach Studien in der Nordsee.* Frankfurt am Main: Kramer.

Schenk, P. E. 1967. The Macumber Formation of the Maritime Province, Canada—a Mississippian analogue to Recent strand-line carbonates of the Persian Gulf. *J. Sediment. Petrol.* 37:365–76.

Schönleber, K. 1936. Scytonema Julianum. Beiträge zur normalen und pathologischen Cytologie und Cytogenese der Blaualgen. *Arch. Protistenk.* 88:36–68.

Schwartz, M. L. 1971. The multiple causality of barrier islands. *J. Geol.* 79:91–94.

Scoffin, T. P. 1970. The trapping and binding of subtidal carbonate sediments by marine vegetation in Bimini Lagoon, Bahamas. *J. Sediment. Petrol.* 40:249–73.

Seilacher, A. 1964. Biogenic sedimentary structures. *In:* J. Imbrie and N. D. Newell (editors), *Approaches to Paleoecology.* New York: Wiley, pp. 296–316.

Shinn, E. A. 1968. Burrowing in recent lime sediments of Florida and the Bahamas. *J. Paleontol.* 42:879–94.

———. 1969. Submarine lithification of Holocene carbonate sediments in the Persian Gulf. *Sedimentology* 12:109–45.

Shinn, E. A., Ginsburg, R. N., and Lloyd, R. M. 1965. Recent supratidal dolomite from Andros Island, Bahamas. *In:* L. C. Pray and R. C. Murray (editors), *Dolomitization and Limestone Diagenesis: A Symposium. Soc. Econ. Paleontologists Mineralogists*, Spec. Publ. 13:112–23.

Shinn, E. A., Lloyd, R. M., and Ginsburg, R. N. 1969. Anatomy of a modern carbonate tidal-flat, Andros Island, Bahamas. *J. Sediment. Petrol.* 39:1202–28.

Smith, C. L. 1940. The Great Bahama Bank. 1. General hydrographic and chemical factors. 2. Calcium carbonate precipitation. *J. Marine Res.* 3, 1–31; 147–89.

Stanley, S. M. 1970. Relation of shell form to life habits of the Bivalvia (Mollusca). *Geol. Soc. Am.*, Mem. 125.

Streeter, S. S. 1963. Foraminifera in the sediments of the northwestern Great Bahama Bank. Ph.D. dissertation, Columbia University, New York.

Sundborg, A. 1956. The river Klarälven, a study of fluvial processes. *Geog. Annaler* 38 (2):127–316.

Swinchatt, J. P. 1969. Algal boring, a possible depth indicator in carbonate rocks and sediments. *Geol. Soc. Am.*, Bull. 80:1391–96.

Tannehill, I. R. 1938. *Hurricanes, their nature and history, particularly those of the West Indies and southern coasts of the United States.* Princeton: Princeton University Press.

Taylor, J. D. M., and Illing, L. V. 1969. Holocene intertidal calcium carbonate cementation, Qatar, Persian Gulf. *Sedimentology* 12:69–109.

Teal, J., and Teal, M. 1969. *Life and Death of the Salt Marsh.* Boston: Little Brown.

Thompson, R. W. 1968. Tidal flat sedimentation on the Colorado River delta, northwestern Gulf of California. *Geol. Soc. Am.*, Mem. 107.

Thunmark, S. 1926. Bidrag till Kännedomen om recenta Kalktuffer. *Geol. Fören. Stockholm Forhandl.* 48:541–83.

Todd, R., and Low, D. 1971. Foraminifera from the Bahama Bank west of Andros Island. *U.S. Geol. Surv.*, Prof. Paper 683-C.

Tourek, T. J. 1968. The tidal creek system on a portion of western Andros Island. Unpublished manuscript.

———. 1970. *The depositional environments and sediment accumulation models for the Upper Silurian Wills Creek Shale and Tonoloway Limestone, central Appalachians.* Ph.D. dissertation, The Johns Hopkins University, Baltimore, Maryland.

Traverse, A., and Ginsburg, R. N. 1966. Palynology of the surface sediments of Great Bahama Bank, as related to water movement and sedimentation. *Marine Geol.* 4:417–59.

———. 1967. Pollen and associated microfossils in the marine surface sediments of the Great Bahama Bank. *Rev. Paleobotan. Palynol.* 3:243–54.

van Straaten, L. M. J. U. 1954. Composition and structure of recent marine sediments in the Netherlands. *Leidse Geol. Mededel.* 19:1–110.

———. 1961. Sedimentation in tidal flat areas. *Alberta Soc. Petrol. Geologists J.* 9:203–26.

Verwey, J. 1952. On the ecology of distribution of cockle and mussel in the Dutch Waddensea, their role in sedimentation and the source of their food supply. *Arch. Neerl. Zool.* 10:171–239.

Vincent, J. 1967. Effect of bedload movement on the roughness coefficient value. Proc. 12th Cong., *Int. Assoc. Hydraulic Res., vol. 1,* Colorado State University, Fort Collins, Colorado,: A20.1–A20.10.

Walker, K. R., and Laporte, L. F. 1970. Congruent fossil communities from Ordovician and Devonian carbonates of New York. *J. Paleo.* 44:928–44.

Wanless, H. R. 1969. Sedimentary structure zonation on tidal levees, Andros Island, Bahamas. *Am. Assoc. Petrol. Geologists,* Bull. 53:748, abstract.

———. 1971. Carbonate tidal flats of the Grand Canyon Cambrian. *Geol. Soc. Am.,* Abs. with Prog., Ann. meetings, Washington, D.C., p. 743, abstract.

———. 1973. Cambrian of the Grand Canyon—a re-evaluation of the depositional environment. Ph.D. dissertation, The Johns Hopkins University, Baltimore, Maryland.

Weimer, R. J., and Hoyt, J. H. 1964. Burrows of *Callianassa major* Say, geologic indicators of littoral and shallow neritic environments. *J. Paleo.* 38:761–67.

Welander, P. 1961. Numerical prediction of storm surges. *Advances in Geophys.* 8:315–79. N.Y.: Academic Press.

Winland, H. D. 1969. Stability of calcium carbonate polymorphs in warm, shallow sea water. *J. Sediment. Petrol.* 39:1579–87.

Wolman, R. G., and Brush, L. M. 1961. Factors controlling the size and shape of stream channels in coarse, non-cohesive sands. *U.S. Geol. Surv.,* Prof. Paper 282-G:183–210.

Woods, P. J., and Brown, R. G. 1975. Carbonate sedimentation in an arid-zone tidal flat, Nilemah Embayment, Shark Bay, Western Australia. *In:* R. N. Ginsburg (editor), *Tidal Deposits.* N.Y.: Springer-Verlag.

INDEXES

Author Index

Ahnert, F., 44
Ahr, W. M., 177
Allen, R. P., 139, 155

Ball, M. M., 103, 187
Bathurst, R.G.C., 106, 131
Beerbower, J. R., 121
Bernoulli, D., 52
Black, M., 1, 26, 29, 49, 51, 68, 69, 74, 95, 106, 110, 114, 116, 117, 159, 165, 166, 169
Bosellini, A., 1, 52, 189
Bradley, W. H., 171
Bricker, O. P., 19
Broeker, W. S., 186
Brown, R. G., 7
Brush, L. M., 107
Bucher, W. H., 50
Burkholder, P. R., 169
Butler, G. P., 121

Calvin, J., 185
Cebulski, D. E., 101
Cloud, P. E., 1, 12, 34, 125, 130, 132, 133, 146, 186

Davies, G. R., 7, 51, 106, 112, 113, 184, 187
Davis, J. H., 156
Deffeyes, K. S., 159
Doty, M. S., 7, 8, 167, 168, 169
Drouet, F., 22
Dunn, G. E., 17

Eardley, A. J., 171
Elenkin, A. A., 171
Enos, P., 103, 108, 186
Epstein, S., 186
Eugster, H. P., 189
Evans, G., 50

Fischer, A. G., 1, 7, 52, 71, 106, 122, 185
Folk, R. L., 84
Fritsch, F. E., 169, 171
Füchtbauer, H., 172, 173, 174

Garrett, P., 7, 12, 53, 55, 57, 68, 114, 118, 119, 120, 124, 136, 139, 157, 186
Gebelein, C. D., 106, 110, 141, 145, 172, 173
Ginsburg, R. N., 1, 2, 6, 7, 8, 12, 13, 15, 19, 20, 21, 51, 92, 103, 106, 121, 122, 131, 139, 144, 146, 159, 160, 169, 181, 182, 184, 186
Graf, W. H., 108
Guy, H. P., 101

Häntzschel, W., 146
Hardie, L. A., 1, 7, 15, 29, 53, 68, 72, 124, 131, 139, 144, 146, 157, 159, 160, 172, 173, 174, 189
Harms, J. C., 102

Harris, D. L., 97, 100, 102
Hedgpeth, J. W., 185
Hjulström, F., 101, 102
Hoffman, P., 55, 85, 106, 116, 172, 173, 176
Hoffmeister, J. E., 156
Hoyt, J. H., 146, 149, 155
Hsü, K. J., 174

Illing, L. V., 1, 51, 132, 146, 159, 184
Irion, G., 171

Jackson, J.B.C., 187
Johnson, J. H., 159, 177
Johnson, R. G., 124

Katz, A., 172
Kendall, C.G.St.C., 1, 51, 115, 187
Kornicker, L. S., 146

Laporte, L. F., 1, 52, 106, 124, 185
Leopold, L. B., 108
Lloyd, R. M., 2, 6, 12
Lochman, C., 185
Logan, B. W., 1, 51, 75, 101, 106, 114, 116, 186
Low, D., 125, 132
Lowenstam, H. A., 186
Lucchi, F. Ricci, 113
Lucia, F. J., 1, 7, 159
Ludlam, S. D., 113

McGugan, A., 52
McKee, E. D., 53
Makurath, J. H., 189
Matter, A., 1, 52, 106, 176
Matthews, R. K., 133
Miller, B. I., 17
Monty, C., 26, 29, 37, 49, 68, 71, 74, 82, 106, 139, 159, 164, 166, 167, 172
Moore, H. B., 146
Moore, R. C., 132
Müller, G., 171
Multer, H. G., 156

Neumann, A. C., 31, 101, 102, 141, 145
Newell, N. D., 1, 34, 124, 125

Perkins, R. D., 103, 108, 186
Phleger, F. B., 184
Pray, L. C., 103, 184
Purdy, E. G., 1, 12, 34, 132, 133, 146

Rabbi, E., 113
Read, J. F., 187
Reineck, H. E., 50, 124, 184
Rhoads, D. C., 137, 156
Ricketts, E. F., 185

197

Rigby, J. K., 34
Roehl, P. O., 1, 52, 106

Sanders, H. L., 138
Schäfer, W., 124
Schenk, P. E., 1, 106, 113
Schönleber, K., 171, 172
Schwartz, M. L., 121
Scoffin, T. P., 101, 141, 145, 156
Seilacher, A., 138, 147, 156
Shinn, E. A., 1, 2, 7, 12, 13, 15, 29, 42, 43, 49, 51, 72, 79, 85, 88, 90, 91, 103, 120, 121, 123, 145 147, 149, 151, 155, 156, 157, 159, 161, 163, 165 173, 175
Siegenthaler, Ch., 174
Singh I. B., 50
Skipwith, P.A.d'E., 1, 51, 115, 187
Stanley, S. M., 136, 182
Streeter, S. S., 125, 132
Sundborg, A., 102
Swinchatt, J., 139

Takahashi, T., 186
Tannehill, I. R., 18
Taylor, J.D.M., 51, 159

Teal, J., 185
Teal, M., 185
Thompson, R. W., 113, 121, 184
Thunmark, S., 171
Todd, R., 125, 132
Tourek, T. J., 43, 44, 106
Traverse, A., 13, 121, 182, 186

van Straaten, L.M.J.U., 50
Verwey, J., 182
Vincent, J., 107

Wagner, C. W., 52
Walker, K. R., 124, 185
Wanless, H. L., 20, 113
Weimer, R. J., 146, 149, 155
Weir, G. W., 53
Welander, P., 97, 100
Wells, A. J., 51
Winland, H. D., 173
Wolman, R. G., 107
Woods, P. J., 7
Wunderlich, F., 50

Young, D. K., 137

Subject Index

Acetabularia, 34, 125
Aequipecten, 125
Algal crusts. See Crusts, algal
Algal heads. See also Stromatolite; Proto-stromatolite
　Black's type C, 49, 68, 114, 116–17, 183
　lithified, 75, 113, 160, 162, 172, 175
Algal marsh
　beach ridge, subenvironment, 41–42
　deposition of hurricane layers in, 102, 105
　high, layering of sediment in, 53, 68–71, 72–78
　high, structures and textures in sediment of, 180
　high, subenvironment, 26–29
　inland, layering of sediment in, 53, 68–71, 72–77, 78–82, 88
　inland, structures and textures in sediment of, 181
　inland, subenvironment, 46–49
　low, layering of sediment in, 53, 82–84
　low, structures and textures in sediment of, 180
　low, subenvironment, 29
Algal mats. See also *Schizothrix*; *Scytonema*
　agglutination of sediment by, 57, 106, 109–10, 112, 115. See also Stromatolite; Proto-stromatolite
　baffling effect of, 103, 105, 109
　binding of sediment against erosion by, 102, 110
　browsing of, by invertebrates, 114, 141–42, 154, 155
　desiccation cracking of, 29, 37, 42, 49, 68, 84
　entrainment of sediment by, 71, 105, 109
　zonation of, 165–70
Algal stromatolite. See Stromatolite; Proto-stromatolite
Algal tufa, 71, 74, 75, 81, 82, 88, 103, 113, 114, 115, 159, 160, 162, 163, 164, 165, 171–73, 175, 177, 183
Alpheus, 35, 45, 126, 145, 149, 150, 151, 155
Anacystis, 29, 31, 167
Andros Island
　climate and weather of, 16–18
　location of study areas on, 12
　physiographic setting of, 13–15
　sediments of, 12–13
　tides of, 18–19
Apseudes, 120, 126, 142, 143, 144, 146, 148, 155
Aragonite
　cement, 29, 76, 165, 170, 173, 174
　sediment, 12, 90, 130, 157, 174
Armandia, 125
Atys, 125
Avicennia, 126

Batillaria, 35, 37, 125, 126, 134, 136, 141, 146, 162, 167
Batophora, 31, 34, 35, 44, 125, 126
Beach
　layering in sediment of, 53, 85, 88, 90
　mound-and-hollow structure, 35
　planar, 35–37
　structures and textures in sediment of, 180
　subenvironment, 34–37
Beach ridge backslope
　layering in sediment of, 53, 63–68
　structures and textures in sediment of, 181
　subenvironment, 41–42
Beach ridge hummock
　layering in sediment of, 53, 90
　structures and textures in sediment of, 180
　subenvironment, 42

Beach ridge washover
　layering in sediment of, 53, 57–63, 64, 85–87, 90
　structures and textures in sediment of, 180
　subenvironment, 37–41
Beach terrace
　layering of sediment of, 53, 57–63, 64, 71, 85–87
　structures and textures in sediment of, 180
　subenvironment, 37
Bedding. See Layering
Bioturbation, 31, 34, 35, 53, 64, 91, 114, 118–21, 140, 141–46, 158. See also Burrows
　by *Apseudes*, 142–43
　by *Marphysa*, rate of, 143–44
　of pond sediments, 141–44
Bonefish, 131, 139, 145
　feeding pits, 35, 142, 154
Boring, algal, 139, 173
Borrichia, 126
Brachidontes, 126, 134
Breaking of shells, 139
Burrows, 147–56, 158
　ant, 25, 37, 55, 57, 145, 151, 156
　crab, 26, 29, 35, 42, 84, 88, 145, 151–53, 156
　insect, 25, 55, 57
　oligochaete, 37, 55, 57, 64, 67, 68, 76, 145, 151, 156
　polychaete, 31, 35, 90, 143, 144, 145, 147–48, 155, 156
　shrimp, 34, 35, 45, 90, 142, 145, 148–51, 156
Buttonwood, 22

Callianassa, 34, 35, 37, 59, 90, 120, 125, 145, 146, 149, 155, 156
Callinectes, 34, 35, 125, 126, 151, 153, 155
Cantharus, 125
Cardisoma, 42, 49, 145, 151, 152, 156, 157
Casuarina, 37, 42, 128
Caulerpa, 125
Cement
　aragonite 29, 76, 165, 170, 173
　magnesian calcite, 29, 76, 165, 170
　micrite, 165
　sparry, 165, 173
　vadose, 165
Ceratonereis, 126, 147
Cerion, 49, 134, 162
Cerithium, 125
Certhidea, 35, 125, 126, 134, 136, 141, 146, 162, 167
Certhids, 31, 145
　boring of, by algal, 139
　distribution of, 135–36
　rate of grazing by, 141–42
Channel
　bank, layering in sediment of, 53–57, 88, 89
　bank subenvironment, 21
　bar, layering in sediment of, 53, 57–63, 64, 91–92
　bar, structures and textures in sediment of, 181
　bar, subenvironment, 43–46
　mouth bars, 37
Chione, 125
Cladium, 126
Clasts. See Intraclasts; Pebbles
Clypeaster, 125
Communities, invertebrate
　levee, 126
　nearshore, 124–25
　pond, 125–26
　tidal flat, 184–85

Conglomerate, 80, 85, 88. *See also* Gravel; Pebbles
Conocarpus, 126
Cross-bedding (cross-bedded sands), 89–90, 92
Crusts
    algal, 46, 113, 117, 159, 171–73, 175–77. *See also* Lamination with palisade structure
    algal structures in, summary of, 163–64
    algal tufa, 171–73
    caliche mechanism for, 173
    cementation textures in, 165
    cemented, 29, 72, 159
    environmental significance of, 175–77
    non-tufa, 173–75
    origin of, 170–75
    paper-thin, 59, 67, 75, 80, 81, 82, 160, 162, 163
    sub-surface, distribution of, 162–63
    surface, distribution of, 161–62
    types of, 160
Current baffling by vegetation, 103, 105, 108, 109
Current lineations, 22, 40, 57, 71, 107
Cycles, depositional
    in algal marsh bedding, 82
    Andros-type, produced by progradation, 121–23, 186
    Lofer, comparison with Andros-type, 122
Cyclones. *See* Hurricanes
Cymodocea, 34, 125

Dasybranchus, 126, 144, 147, 155, 156
Depression fills, 59, 63, 66, 107, 111
Desiccation cracks. *See* Mud-cracks; Sheet-cracks
Detracia, 126, 134
Diatoms, 125
Dictyosphaeria, 125
Didemnum, 125
Diplantheria, 31, 44, 126
Distichlis, 126
Dolomite. *See* Proto-dolomite

Earthworms. *See* Oligochaetes
Echinaster, 125
Elops, 154
Exposure index, 7–11
    definition of, 8
    determination of, 8–9
    seasonal variation of, 9, 11
    sedimentary features as a function of, 10

Fasciolaria, 125
Fenestral pores, 55, 57, 59, 68, 69, 71, 73, 75, 76, 78, 81, 82, 84, 117, 164, 165, 174
Fiddler-crab. *See Uca*
Filament molds, algal, 56, 59, 62, 67, 68, 69, 71, 74, 75, 81, 82, 84, 112, 115, 117, 139, 158, 159, 164, 165, 171, 172
Fimbristylis, 126
Foraminifera, 29, 31, 55, 59, 80, 84, 90, 91, 125, 126

Gastropods, 29, 31, 45, 55, 80, 84, 88, 90, 91, 125, 126, 134, 141, 154. *See also* Cerithids
Geloina, 35, 126, 134, 136, 142, 153
Girvanella, 82, 114, 115, 159, 177
Graded bedding, 59, 63, 112
    reverse, 66, 112
Grasses, 22, 26, 37, 49, 126
Gravel
    flat-pebble, 37, 80, 85–88
    lag, 44, 88
    round-pebble, 37, 88–89
    shell, 45, 88

Halimeda, 34, 125, 166
Hurricane flooding, 82, 84, 88, 100, 103, 108, 110
Hurricane layers, thin beds as, 103
Hurricanes, 18, 100, 103, 105, 108

Intraclast, 21, 26, 29, 37, 40, 44, 55, 63, 64, 66, 75, 76, 81, 84, 88, 90, 163. *See also* Pebble
Ircinia, 125
Islands, significance of, 186–87

Juncus, 126

Laevicardium, 125
Lamination
    as algal "stick-ons," 109–10, 111, 112, 115
    areal continuity of, 112–13
    conditions required for storm deposition of, 97–102
    crinkled fenestral, description of, 68–71
    disrupted flat, description of, 63–68
    origin of, 93–102
    problem of uniform thickness of, 106–10
    review of criteria for recognizing Andros-type, 111–13
    role of blue-green algae in origin of, 106, 109–10
    smooth domal, description of, 53–57
    smooth flat, description of, 57–63
    as storm layers, evidence for, 95–97
    time value of, 93
    wavy fenestral, description of, 71
    with palisade structure, description of, 71–78
Laminite, 106, 109, 111, 112, 113
"Laminite cap," 118, 121–23
Laurencia, 44
Layering. *See also* Lamination; Thin bedding; Thick bedding
    method of determining time value of, 92
    origin of, 93–105
    preservation and destruction of, 118–21
    previous work on, in modern carbonate sediments, 50–52
    review of criteria for recognizing Andros-type, 111–14
    as "settle-outs," 109
    storm deposition of, 95–97
    time value of, 92–93
    types of, in Andros Island tidal flat sediments, 53
    vertical and lateral distribution of, 118–23
    vertical succession of, produced by progradation, 121–23
Levee
    backslope, layering in sediment of, 53, 63–68
    backslope, structures and textures in sediment of, 180
    backslope, subenvironment, 25–29
    community, 126
    crest, layering in sediment of, 53, 57–63, 64
    crest, structures and textures in sediment of, 180
    crest, subenvironment, 21–25

Magnesian calcite
    cement, 29, 46, 76, 159, 162, 165, 170, 174, 175
    filament molds, 13, 46, 73, 74, 82, 113, 139, 158, 164, 165, 171, 172
    Mg distribution coefficient for, 172, 173
    sediment, 132
Mangrove
    black, 22, 29, 37, 126
    red, 22, 26, 29, 31, 49, 84, 125, 126, 156
Manicina, 125

Marphysa, 31, 35, 45, 91, 126, 141, 143, 144, 145, 146, 147, 155, 156, 157
Marsh. *See* Algal marsh
Micrite, 55, 56, 59, 63, 66, 71, 73, 84, 113, 115, 165, 173
Microcoleus, 31
Microspar, 165, 173
Monanthochloe, 126
Mud
  clotted, 55, 56, 66, 71, 76, 78, 82, 84, 90, 91, 146
  pelleted, 12, 13, 31, 59, 66, 91
  peloidal, 55, 59, 63, 76, 89, 90, 162, 164
Mud chips. *See* Intraclast
Mud-cracks, 25, 63, 64, 67, 68, 80, 82, 84, 103, 113, 117, 164, 176
Myriochele, 125

Nematodes, 125, 126

Offshore (nearshore)
  community, 124–25
  layering in sediment of, 53, 90
  structures and textures in sediment of, 180
  subenvironment, 34
Oligochaetes, 57, 64, 126, 145, 151, 156
Ostracods, 125, 126, 132
Ottonosia, 159

Palisade structure. *See* Lamination with palisade structure; Crusts, algal
Palm hammocks, 13, 161, 163, 175
Panopeus, 139, 151, 155
Peat, freshwater
  used to date sediments, 13, 121
Pebble, 25, 37, 38, 39, 80, 85–89, 113. *See also* Intraclast
  cemented crust, 38, 81, 82, 85, 88, 162
  curled, 85, 88
  welded to surface by algal mats, 37, 85
Pellet, fecal, 31, 37, 45, 55, 59, 64, 68, 130, 132, 133, 142, 143, 145, 146–47, 148, 151, 158
Peloid, 12, 13, 40, 55, 57, 59, 62, 66, 71, 78, 80, 81, 82, 84, 90, 91, 107, 113, 132, 133, 164, 165, 173, 174
Peneroplis, 35, 45, 55, 126, 132, 145, 157
Penicillus, 31, 34, 35, 36, 37, 44, 125, 131, 157, 166
Perinereis, 126, 147
Phormidium, 22, 166, 169
Pitar, 125
Plectonema, 166
Polychaete, 29, 37, 125, 126, 144, 147, 154
Pond
  bioturbation of, 141–44
  community, 125–26
  erosion of, 110
  layering in sediment of, 53, 88, 90–91
  structures and textures in sediment of, 180
  subenvironment, 29–33
Pontodrilus, 57, 64, 126, 145, 151
Precambrian Bahama-type tidal flat features, 185–86
Prism-cracks, 31, 45
Proto-dolomite, 29, 159, 162, 163, 171, 173
Proto-stromatolite
  criteria for recognizing Andros-type, 114–18
  as current direction indicators, 55, 116
  definition of, 21
  description of, 53–57
  origin of, 57, 85

Prunum, 126, 134
Rachicallis, 126
Rainfall
  on Andros Island, 16–17
  influence on tidal flooding, 19
Rate of deposition experiments, 92–93
Rhipocephalus, 34, 125, 131, 157
Rhizophora, 125, 126, 156, 157
Ripples, 22, 37, 39, 40, 45, 57, 59, 66, 71, 107, 111
Rivularia, 37, 53, 71, 115, 117
Roots and root-molds, 57, 59, 79, 80, 84, 85, 90, 92, 156–57, 158

Salicornia, 126
Salinity
  relation to exposure index, 130
  of tidalwaters in study area, 19
Samphospongia, 159
Sand
  depression fills of, 59, 63, 66, 107, 111
  peloidal, 39, 40, 59, 63, 64, 66, 75, 82, 84, 107, 111
  skeletal, 39, 40, 42, 85, 88, 90
Sayella, 126, 134
Schizoporella, 125
Schizothrix, 21, 22, 26, 28, 35, 37, 39, 41, 45, 53, 56, 57, 63, 71, 75, 82, 108, 109, 110, 116, 117, 118, 122, 125, 126, 128, 139, 140, 158, 162, 166, 167, 169, 170, 185, 186
Scour, erosion, 67
Scytonema, 14, 20, 22, 26, 28, 37, 41, 46, 49, 53, 55, 56, 63, 64, 68, 69, 71, 73, 75, 78, 81, 82, 84, 88, 103, 105, 109, 112, 113, 114, 115, 117, 126, 128, 135, 139, 158, 159, 162, 163, 166, 167, 168, 169, 170, 171, 172, 175, 176, 177
  algal tufa, 171–73
  factors influencing distribution of, 165–70
Scytonema androsense, 166, 167
Scytonema crustaceum, 166, 167
Scytonema Julianum, 171
Scytonema myochrous, 166, 167
Sea-grass, 37, 92, 125
Sediment, produced by algae, 82, 130–31, 187. *See also* Algal tufa
Sedimentary record of environmental elements, 181–83
Sedimentary structures
  catalog of, 178–79
  summary of, in each subenvironment, 180–81
Sediment by-passing, 108
Sesarma, 29, 84, 126, 128, 145, 151, 155
Sheet-cracks, 55, 57, 59, 64, 67, 68, 78, 81, 84, 117, 165, 174
Skeletal record
  of codiacean algae, 130–32
  of foraminifera, 132–33
  of molluscs, 134–37
Spartina, 126
Sponges, 125
Storm deposition as a widespread phenomenon, 184
Storm flooding, measurements of, 98–99
Storm layers, lamination as, 95–97
Storm surge, 97–100
Stratigraphic relations, 121–23, 185–86
Stromatolite, 21, 29, 46, 57, 75, 82, 85, 111, 114, 140, 159, 171, 172, 175, 183, 186. *See also* Proto-stromatolite
  criteria for recognizing Andros-type, 114–18
Strombus, 125

Subenvironment
    channel, surface features of, 43–46
    definition of, 3
    inland algal marsh, surface features of, 46–49
    levee-pond, surface features of, 20–33
    shoreline, surface features of, 34–43
Suspended sediment, measurement of, 95
Symploca, 166
Syncera, 126, 134, 142

Thalassia, 31, 34, 35, 44, 92, 125, 187
Thick bedding (including very thick bedding), 90–92
Thin beds (thin bedding)
    algal tufa-peloidal mud interbeds, description of, 78–82
    disrupted fenestral, description of, 82–84
    flat-pebble gravel, description of, 85–88
    as hurricane layers, 103
    origin of, 102–5
    round-pebble gravel, description of, 88–89
    as "settle-outs," 109
Tidal
    channel. *See* Channel
    currents, 19
    range. *See* Tides

Tides
    broad definitions used, 1, 18
    measurements of, in study areas, 18–19
Tonna, 125
Tornatina, 126, 134, 142
Tracks and trails
    bird footprints, 154–55
    crab grazing scratches, 155
    gastropod grazing trails, 154
    polychaete trails, 154
Tufa, algal. *See* Algal tufa

Uca, 21, 29, 57, 84, 88, 116, 126, 128, 145, 151, 155, 156
Udotea, 166

Vermicularia, 125
Verongia, 125

"Whitings," 131, 139

Zonation of algae, 165–70
Zonation of surface features
    in levee-pond subenvironments, 20
    in shoreline subenvironments, 34

The Johns Hopkins University Press

This book was composed in Baskerville text and Permanent display type by Maryland Linotype Composition Company, Inc., from a design by Patrick Turner. It was printed on 70-lb. Patina Coated Matte paper and bound in Holliston Roxite cloth by The Maple Press Company.

Library of Congress Cataloging in Publication Data

Main entry under title:

Sedimentation on the modern carbonate tidal flats of
    northwest Andros Island, Bahamas.

  (The Johns Hopkins University studies in geology;
no. 22)
    Bibliography: pp. 191–96
    Includes index.
    1. Marine sediments—Bahamas—Andros (Island)
2. Rocks, Carbonate. 3. Tidal flats—Bahamas—Andros
(Island) I. Hardie, Lawrence A. II. Series.
GC383.2.S4   551.4′63′3   76–47389
ISBN 0–8018–1895–8